T0338297

Autoecology and ecophysiology of woody shrubs and trees

Autoecology and ecophysiology of woody shrubs and trees

Concepts and applications

EDITED BY

Ratikanta Maiti
Universidad Autónoma de Nuevo León
Nuevo León, Mexico

Humberto González Rodríguez
Universidad Autónoma de Nuevo León
Nuevo León, Mexico

Natalya Sergeevna Ivanova
Botanical Garden of Ural Branch of the Russian Academy of Sciences
Yekaterinburg, Russia

WILEY Blackwell

Registered office: John Wiley & Sons, Ltd, The Atrium, Southern Gate, Chichester, West Sussex, PO19 8SQ, UK

Editorial offices: 9600 Garsington Road, Oxford, OX4 2DQ, UK
The Atrium, Southern Gate, Chichester, West Sussex, PO19 8SQ, UK
111 River Street, Hoboken, NJ 07030-5774, USA

For details of our global editorial offices, for customer services and for information about how to apply for permission to reuse the copyright material in this book please see our website at
www.wiley.com/wiley-blackwell.

Library of Congress Cataloging-in-Publication Data is applied for

ISBN: 9781119104445

A catalogue record for this book is available from the British Library.

Wiley also publishes its books in a variety of electronic formats. Some content that appears in print may not be available in electronic books.

Set in 9.5/13pt, MeridienLTStd by SPi Global, Chennai, India.
Printed and bound in Malaysia by Vivar Printing Sdn Bhd

1   2016

Contents

Preface

Forest trees and shrubs play a vital role as important sources of our daily needs; and they also serve as a life-saver by their capture of CO_2, thereby reducing the atmospheric carbon load. At the same time, they use CO_2 in the process of photosynthesis and they store carbon in their biomass and wood as a source of energy. However, adverse climatic conditions, increasing global warming associated with constant emission of greenhouse gases, illegal logging, human activities, burning fossil fuels, expanding agriculture, high temperatures, global warming, cyclones, water and cold stress affect the growth of trees and shrubs in the forests prevailing in different regions of the world.

Forests save our lives by absorbing and reducing the CO_2 load liberated in the lower horizon of the atmosphere through the process of photosynthesis, liberating oxygen for our respiration and storing carbon in their wood, important for the wood industry. This situation in turn endangers the security of mankind and animals, reduces crop productivity and enhances hunger and poverty in the world. There is a great necessity for protecting forest resources against all these untoward menaces. Concerted research activities in multidisciplinary facets need to be directed to protect our forest treasures.

At this juncture, there is a great necessity to study the autoecology and ecophysiology of tree species in the forest ecosystem.

Autoecology deals with all aspects of the dynamism of populations, the physiological traits of trees, their light requirements, their life history pattern and their physiological and morphological characters (Fetcher et al., 1994). On the other hand, ecophysiology is defined by growth in terms of various plant parameters, such as leaf traits, xylem water potential, plant height, basal diameter, crown architecture and so on, which are influenced by several physiological traits and the environmental conditions of the forest ecosystem. This also involves various functional, physiological, biochemical and biophysical aspects of woody trees for plant productivity. In this regard, there is a great necessity to determine the variability of physiological functions among tree species and their adaptation to the prevailing environmental conditions. A clear understanding of different facets of autoecology and ecophysiology will help forest scientists and foresters in the efficient management and protection of trees in a forest ecosystem.

The main objectives are to provide various aspects of autoecology and ecophysiology. In both cases, we discuss and present research advances obtained in all of the major forests of the Planet: temperate, tropical and alpine regions of the world, with supporting literature on all aspects of autoecology and ecophysiology mentioned herein. With respect to autoecology, we describe the branching patterns and crown architecture necessary for the capture of solar radiation, the variability in leaf traits, leaf anatomy, wood anatomy, phenological events related to the growth, development and maintenance of the life cycle and the seed characteristics and plant traits related to plant productivity.

With respect to ecophysiology we report results on leaf pigments, leaf nutrients, carbon sequestration (carbon fixation), plant–water relations and factors affecting plant productivity. As mentioned earlier, we provide published research advances and original research results in various topics. These used different approaches in their research: physiological, population and ecosystem approaches. A global integrated approach is complemented by modern methods and statistical data analysis. In a nutshell, this book contains research advances on various aspects of woody plants required by forest scientists and foresters to manage and protect forest trees and to plan future research. Besides, it is expected that this book will find a place among researchers, academic scholars, undergraduate and graduate students in disciplines related to agronomy, biology, forest science and range management. It will also be useful for policy makers as a ready reference as well as for retrieving detailed information on the plant species documented which support forest and range ecosystems for different uses and purposes such as forage, timber, charcoal, shelter, reforestation and soil erosion prevention.

Bibliography

Fetcher, N., Oberbauer S.T., Chazdon, R.L. **1994**. Physiological ecology of plants at La Selva. In: Lucinda A. McDade, Kamaljit S. Bawa, Henry A. Hespenheide and Gary S. (eds.). *La Selva: Ecology and Natural History of a Neotropical Rainforest*. University of Chicago Press, Chicago. pp. 128–141.

List of contributors

Humberto Gonzalez-Rodriguez

Facultad de Ciencias Forestales, Universidad Autonoma de Nuevo Leon, Linares, Mexico

Natalya Sergeevna Ivanova

Botanical Garden of Ural Branch of the Russian Academy of Sciences, Yekaterinburg, Russia

Theodore Karfakis

Terra Sylvestris Non-Governmental Organization, Kalamos Lefkados, Greece

Ratikanta Maiti

Facultad de Ciencias Forestales, Universidad Autonoma de Nuevo Leon, Linares, Mexico

Ek Raj Ojha

Climate Change and Sustainable Development Program, Tribhuvan University, Kathmandu, Nepal

Artemio Carrillo Parra

Facultad de Ciencias Forestales, Universidad Autonoma de Nuevo Leon, Linares, Mexico

Maria Victorovna Yermakova

Botanical Garden of Ural Branch of the Russian Academy of Sciences, Yekaterinburg, Russia

Ekaterina Sergeevna Zolotova

Botanical Garden of Ural Branch of the Russian Academy of Sciences, Yekaterinburg, Russia

CHAPTER 1

Background

In a forest different species of trees, shrubs and herbs grow together mutually, share solar radiation for photosynthesis and absorb soil moisture and nutrients from the soil horizon with their efficient root systems distributed at different soil profiles depending on the species. They grow together and maintain their dynamic growth if not disturbed by environmental stresses, such as cyclones, illegal logging, over-grazing by animals, expansion of agriculture other human activities and so on. The plants absorb soil moisture and nutrients for their growth and development. The mature leaves finally fall to the ground, mix with the soil and thereby recycle nutrients. Adverse climatic conditions, high temperatures, global warming, cyclones, water and cold stress affect the growth of the trees and shrubs in the forests prevailing in different regions of the world. Some species have a capacity of adaptation to these environmental stresses, others are susceptible and fail to survive. Forests give us timber, fire wood, medicinal plants, other products for our livelihood and forage for animals. There exists a great diversity in the leaf canopy and crown architecture, branching patterns, leaf morphology and floral structure. Forests provide different types of soft and hard woods for the wood and paper industries depending on the structural organisation of wood. Different species flower in different seasons, produce fruits and disperse seeds for germination, regeneration of seedlings and maintaining their life cycles. Forests save our lives by absorbing and reducing the carbon dioxide load liberated in the lower horizon of the atmosphere through the process of photosynthesis, liberating oxygen for our respiration and storing carbon in the wood, important for the wood industry. Most of them are perennial, few are annual. Different species flower at different times of the year depending on their photoperiod requirement. They flower, produce fruits and finally disperse seeds on the ground which germinate and emerge on the advent of favorable weather, thereby maintaining the life cycle. It is necessary to understand the dynamic activities of the species present in a forest varying widely in different climatic zones of the world, in arid, semiarid, temperate or alpine zones. The authors have original findings of these activities.

A knowledge of the autoecology and ecophysiology of the trees in a forest is an essential pre-requisite to understand

Autoecology and Ecophysiology of Woody Shrubs and Trees: Concepts and Applications, First Edition.
Edited by Ratikanta Maiti, Humbero Gonzalez Rodriguez and Natalya Sergeevna Ivanova.
© 2016 John Wiley & Sons, Ltd. Published 2016 by John Wiley & Sons, Ltd.

the growth of trees required for efficient forest management by forest managers and rangers. They should know the phenology (flowering, fruiting, seed dispersal) and the effects of environments on the growth and development of each species and globally for taking effective measures for maintaining forest health. We present here up to date literature on various aspects of autoecology and ecophysiology from tropical, temperate, arid–semiarid and a few alpine forests globally. We present here our research advances on various aspects of autoecology and ecophysiology of semiarid Tamaulipan thorns which may serve as a model to study the autoecology of trees in other regions. In addition, we provide an extensive review on various aspects of autoecology globally.

1.1 A definition of autoecology

Autoecology involves all aspects of the dynamism of populations and the physiological traits of trees, their light requirements and also their life history pattern and physiological and morphological characters (Fetcher et al., 1994).

1.2 A definition of ecophysiology

Ecophysiology (from Greek οῖκος, *oikos*, "house(hold)"; φύσις, *physis*, "nature, origin"; and -λογία, *-logia*, "discussion"), environmental physiology or physiological ecology is a biological discipline that studies the adaptation of an organism's physiology to environmental conditions. It is closely related to comparative physiology and evolutionary physiology. See: *ecophysiology of tree growth* onWikipedia, the free online encyclopedia.

Ecophysiology of tree growth can be defined in terms of an increase in the size of an individual or a stand. Growth is usually expressed as a change in size per unit of time and area. The growth of trees is influenced by several physiological traits and environmental conditions of the forest ecosystem. This also deals with various functional, physiological and biophysical aspects of woody trees to plant productivity. There is a great necessity to determine the variability of physiological functions among tree species and their adaptation to the environment.

1.3 Environment

Forest environment plays an important role in the growth of a tree. Environment can be classified as climatic, edaphic, physiographic and biotic. The climatic factors prevailing in different semiarid, tropical and temperate regions affect the growth and adaptation of trees. The climatic factors related to atmospheric conditions are solar radiation, light, air composition, wind, temperature, precipitation, relative humidity and intensity of light. These climatic factors determine the distribution of vegetation.

We want to mention here a brief account of climate change and its effect on forests. A change in climate due to an increase of carbon dioxide concentration has a direct impact on the productivity of forests. Climate determines the distribution of vegetation in a forest ecosystem. There exists a good relation of the climate with the conservation and development of forest. It is essential for foresters to have good understanding of the climate changes and its impact on forest productivity and to take the necessary measures to protect it.

We are very much concerned about how human activities such as the burning of fossil fuels, conversion of forests to agricultural lands and other illegal activities cause a significant increase of carbon dioxide and other green house gases (GHG) in the atmosphere. On the other hand, both forests and human uses of forest products contribute to a gradual increase of GHG in the atmosphere. Fortunately the trees and forests with their ability to absorb CO_2 and carbon have an opportunity to mitigate climate change. GHG cause the retention of heat in the lower atmosphere due to absorption of light and its reflection by clouds and other gases.

The earth receives radiant energy from solar radiation for its utilisation by the plants for photosynthesis and by humans for other activities. Short-wave solar energy (visible) received from the sun passes through the atmosphere, thereby warming the earth's surface. Long-wave thermal radiation is absorbed by a number of GHG. These GHG accumulate in small amounts in the lower layer of the atmosphere and reflect long-wave thermal radiation in all directions. Some of the radiation is directed towards the earth's surface. The amount of GHG in the atmosphere influences global temperature.

Green house gases present in the earth's surface are water vapour (H_2O), carbon dioxide (CO_2), nitrous oxide (NO_2), ozone (O_3), carbon monoxide (CO) and chlorofluorocarbons (CFC). The concentrations of these gases have changed in the earth over geological time scales. The increase of agriculture, animal husbandry, grazing and an increase in human population has indirectly increased the levels of these GHG, enhancing global warming, thereby threatening the security of human life, animals and so on. Over and above this, an incessant logging of trees for timber has a direct impact in the increase of GHG. Constant emissions of carbon dioxide and its accumulation in the lower profile of the atmosphere enhances contamination, but thanks to the forest ecosystem of trees and shrubs, even the lower organisms capture carbon dioxide and utilise it in the process of photosynthesis and the accumulation of carbon in wood and biomass, thereby reducing the carbon dioxide load and at the same time liberating oxygen necessary for the respiration of living organisms. Now we can present a brief account on the effects of environmental components on forest growth and productivity.

1.4   Solar radiation

Solar radiation is the main source of energy for our life and all living organisms. It provides light, temperature and energy for growth and development. Although a large part of radiation is absorbed by the atmosphere, a small part reaches the earth's surface and is captured by leaves through chlorophylls and converted into chemical energy by the process of photosynthesis. This energy is stored in the grains and fruits serving as a source of energy for us. Not all solar radiation fully reaches the earth's surface. A large part of the ultraviolet rays (wavelength 0.12–0.40 µm) are absorbed by oxygen, nitrogen and ozone present in the atmosphere. The visible portion of solar radiation (wave length 0.40–0.71 μm) is called visible light.

1.5 Solar radiation and vegetation

Solar radiation supplies heat, illumination and chemicals and produces electrical effects on the earth's surface. Solar radiation contributes with light and temperature for the growth of the ecosystem: a proper temperature range is necessary for various physiological activities, viz. transpiration, photosynthesis and respiration. For proper growth of plants a suitable range of temperature is very essential. An optimum air temperature is necessary for the germination of seeds. Higher air temperature increases microbial activity on the soil surface, leading to a rapid decomposition of organic matter, release of nutrients and formation of humus. Temperature affects the activities of enzymes. Plants are adapted to different regions, such as arid, temperate and alpine climates, thereby temperature requirements vary in the different regions.

1.6 Light requirement of tree species

Plants require light for photosynthesis and growth. The species require abundant light for their optimum growth and development. Shade plants are capable of growing under shade. These plants require shade at least in the early stages for optimum growth and development. Shade-tolerant plants possess a range of adaptations to

help them to survive and alter the quantity and quality of light typical of shade environments. Shade leaves and shade-tolerant species have a higher photosynthetic efficiency per unit of leaf area under low-light conditions than sun leaves and intolerant species. Under high light conditions, the reverse is true. This is because shade leaves of intolerant species lose photosynthetic efficiency when they are suddenly exposed to high light intensities. Excess light incident on a leaf can cause photoinhibition and photodestruction. Plants adapted to high light environments have a range of adaptations to avoid or dissipate the excess light energy, as well as mechanisms that reduce the amount of injury caused.

1.7 Photomorphogenesis and photoperiodism

Light intensity is an important component in influencing the temperature of plant organs (energy budget). Light directly affects tree growth through its intensity, quality and duration.

1.8 Photosynthesis

Photosynthetic activity contributes greatly to the production of biomass. This is also influenced by the intensity of leaves and canopy orientation. It is expected that trees with an open leaf canopy are more efficient in photosynthesis than those with a closed canopy. The rate of photosynthesis is influenced by both plant and environmental factors. Light quality is of interest in relation to considerations of the merits of uneven- versus even-aged stands. Light plays an important role for the natural regeneration and maximum production of high quality wood.

1.9   Temperature

Temperature is a very important factor for photosynthesis. Both air and soil temperature influence plant growth and thus affect forest vegetation. The proper temperature range is necessary for various physiological activities. Each plant species requires an optimum temperature for plant growth. A temperature lower than the optimum required for each species affects its growth. Some species are well adapted to a xeric habitat, some to temperate and others to alpine zones. With an increase in temperature plant growth increases up to its optimum temperature. The low temperatures prevailing in winter affect seed germination.

1.10   Water relations

Water is one of the most important factors influencing the distribution, growth and development of vegetation and is essential for all vital processes such as photosynthesis. It forms the essential constituents of cell sap and cell vacuoles. It works as a medium for the absorption of plant nutrients and plant metabolism. It is necessary for the germination of seeds; and it is necessary for all plant movement. A plant cell requires about 90% water to maintain its vital activity. A substantial decrease in water in the cell causes plasmolysis, thereby inhibiting all metabolic activities. Water potential plays an important role to maintain

a pressure gradient in the plant cells required for the transport of water from one to another. Water is absorbed by roots and is transported upwards through xylem vessels to the leaves and other organs to help in metabolic activities. Excess water is lost through transpiration via the stomatas, thereby maintaining a water balance in the plant cells. Water deficiency causes a lowering in plant water potential, while an excess of water due to flooding affects respiration and plant growth. The sequential process of water absorption, its translocation and loss by the transpirational flux is discussed in brief below.

Water is absorbed by roots and then is transported through the mostly dead xylem tissue from the soil to the leaves and finally transpired to the atmosphere. The loss of water through transpiration creates a vacuum pressure in the leaf mesophyll due to which water in the xylem is under negative pressure and creates cohesive forces in xylem vessels between water molecules to maintain the water columns intact. Roots absorb water from soils owing to the difference of water potential between the soil and cells. There is always a gradient of water potential from the peripheral cortical cells to the interior cells, finally reaching the endodermis, due to water entering the xylem vessels. Once it reaches the xylem vessels, water moves up by a suction force called the ascent of sap. The adjacent phloem tissue is under positive pressure that is maintained osmotically with assimilated sugars and dissolved minerals. Variability in soil and atmospheric conditions influences the interaction between the pressures and structural properties, determining the tissue resistivity against embolism formation under high negative pressures in xylem tissue that threatens the integrity of xylem transport. Only a small amount is consumed in the process of photosynthesis. The main effect of water on photosynthesis is indirect, through the hydration of protoplasm and stomatal closure.

The deficiency of water called water stress affects the growth and development of plants. Each species requires an optimum amount of water below which the growth of this species is reduced. The species adapted to drought is called drought-resistant. Drought-resistant plants have several morphological, anatomical and biochemical mechanisms of resistance, similar to species tolerant to low temperatures. The density of trichomes, leaf surfaces with a waxy coating, thick cuticles, compact palisade cells and a few biochemical components, such a proline, sugars and ABA are related to drought resistance. In the semiarid tropics, several species are tolerant to drought, others not so much. We need to identify them in a forest.

1.11   Plant nutrients

The photosynthetic efficiency of leaves depends on the soil nutrient supplies. A good amount of nutrient improves the photosynthetic capacity of trees. Photosynthesis involves physicochemical processes that occur in the presence of light energy and enzymatic processes that occur in dark reactions. It involves mainly two processes: Photosystem I, during which radiant energy is absorbed by chloroplasts, leading to the photolysis of water. The chemical process during which CO_2 is fixed occurs in dark reactions. This is influenced by the quantity of CO_2 fixed by each gram of foliage and indirectly by the increased size of individual leaves and the total size of crown and root system. An increase in leaf area contributes to a high photosynthetic capacity. The photochemical process starts when the chloroplasts capture radiant energy and transfer the energy to a chemical process during the dark phase. Two types of chlorophylls are involved in this process: chlorophyll *a* and chlorophyll *b*. The chloroplasts are composed of stacks of thylakoids (like coins) and stroma between the chloroplasts. The photochemical process occurs in the thylakoids while the chemical reaction combining the CO_2 with H_2O occurs in the stroma. The light energy absorbed by chlorophyll in the photochemical reaction is transmitted to the dark reaction in the absence of light for the chemical combination of CO_2 and H^+ liberated during the photolysis of water to form the first product of photosynthesis, phosphoglyceraldehyde. Then glucose-1-phosphate, a three-carbon compound, enters the phosphate pentose pathway (Calvin Cycle) and phosphate, a high energy compound, helps in the transfer of energy from one compound to another. The Calvin Cycle ultimately forms glucose, the final product; and the phosphate is liberated to function again in the transfer of energy. Thereby 674 calories of radiant energy absorbed by the chlorophyll pigments are stored in the glucose; and the glucose in turn is stored as a carbohydrate in insoluble starch. The phosynthates and glucose are transmitted and finally stored in plant biomass and wood as sources of energy liberated during the process of respiration for various metabolic processes in the plant.

$$6CO_2 + 12H_2O + 674 \text{ calories}$$

$$\rightarrow C_6H_{12}O_6 + 6O_2$$

1.12   Role of nutrients in plant life

Several nutrients take part in the growth and development of plants, such as carbon, oxygen and hydrogen from the air or from water. Although plants absorb more than 40 elements during their growth, not all of them are essential for growth and development.

Elements which have been proved to be essential for the growth and development of plants are called essential elements. The nutrients which are required in larger proportion are called major nutrients, for example carbon (C), hydrogen (H), nitrogen (N), calcium (Ca), phosphorus (P), magnesium (Mg), potassium (P), oxygen (O) and sulfur (S). Those required in smaller amounts are called minor nutrients, for example iron (Fe), zinc (Zn), chlorine (Cl), copper (Cu), molybdenum (Mo), boron (B) and manganese (Mn). Deficiency in these micronutrients affects plant growth.

Nitrogen (N) is the most common nutrient that limits forest growth. It forms the skeleton of protein and help in enzyme action. Phosphorus (P) is for energy transformation. Sulphur acts as a milder substitute for oxygen although it is required to form two aminoacids: methionine and cysteine. Potassium (K) activates many enzymes. Calcium (Ca)

connects organic molecules. Magnesium (Mg) is present at the center of chlorophyll. Iron (Fe) is a key element for respiration and photosynthesis processes. Manganese (Mn) plays a significant role during respiration and photolysis of water. Copper (Cu) is involved in oxidation–reduction reactions.

The photosynthetic efficiency of foliage depends decisively on soil nutrient supplies. By improving the nutrient status of a site we also improve the photosynthetic capacity of trees. The effect is both direct (quantity of CO_2 fixed by each gram of foliage) and indirect by increasing the size of individual leaves, total size of crown and root system.

1.13   Plant factors

In plants, net photosynthesis is dependent on leaf age position within crown. The fully expanded leaves in a conifer are the most efficient of all age classes owing to varying rates of respiration and insect or disease damage. The upper crown leaves are the most productive, exposed directly to sunlight. The leaves present in the lowest whorls contribute little to net photosynthesis as most of the leaves do not receive direct sunlight. Variation in photosynthesis occurs between crown classes and species (Gholz et al., 1979). We have developed a hypothesis that trees with an open canopy have a great capacity of capturing solar radiation and photosynthesising, compared to those with close, overlapping leaves. Few of these trees with an open canopy have the capacity to capture 50% carbon, as discussed later. Plants with an open canopy have highly branched stems while those with a close canopy possess a stout basal stem, yet to be confirmed. Leaf surface structures containing high stomatal frequency, trichome density, silica contents, wax contents and so on could be related to the adaptation of trees to xeric and biotic environments, as discussed later.

1.14   Respiration

Respiration is the process by which energy fixed by photosynthesis in plants is made available for metabolic processes. This energy is stored in wood as potential energy which is released by combustion. The implications are obvious. Per unit area, having fewer but larger diameter trees is better than keeping more but smaller diameter trees from the standpoint of wood production.

1.15   Phenology and ecology

Variations in tree phenology are good indicators of the plant species' response to climate change. In a tropical region, Borchert (1980) investigated the phenology and ecophysiology of a tropical tree: *Erythrina poeppigiana* O. F. Cook. Under the climatic conditions of San Jose in Costa Rica, leaf fall, flowering and shoot emergence of *E. poeppigiana* were markedly asynchronous among trees of the same population, suggesting strong endogenous control of the tree's development. During one year, trees showed two cycles of leaf shedding and shoot emergence. The endogenous periodicity of leaf shedding appeared to be primarily the result of leaf senescence. During the dry season, leaf senescence and hence leaf shedding were enhanced owing to tree severe water deficits. As a consequence of leaf shedding, water stress was reduced and shoot emergence started under continued drought. The degree of water stress, causing stem shrinkage, was affected by tree size, soil moisture availability and seasonal changes in evapotranspiration. The phenology of *Erythrina* along an altitudinal gradient of increasing atmospheric water stress showed a transition from an evergreen to a deciduous habit. With increasing drought, consecutive developmental stages tended to be more separate in time and more synchronised.

1.16   Effect of drought stress

Drought stress greatly affects the growth of trees, frequently causing fires during summer. Bréda et al. (2006) made a review of the pysiological responses, adaptation processes and long-term consequences under a drough event that occurred in temperate Western Europe during 2003. They emphasised the need to understand the key processes that may allow trees and stands to overcome such severe water shortages. They reviewed the impact of drought on exchanges at soil–root and canopy–atmosphere. They quantified and modelled CO_2 flux, the decline in transpiration, water uptake and in net carbon assimilation due to stomatal closure. Estimates of soil water deficit gave a quantitative index of soil water shortage. They reported the irreversible damage imposed on water transfer within trees and particularly within xylem and also the inter-specific variability of these properties among a wide range of tree species. There occurred inter-specific diversity of hydraulic and stomatal responses and a large diversity in traits potentially related to drought tolerance. The potential involvement of hydraulic dysfunctions or of deficits in carbon storage under long-term decline of tree growth and development and for the onset of tree dieback was emphasised. Bréda et al., 2006 reported that the starch content in stem tissues recorded at the end of summer 2003 was used to predict crown conditions of oak trees. Similarly, Bréda et al. (2006) made a review of ecophysio-logical responses, adaptation processes of temperate forest trees and stands under severe drought and long-term conse-quences that occurred in Western Europe during 2003. They highlighted the need to understand the key processes that may allow trees and stands to overcome such severe water shortages. They reviewed the current knowledge available about such processes. They explained the impact of drought on exchanges at soil–root and canopy–atmosphere interfaces and illus-trated this with examples from water and CO_2 flux measurements following spring: low starch contents were correlated with large twig and branch decline in the tree crowns.

1.17   Ecological plasticity

Man's broad diffusion and active man-agement of the sweet chestnut in past centuries established the species at the limits of its potential ecological range in many European mountain areas. It is hypothesised that the chestnut tree has considerable plasticity, enabling the species to adapt to very different sites and conditions. In order to test this hypothesis, Pezzatti et al. (2009) applied the pipe-model approach, postulating the constancy of the leaf area/sapwood area (LA/SA) relationship for single portions of a tree. They analysed the variation of the LA/SA coefficient among chestnut trees subjected to different sites conditions (e.g. convex vs concave sites) in order to gain an insight into the plasticity of the species. The results established that the sweet chestnut is able to greatly vary the allocation of resources with respect to environmental conditions. In particular, the LA/SA coef-ficient was high when trees were growing in sites with a good water supply.

1.18   Productivity

The productivity of a tree is influenced by various climatic conditions, solar energy, temperature, pigments contributing to the productivity of biomass, dry matter and timber production, radiation, tempera-ture and energy budget. Among them the following factors are important.

1 Carbon utilisation
2 Utilisation and cycling of mineral elements
3 Water relations.

In a forest ecosystem, individual species compete for the capture of solar radiation by plant pigments, chlorophyll and other pigments by the process of photosynthe-sis, thereby leading to carbon fixation (carbon sequestration) and dry matter production. Several physico-chemical and biochemical process occur in this vital process. Solar energy passes through the atmosphere, finally reaches the plant cover and is distributed at different profiles of leaf canopy, vertically, horizontally, then captured by leaves for the process of CO_2 assimilation during the process of photosynthesis; this finally leads to the production of dry matter. The leaves exposed directly to solar energy have a greater capacity for photosynthesis than those inside the plant cover. It is expected that the photosynthetic capacity of leaves present at different profiles of the tree varies widely. Variation in the contents of plant pigments and leaf canopy cover is expected to cause variation in carbon sequestration among different plant species. Therefore, there is a necessity to make comparative studies in the pigment contents and carbon fixation capacity.

Bibliography

Adams J.M., Piovesan G. **2002**. Uncertainties in the role of land vegetation in the carbon cycle. *Chemosphere* **49**(8):805–819.

Adams J.M., Piovesan G. **2005**. Long series relationships between global interannual CO_2 increment and climate: Evidence for stability and change in role of the tropical and boreal-temperate zones. *Chemosphere* **59**(11):1595–1612.

Adams J.M., Piovesan G., Strauss S., Brown S. **2002**. The case for genetic engineering of native and landscape trees against introduced pest and diseases. *Conservation Biology* **16**(4):874–879.

Alessandrini A., Biondi F., Di Filippo A., Ziaco E., Piovesan G. **2011**. Tree size distribution at increasing spatial scales converges to the rotated sigmoid curve in two old-growth beech stands of the Italian Apennines. *Forest Ecology and Management* **262**:1950–1962.

Alessandrini A., Vessella F., Di Filippo A., Salis A., Schirone B., Piovesan G. **2010**. Combined dendroecological and NDVI analysis to detect regions of provenance in forest species. *Scandinavian Journal of Forest Research* **25**(1):121–125.

Bellarosa R., Codipietro P., Piovesan G., Schirone B. **1996**. Degradation, rehabilitation and sustainable management of a dunal ecosystem in central Italy. *Land Degradation and Development* **7**(4):297–311.

Bernabei M., Lo Monaco A., Piovesan G., Romagnoli M. **1996**. Dendrocronologia del faggio (*Fagus sylvatica* L.) sui Monti Sabini. *Dendrochronologia* **14**:59–70.

Bernetti G. **1995**. *Selvicoltura speciale*. UTET, Torino.

Bernetti G. **2007**. *Botanica e selvicoltura*. Accademia Italiana di Scienze Forestali. Coppini, Firenze.

Biondi F., Strachan S.D.J., Mensing S., Piovesan G. **2007**. Radiocarbon analysis confirms the annual nature of sagebrush growth rings. *Radiocarbon* **49**(3):1231–1240.

Blasi S., Menta C., Balducci L., Conti F.D., Petrini E., Piovesan G. **2013**. Soil microarthropod communities from Mediterranean forest ecosystems in Central Italy under different disturbances. *Environmental Monitoring and Assessment* **185**(2):1637–1655.

Bonsen K.J.M. **1996**. Architecture, growth dynamics and autoecology of the Sycamore (*Acer pseudoplatanus* L.). *International Journal of Urban Forestry* **20**(3):339–354.

Borchert R. **1980**. Phenology and ecophysiology of tropical trees: *Erythrina poeppigiana* O. F. Cook. *Ecology* **61**:1065–1074.

Bréda N., Huc R., Granier A., Dreyer E. **2006**. Temperate forest trees and stands under severe drought: a review of ecophysiological responses, adaptation processes and long-term consequences. *Annals of Forest Science* **63**(6):625–644.

Brokav N.V.I. **1982**. The definition tree fall gap and its effects on measures of forest community. *Biotropica* **14**:158–160.

Clark D.A., Clark D.B. **1992**. Life history diversity of canopy and emergent trees in a neotropical rain forest. *Ecological Monographs* **62**(3):315–344.

de Luis M., Cufar K., Di Filippo A., Novak K., Papadopoulos A., Piovesan G., Rathgeber C.B.K., Raventos J. **2013**. Plasticity in dendroclimatic response across the distribution of Aleppo pine (*Pinus halepensis*). *Plos One* **8**(12):e83550.

Denslow, J.S., Hartshorn, G.S. **1994**. Tree-fall gap environment and forest dynamic process. In: McDalr L.A., Bawa K, Hespenhede. H.A. (eds.). *La Selva – Ecology and Natural History of a Neotropical Rain Forest*. University of Chicago Press, Chicago. Pp.120–127.

Di Filippo A., Alessandrini A., Biondi F., Blasi S., Portoghesi L., Piovesan G. **2010**. Climate change and oak decline: Dendroecology and stand productivity of a Turkey oak (*Quercus cerris* L.) old stored coppice in Central Italy. *Annals of Forest Science* **67**(7):706

Di Filippo A., Biondi F., Cufar K., De Luis M., Grabner M., Maugeri M., Presutti Saba E., Schirone B., Piovesan G. **2007**. Bioclimatology of beech (*Fagus sylvatica* L.) in the Eastern Alps: spatial and altitudinal climatic signals identified through a tree-ring network. *Journal of Biogeography* **34**(11):1873–1892.

Di Filippo A., Biondi F., Maugeri M., Schirone B., Piovesan G. **2012**. Bioclimate and growth history affect beech lifespan in the Italian Alps and Apennines. *Global Change Biology* **18**(3):960–972.

Donovan D.G., Puri R.K. **2004**. Learning from traditional knowledge of non-timber forest products: Penan Benalui and the autecology of *Aquilaria* in Indonesian Borneo. *Ecology and Society* **9**(3):1–23.

Fetcher, N., Oberbauer S.T., Chazdon, R.L. **1994**. Physiological ecology of plants at La Selva. In: L.A. McDade, K.S. Bawa, H.A. Hespenheide and G.S. Gale (eds.). *La Selva: Ecology and Natural History of a Neotropical Rainforest*. University of Chicago Press, Chicago. Pp. 128–141.

Gale M.R., Grigal D.F. **1987**. Vertical root distributions of northern tree species in relation to successional status. *Canadian Journal of Forest Research* **17**(8):829–834.

Gholz H.L., Grier C.C., Campbell A.G., Brown A.T. **1979**. Equations for estimating biomass and leaf area of plants in the Pacific Northwest. *Research Paper* **41**, Oregon State University, Corvallis, 39 pp.

Givnish T.J. **1998**. Leaf and canopy adaptations in tropical forests. In: Medina E., Mooney H.A., Vázquez-Yáñanes C. (eds.). *Physiological Ecology of Plants of the West Tropics. Tasks for Vegetation Science 12*. Springer, Netherlands. Pp. 51–84.

Grubb P.J. **2008**. The maintenance of species-richness in plant communities: the importance of the regeneration niche. *Biological Reviews* **52**(1):107–145.

Kamnesheidt L. **2000**. Some autoecological characteristics of early to late successional tree species in Venezuela. *Acta Oecologica* **21**(1):37–48.

Oldeman R.A.A., van Dijk J. **1991**. Diagnosis of the temperament of tropical rainforest trees. In: Rainforest Regeration and Management. A. Gomez-Pompa et al, (eds.). *Man and the biosphere series, Vol* **6**, UNESCO Parijs and Parthenon Publ. Group, London. Pp. 21–65.

Pezzatti G.B., Krebs P., Gehring E., Fedele G., Conedera M., Mazzoleni S., Monaco E., Giannino F. **2009**. Using the leaf area/sapwood area (LA/SA) relationship to assess the ecological plasticity of the Chestnut tree (*Castanea sativa* MILL.). In: I European Congress on Chestnut – Castanea 2009. *Acta Horticulturae* **866**.

Piovesan G., Adams J.M. **2000**. Carbon balance gradient in European forests: Interpreting EUROFLUX. *Journal of Vegetation Science* **11**(6):923–926.

Piovesan G., Adams J.M. **2001**. Masting behaviour in beech: linking reproduction and climatic variation. *Canadian Journal of Botany* **79**(9):1039–1047.

Piovesan G., Adams J.M. **2005**. The evolutionary ecology of masting: does the environmental prediction hypothesis also have a role in mesic temperate forests? *Ecological Research* **20**(6):739–743.

Piovesan G., Alessandrini A., Baliva M., Chiti T., D'Andrea E., De Cinti B., Di Filippo A., Hermanin L., Lauteri M., Mugnozza G.S., Schirone B., Ziaco E., Matteucci G. **2010**. Structural patterns, growth processes, carbon stocks in an Italian network of old-growth beech forests. *Italian Journal of Forest and Mountain Environments* **65**(5):557–590.

Piovesan G., Bernabei M., Di Filippo A., Romagnoli M., Schirone B. **2003**. A long-term tree ring beech chronology from a high-elevation old-growth forest of Central Italy. *Dendrochronologia* **21**(1):13–22.

Piovesan G., Biondi F., Bernabei M., Di Filippo A., Schirone B. **2005**. Spatial and altitudinal bioclimatic zones of the Italian peninsula identified from a beech (*Fagus sylvatica* L.) tree-ring network. *Acta Oecologica* **27**(3):197–210.

Piovesan G., Biondi F., Di Filippo A., Alessandrini A., Maugeri M. **2008**. Drought-driven growth reduction in old beech (*Fagus sylvatica* L.) forests of the central Apennines, Italy. *Global Change Biology* **14**(6):1265–1281.

Piovesan G., Di Filippo A., Alessandrini A., Biondi F., Schirone B. **2005**. Structure, dynamics and dendroecology of an old-growth Fagus forest in the Apennines. *Journal of Vegetation Science* **16**(1):13–28.

Piovesan G., Pelosi C., Schirone A., Schirone B. **1993**. Taxonomic evalutations of the genus Pinus (Pinaceae) based on electrophoretic data of salt soluble and insoluble seed storage proteins. *Plant Systematic and Evolution* **186**(1/2):7–68.

Piovesan G., Saba E.P., Biondi F., Alessandrini A., Di Filippo A., Schirone B. **2009**. Population ecology of yew (*Taxus baccata* L.) in the Central Apennines: spatial patterns and their relevance for conservation strategies. *Plant Ecology:* **205**(1): 23–46.

Piovesan G., Schirone B. **2000**. Winter North Atlantic oscillation effects on the tree rings of the Italian beech (*Fagus sylvatica* L.). *International Journal of Bio-Meteorology* **44**(3): 121–127.

Piraino S., Camiz S., Di Filippo A., Piovesan G., Spada F. **2012**. A dendrochronological analysis of *Pinus pinea* L. on the Italian mid-Tyrrhenian coast. *Geochronometria* **40**(1):77–89.

Schirone B., Pedrotti F., Spada F., Bernabei M., Di Filippo A., Piovesan G. **2005**. L'hêtraie de plusieurs siècles de la Vallée Cervara (Parc National des Abruzzes, Italie). *Acta Botanica Gallica* **152**(4):519–528.

Schirone B., Piovesan G. **1995**. L'approccio dendrologico nello studio del dinamismo della vegetazione forestale. *Colloques Phytosociologiques* **24**:265–271.

Schirone B., Piovesan G., Bellarosa R., Pelosi C. **1991**. A taxonomic analysis of seed proteins in *Pinus* spp. (Pinaceae). *Plant Systematic and Evolution* **178**(1/2):43–53.

Selås V., Piovesan G., Adams J.M., Bernabei M. **2002**. Climatic factors controlling reproduction and growth of Norway spruce in southern Norway. *Canadian Journal of Forest Research* **32**(2):217–225.

Wullschleger S.D. Tuskan G.A., Difazio S.P. **2002**. Genomics and the tree physiologist. *Tree Physiology* **22**:1273–1276.

Yoder B., Ryan M.G., Waring R.H., Schettle A.W., Auffmann M.R. **1994**. Evidence of reduced photosynthetic rates in old trees. *Forest Science* **40**:513–527.

Zerhun N.A., Monontagu K.D. **2004**. Belowground to aboveground biomass ratio and vertical root distributionresponses of mature *Pinus radiata* stands to phosphorusfertilization at planting. *Canadian Journal of Forest Research* **34**(9):1883–1894.

Ziaco E., Alessandrini A., Blasi S., Di Filippo A., Dennis S., Piovesan G. **2012**. Communicating old-growth forest through an educational trail. *Biodiversity and Conservation* **21**(1):131–144.

Ziaco E., Biondi F., Di Filippo A., Piovesan G. **2012**. Biogeoclimatic influences on tree growth releases identified by the boundary 7 line method in beech (*Fagus sylvatica* L.) populations of southern. *Europe Forest Ecology and Management* **286**:28–37.

Ziaco E., Di Filippo A., Alessandrini A., Baliva M., D'Andrea E., Piovesan G. **2012**. Old-growth attributes in a network of Apennines (Italy) beech forests: disentangling the role past human interferences and biogeoclimate. *Plant Biosystems* **146**(1):153–166.

Zobel B.J., Talbert J.T. **1984**. *Applied forest tree improvement*. The Blackburn Press, Wiley & Sons, Inc., New York. 524 p.

PART I

CHAPTER 2

Autoecology

2.1 Background

Autoecology deals with various aspects of trees/plants related to the growth and productivity of trees in a forest. We cite a few research advances on various aspects of autoecology of trees in temperate, tropical, semiarid and alpine regions of the northern hemisphere.

The autoecological role of the crown architecture and branching patterns of trees have received little attention. In this context, a reiteration capacity (Denslow and Hartshorn, 1994) comprising all growth responses is outside the deterministic architectural model. Being induced by environmental changes is considered an adaptative feature (Kamnesheidt, 2000). It means that both the crown architecture and branching pattern are adaptive measures of tree species in a forest ecosystem depending on the space available. Oldeman and van Dijk (1991) postulated that organised leaves, flowers, fruits, shoots, branching and reiteration patterns are all adaptive tools in determining ecological groups. The following autoecological parameters were considered:

1 Regeneration pattern
2 Distribution along drainage gradient
3 Gap association, crown architecture and reiteration

4 Leaf characteristics
5 Light compensation.

Early successional species had a monolayered leaf arrangement at all stages, while most shade tolerant species showed a mulilayered leaf arrangement. Early successional species in tropical rainforest were extensively adapted to a rapid height growth without wasting biomass to build a multilayered leaf arrangement (Oldeman and van Dijk, 1991). In shade-tolerant canopy species a multilayered leaf arrangement seems to be adavantageous. Compound leaves are generally found in gap-adapted camopy species. In our study, we classified trees in three classes:

1 Open canopy
2 Semiclosed
3 Closed canopy.

Trees with an open canopy have more capacity for the capture of solar radiation and photosynthesis. We selected a few open canopy species with 50% carbon dioxide fixation and a high wood density, while the closed canopy ones have less capacity; this needs further confirmation (discussed in a later section). A study has been made on the architecture and development of the sycamore (*Acer pseudoplatanus* L.) in the Netherlands and the behaviour of sycamore in these

Autoecology and Ecophysiology of Woody Shrubs and Trees: Concepts and Applications, First Edition.
Edited by Ratikanta Maiti, Humbero Gonzalez Rodriguez and Natalya Sergeevna Ivanova.
© 2016 John Wiley & Sons, Ltd. Published 2016 by John Wiley & Sons, Ltd.

systems, using specimens growing under favourable conditions. Scarrone's model was used to explain the branching system with different axes distinguished into sub-axes, capital axes and cardinal axes or explained in terms of reiteration. Sycamore in the Netherlands was compared with sycamore growing in Denmark and Great Britain, exhibiting a correlation with growth and flowering, age of seed production, the sub-axes, the architectural model, leaf size, architecture and silviculture (Bonsen, 1996). With respect to the role of canopy structure, a long-term demographic study undertaken in neotropical rainforest showed a great interspecific variation in growth and mortality among different microsites. In general, canopy species grow through different light conditions within the forest to attain maturity, thereby showing phosynthetic plasticity and acclimatation of their autoecological features (Kamnesheidt, 2000).

Donovan and Puri (2004) reported traditional knowledge of non-timber forest products in Penan Benalui and the autoecology of *Aquilaria* in Indonesia. *Aquilaria* is a tropical forest tree of South and Southeast Asia. Attempts at cultivating the valuable aromatic resin, *gaharu*, have been uneven at best. Thus, *gaharu* remains largely a natural forest product, increasingly under threat as the trees are overexploited and the forest is cleared. In this paper, they compared scientific knowledge and traditional knowledge of the Penan Benalui and other forest product collectors of Indonesian Borneo. They found that the Penan recognise the complex ecology of resin formation involving two, or maybe three living organisms: the tree, one or more fungi and possibly an insect intermediary.

Developing a sustainable production system for this resource will require a clear understanding of how these various natural elements function, separately and synergistically. Traditional knowledge can provide an information base and identify promising areas for future research. This knowledge supports the call for a greater role for ethnobiological research and interdisciplinary cooperation, especially between ethnobiologists and foresters, in developing sustainable management systems for this traditional resource and its natural habitat.

Regeneration of species is an important issue for determining the productive capacity of a tree species. In a forest ecosystem, Grubb (2008) discussed the importance of regeneration and species richness in a plant community. According to "Gause's hypothesis" during the process of evolution by natural selection, in a community at equilibrium every species must occupy a different niche. Most plant communities live longer than their constituent individual plants. In the case of an individual death, it may or may not be replaced by an individual of the same species. Several mechanisms also contribute to the maintenance of species-richness: In contrast, there seems to be almost limitless possibilities for differences between species in their requirements. It is emphasised that foresters were the first by a wide margin to appreciate its importance. The regeneration cycle gives potentially important examples of differentiation between species for each of the following stages: (a) production of viable seed (including the sub-stages of flowering, pollination and seed-set), (b) dispersal, in space and time, (c) germination, (d) establishment and

(e) further development of the immature plant. In conclusion, emphasis is placed on the following themes: (a) the kinds of work needed in future to prove or disprove that differentiation in the regeneration niche is the major explanation of the maintenance of species-richness in plant communities, (b) the relation of the present thesis to published ideas on the origin of phenological spread, (c) the relevance of the present thesis to the discussion on the presence of continua in vegetation, (d) the co-incidence of the present thesis and the emerging ideas of evolutionists about differentiation of angiosperm taxa and (e) the importance of regeneration studies for conservation.

2.2 Temperate region

A study was undertaken on the vertical root distributions of northern tree species in relation to successional status. Data were taken on root biomass, number, diameter and length by soil depth for northern tree species from 19 published papers, showing a total of 123 vertical root distributions. Species were classified into three tolerance classes based on successional status. A nonlinear function, $Y = 1 - \beta^d$, was fitted to the data for each excavation, where Y is the cumulative root fraction from the soil surface to depth d in centimetres. The regression coefficient, β, was considered as a measure of vertical root distribution and was used as a response variable to test whether significant differences in vertical root distributions existed among tolerance classes. Early successional or intolerant species demonstrated a significantly greater proportion of roots occurring deeper than did late successional or tolerant species.

Differences in vertical root distributions are considered to be related to the inherent genetic potential of early successional species for deep exploitation of a more homogeneous substrate, owing to either geologic deposition or nutrient and water redistribution following forest disturbance. Early successional species are also able to adapt to sites limiting in water and nutrients because of their ability to exploit larger volumes of soil (Gale and Grigal, 1987).

A study was made by Foster (1988) on species and stand responses to catastrophic wind in Central New England, USA. Relationships between age and average percentage damage of the important tree species in all the landscape after the 1938 storm showed it was actually a much less important form in the proportion of leaning trees and corresponding increase in uprooted trees. Subsequently, Baskin and Baskin (1988) investigated the germination ecophysiology of herbaceous plant species in a temperate region. They collected germination phenology data from 75 winter annuals, 49 summer annuals, 28 monocarpic perennials and 122 polycarpic perennials and undertook experimental investigations of dormancy breaking and germination requirements. Gerhardt and Hytteborn (1992) discussed natural dynamics and regeneration methods in tropical dry forests. Species which occur in these forests differ remarkably in many morphological and autoecological characteristics. Curtis and McIntosh (1991) reported an upland forest continuum in the Prairie–Forest Border Region of Wisconsin. Limited range filling is a problem for European tree species in a temperate environment. Svenning and Skov (2004) reported limited filling of the potential range in European tree species. The relative roles of environment and history in controlling large-scale species distributions are important. They use atlas data to examine the extent to which 55 tree species fill their climatically determined potential ranges in Europe. By quantifying range filling (R/P), they observe mean R/P = 38.3% (±30.3% SD). Many European tree species are naturally adapted outside their native ranges, thereby providing support for interpreting the many low R/P as a primary demonstration of dispersal limitation. R/P increases strongly with latitudinal range centroid and secondarily with hardiness and decreases weakly with longitudinal range centroid. Therefore, European tree species appear strongly controlled by the geographical dispersal constraints of post-glacial expansion as well as climate. They expect European tree species to show only limited tracking of near-future climate changes.

A study was made by Kamnesheidt (2000) on some autoecological characteristics of early to late successional tree species in Venezuela. The validity of strategy with respect to succession was tested on 21 tree and shrub species common in either unlogged or logged stands, respectively, in the Forest Reserve of Caparo, Venezuela, by taking data on morphological, physiological and population characteristics. Based on a preliminary abundance analysis, "early", "mid" and "late" successional species as well as "generalists" were distinguished. Early successional species, that is *Ochroma lagopus*, *Heliocarpus popayanensis* and *Cecropia peltata*, were similar in many autoecological aspects, for example monolayered leaf arrangement, orthotropic architectural models, no adaptive reiteration and clumped distribution, but they showed difference in gap association and distribution along a drainage gradient. They showed more diverse crown and leaf characteristics than early successional species. Late successional species established themselves only in small gaps and understorey and demonstrated a regular spatial pattern in undisturbed areas. All late successional species demonstrated architectural models with plagiotropic lateral axes and showed a multilayered leaf arrangement. Adaptive reiteration was a common feature of late successional species, which could be further subdivided into large, medium-sized and small trees, indicating the necessity of different light requirements at maturity. Crown illuminance is an important feature of the productive capacity of the tree species. The light compensation point (LCP) of an individual plant was strongly influenced by its crown illuminance. Large late successional species showed the widest range of LCP values, indicating the increasing light availability with increasing height in mature forest. On the basis of many autoecological characteristics, it was concluded (a) that there is in fact a continuum of species strategies with respect to succession even among early and mid-successional species and (b) that the latter group of species showed the widest breadth of autoecological traits, reflecting the heterogeneous environment in which they establish and mature.

2.3 Tropical rainforest

Very little is known about the autoecology of tropical tree species.

A review was made by West (1990) on the structure, function, environmental response, interactions with other ecosystem components and indicator values of microphytic crust on non-tilled, extensively managed land in arid to semiarid regions. A good knowledge on the characteristics and roles of these crusts in such ecosystems can help ensure that they are observed and accounted for monitoring changes, basic field ecology research and land management decisions. The soil surface is covered with microphytes. However, microphytes are often associated with other kinds of soil surface and near-surface features, variously called crusts, caps, films, veils, pans, mats, skins or scaly micro-horizons that can interact with microphytes. The kinds of organisms that can be involved in microphytic crusts are very diverse. These are commonly called thallophytic crusts and those parts of crusts whose individuals are visible only microscopically are known as microscopic crusts.

A comparative study of germination characteristics was undertaken in a local flora using a standardised procedure using seeds collected from a wide range of habitats in the Sheffield region. Measurements were taken on freshly collected seeds and on samples subjected to different techniques (Grime et al., 1981).

Fetcher et al. (1983) studied autoecological characteristics, including the effects of light regime on the growth, leaf morphology and water relations of seedlings of two species of tropical trees.

Some studies have been undertaken on germination ecology in tropical forests. In this respect, a comparative study was done by Washitani and Masuda (1990) on the germination characteristics of seeds from a moist tall grassland community. They studied the patterns of emergence in the warm temperate grasslands of Japan on seed germination of several dozen species, following various types of thermal pre-treatment. An enormous variety of responses on germination was found among the species.

Baskin and Baskin (1989) studied the seed germination ecophysiology of *Jeffersonia diphylla*, a perennial herb of mesic deciduous forests. Freshly matured seeds of the mesic deciduous woodland herb *Jeffersonia diphylla* (L.) Pers. (Berberidaceae) contained underdeveloped embryos (ca. 0.6 mm in length) and showed deep, simple morphophysiological dormancy (MPD).

Environments prevailing in a forest ecosystem influence the dormancy and germination of seeds of tree species. In this respect, Bouwmeester and Karssen (1993) studied the effect of environmental conditions on changes in dormancy and germination of seeds of *Sisymbrium officinale* (L.) Scop by burying tree seeds in soils exposed to different environments. Seeds were buried in soil in the field or in incubators; and at regular intervals, the germination of exhumed seeds was tested over a range of conditions. Seeds that were buried in the field showed clear seasonal changes in dormancy. Fluctuations in soil moisture and nitrate content were not required for the changes in dormancy. Temperature appeared to be the only factor regulating these changes. Dormancy was broken in periods of low temperatures and induced in periods of high temperatures. Fresh seeds and seeds buried for only a few months germinated best at high temperatures whereas seeds buried for a longer time germinated best at low temperatures. Light, nitrate and desiccation induced germination of exhumed seeds. As a consequence, these treatments extended germination over a much longer period of the year. Germination of exhumed seeds in Petri dishes at field temperature was reasonably described by a model based on the dual role of temperature, in the regulation of both dormancy and subsequent germination.

With respect to the succession of species, Reiners et al. (1994) in the Atlantic Zone of Costa Rica studied the effects of converting lowland tropical rainforest to pasture and of subsequent succession of pasture lands to secondary forest. Three replicate sites of each of four land-use types representing this disturbance-recovery sequence were sampled for studying changes in vegetation, pedological properties and potential nitrogen mineralisation and nitrification. The four land-use types included (a) primary forest, (b) actively grazed pasture (10–36 years old), (c) abandoned pasture (abandoned 4–10 years) (d) and secondary forest (abandoned 10–20 years). Conversion and succession showed significant effects on canopy cover, canopy height, species composition and species richness. It was assumed that the succession of secondary forests was proceeding toward a floristic composition like that of the primary forests. Significant changes in soil properties were associated with conversion of forest to pasture including:

1 A decrease in acidity and increase in some base exchange properties
2 An increase in bulk density and a concomitant decrease in porosity
3 Higher concentrations of NH_4^+
4 Lower concentrations of NO_3^-
5 Lower rates of N-mineralisation
6 In some cases, lower rates of nitrification.

Chemical changes involving cations associated with conversion from forest to pasture indicated increases in soil fertility under the pasture regimes, while changes associated with nitrogen indicated decreases in fertility.

It is necessary to study the dynamics and regeneration of tree species in tropical rain forest. Gerhardt and Hytteborn (1992) studied natural dynamics and regeneration methods in tropical dry forest species and demonstrated differences in many morphological and autoecological characteristics. Janzen (1967) discussed the autoecology, especially the reproductive. In this respect, dry forest ecosystems have been little studied as compared with tropical rain forests and savannas. Vetaas (1992) studied microsite effects on trees and shrubs in dry savannas. The physiognomy of dry savannas can be described as a combination of discontinuous woody perennials and a continuous grassland matrix. Interactions between these two components are very important features for the persistence of a savanna landscape. Studies have argued that small-scale facilitating interactions between woody perennials and the herbaceous understorey are also important. They put forward some of the evidence for microsite effects on trees and shrubs and attempted to integrate their interactions with the surrounding open grassland. Woody perennials modify the microclimate by intercepting solar radiation and rainfall.

Their root systems absorb nutrients horizontally and vertically, which are concentrated in the sub-soil from litter decomposition and root turnover. Legumes are abundant in dry savannas, showing symbiotic relationships with *Rhizobium* bacteria. This symbiosis increases depending on the availability of nitrogen in the soil. Isolated trees and shrubs possess feedback mechanisms in their interactions with other organisms and contribute to an uneven distribution of water and nutrients in dry savanna. This can influence the species composition and community diversity, thereby facilitating interaction between the woody and herbaceous components and competitive interaction on larger scales; these are complementary processes which together explain a dynamic coexistence.

Subsequently, Swaine et al. (1997) made a review on the dynamics of tree populations in tropical forest. In most of the forests studied, annual mortality is between 1 and 2% and is independent of size class in trees >10 cm dbh; mortality shows a negative correlation with growth rate and crown illumination. Growth rate is highly variable between individual trees, but shows a strong autocorrelation between successive measurements on the same tree. Differences in the rate of dynamic processes can be observed between some species at the same site. It is concluded that none of the studies discussed are of sufficient duration to draw any conclusions about the equilibrium or non-equilibrium of floristic composition.

Webb (2000) explored the phylogenetic structure of ecological communities for rain forest trees. Webb (2000) in *The American Naturalist* mentioned that species are partitioning habitat according to their autoecology in tropical rainforests.

2.4   Semiarid and arid lands

Very little information is available on the autoecology of tree species in arid and semiarid regions. Hellmuth (1968) studied the ecophysiological studies on plants in arid and semiarid regions in Western Australia with respect to the ecology of *Rhagodia baccata* (Labill.) Moq. in semiarid environmental conditions.

Illius and O'Connor (1999) made a critical comment on the relevance of non-equilibrium concepts to arid and semiarid grazing systems. These environments prevailing in the arid and semiarid regions impart extreme and unpredictable variability in rainfall; and they impart non-equilibrium dynamics by continually disrupting the light consumer–resource relations otherwise considered to pull a system towards equilibrium. The livestock grazing in drylands cause degradation and "desertification" through bad management practices leading to overstocking. An article recently published in *Ecological Applications* (Illius and O'Connor, 1999), however, argues that variability in arid and semiarid grazing systems is not the outcome of qualitatively different dynamical behaviour and that livestock may cause negative change through "normal" density-dependent relations. The authors state that these operate primarily in key resource areas and during drought periods, and "desertifications" are endemic in drylands.

Sullivan and Rohde (2002) reported the prevalence of non-equilibrium in arid and semiarid grazing systems.

Bradford and Lauenroth (2006) developed controls over the invasion of *Bromus tectorum* by investigating the effect of climate, soil, disturbance and seed availability. They utilised a soil water model to simulate seasonal soil water dynamics in multiple combinations of climatic and soil properties. In addition, they also utilised a gap dynamics model to simulate the impact of disturbance regime and seed availability on competition between *B. tectorum* and native plants. The results suggest that climate is very important, but soil properties have no significant effect on the probability of observing conditions suitable for *B. tectorum* establishment. The results suggest that frequent disturbance causes more *Bromus tectorum* in invaded areas and higher seed availability causes faster invasion.

2.5   Alpine region

Very little information is available on the autoecology of alpine forest. Kullman (2008) reported early postglacial appearance of tree species in northern Scandinavia. He made a review on megafossil evidence for the first postglacial records of different tree species in northern Scandinavia. *Betula pubescens* Coll. appeared at the Arctic coast of northern Norway by 16 900 years BP. In addition, *B. pubescens* (14 000 years BP), *Pinus sylvestris* (11 700 years BP) and *Picea abies* (11 000 years BP) existed on early ice-free mountain peaks (nunataks) at different locations in the Scandes during the Late Glacial. *Larix sibirica*, currently not native to Fennoscandia and several thermophilous broadleaved tree species were recorded in the earliest part of the Holocene. The conventional interpretation of pollen and macrofossil records from peat and sediment stratigraphies do not consider the occurrence of the species

mentioned above that early at these northern and high altitude sites. This very rapid arrival after the local deglaciation suggests that the traditional model of far distant glacial refugial areas for tree species has to be challenged. The results are more compatible with a situation involving scattered "cryptic" refugia quite close to the ice sheet margin at its full-glacial extension.

Silvicultural practices remarkably disturb the development of a natural ecosystem. In this respect, Franklin et al. (2002) investigated disturbances and structural development of natural forest ecosystems with silvicultural implications, using Douglas fir forests as an example. He asserts that forest managers should have a comprehensive scientific knowledge of natural stand development processes when designing silvicultural systems. This will help to integrate ecological and economic objectives, including a better appreciation of the nature of disturbance regimes and the biological legaciesthat they leave behind, such as live trees, snags and logs. They also argue that most conceptual forest development models do not incorporate current knowledge of: (a) the complexity of structures (including spatial patterns) and developmental processes; (b) the duration of developments in long-lived forests; (c) complex spatial patterns of stands that develop in later stages of seres; and particularly (d) the role of disturbances in creating structural legacies that become key elements of the post-disturbance stands. They elaborated the existing models for stand structural development using the natural stand development of the Douglas fir–western hemlock sere in the Pacific Northwest as a primary example. Most of the principles are broadly applicable while some processes are related to specific species. They discuss the use of principles from disturbance ecology and natural stand development to create silvicultural approaches that are more aligned with natural processes.

In the context of the literature reviewed, it may be stated that, though studies on ecology are very important as a guide for forest managers managing forest, very little information is available on this aspect. Therefore, concerted research input needs to be directed towards this topic.

Bibliography

Baskin C.C., Baskin J.M. **1988**. Germination ecophysiology of herbaceous plant species in a temperate region. *American Journal of Botany* **75**(2):286–305.

Baskin J.M., Baskin C.C. **1989**. Seed germination ecophysiology of *Jeffersonia diphylla*, a perennial herb of mesic deciduous forests. *American Journal of Botany* **76**(7): 1073–1080.

Bazzaz F.A., Pickett S.T.A. **1980**. Physiological ecology of tropical succession: A comparative review. *Annual Review of Ecology and Systematics* **11**:287–310.

Bonsen K.J.M. **1996**. Architecture, growth dynamics and autoecology of the Sycamore (*Acer pseudoplatanus* L.). *Arboricultural Journal: The International Journal of Urban Forestry Volume* **20**(3):339–354.

Bouwmeester J.J., Karssen C.M. **1993**. Annual changes in dormancy and germination in seeds of *Sisymbrium officinale* (L.) Scop. *New Phytologist* **124**(1):179–191.

Bradford J.B., Lauenroth W.K. **2006**. Controls over invasion of *Bromus tectorum*: The importance of climate, soil, disturbance and seed availability. *Journal of Vegetation Science* **1**(6):693–704.

Curtis J.T., McIntosh R.P. **1991**. An upland forest continuum in the prairie–forest border region of Wisconsin. *Ecology* **32**(3):476–496.

Denslow J.S., Hartshorn, G.S. **1994**. Tree-fall gap environment and forest dynamic process. In: McDalr L.A., Bawa K, Hespenhede. H.A. (Eds.). *La Selva – Ecology and Natural History of a Neotropical Rain Forest*. University of Chicago Press, Chicago. Pp. 120–127.

Donovan D.G., Puri R.K. **2004**. Learning from traditional knowledge of non-timber forest products: Penan Benalui and the autecology of *Aquilaria* in Indonesian Borneo. *Ecology and Society* **9**(3):1–23.

Fetcher N., Strain B.R., Oberbauer S.F. **1983**. Effects of light regime on the growth, leaf morphology and water relations of seedlings of two species of tropical trees. *Oecologia* **58**:314–319.

Fischer R.A., Turner N.C. **1978**. Plant productivity in the arid and semiarid zones. *Annual Review of Plant Physiology* **29**:277–317.

Foster D.R. **1988**. Species and stand response to catastrophic wind in Central New England, USA. *Journal of Ecology* **76**(1):135–151.

Foster S.A. **1986**. On the adaptive value of large seeds for tropical moist forest trees: a review and synthesis. *The Botanical Review* **52**(3):260–299.

Franklin J.F., Spies T.A., Pelt R.V., et al **2002**. Disturbances and structural development of natural forest ecosystems with silvicultural implications, using Douglas fir forests as an example. *Forest Ecology and Management* **155**(1/3): 399–423.

Gale M.R., Grigal D.F. **1987**. Vertical root distributions of northern tree species in relation to successional status. *Canadian Journal of Forest Research* **17**(8):829–834.

Gerhardt K., Hytteborn H. **1992**. Natural dynamics and regeneration methods in tropical dry forests: An introduction. *Journal of Vegetation Science* **3**(3):361–364.

Grime J.P., Mason G., Curtis A.V., Rodman J., Band S.R., Mowforth M.A.G., Neal A.M., Shaw S. **1981**. A comparative study of germination characteristics in a local flora. *Journal of Ecology* **69**(3):1017–1059.

Grubb P.J. **2008**. The maintenance of species-richness in plant communities: the importance of the regeneration niche. *Biological Reviews* **52**(1):107–145.

Hellmuth E.O. **1968**. Eco-physiological studies on plants in arid and semi-arid regions in Western Australia: I. Autecology of *Rhagodia baccata* (Labill.) Moq. *Journal of Ecology* **56**(2):319–344.

Illius A.W., O'Connor T.G. **1999**. On the relevance of non-equilibrium concepts to arid and semi-arid grazing systems. *Ecological Applications* **9**(3):798–813.

Janzen D.H. **1967**. Synchronization of sexual reproduction of trees within the dry season in Central America. *Evolution* **21**(3):620–637.

Kamnesheidt L. **2000**. Some autecological characteristics of early to late successional tree species in Venezuela. *Acta Oecologica* **21**(1):37–48.

Kullman L. **2008**. Early postglacial appearance of tree species in northern Scandinavia: review and perspective. *Quaternary Science* **27**(27/28):2467–2472.

Oldeman R.A.A., van Dijk J. **1991**. Diagnosis of the temperament of tropical rainforest trees. In: *Rainforest Regeneration and Management*. A. Gomez-Pompa et al. (Eds.) Man and the biosphere series, Vol 6, UNESCO and Parthenon Publishing, Paris. Pp. 21–65.

Reiners W.A., Bouwman A.F., Parsons W.F.J., Keller M. **1994**. Tropical rain forest conversion to pasture: Changes in vegetation and soil properties. *Ecological Applications* **4**(2):363–377.

Schulze E.D. **1982**. Plant Life Forms and Their Carbon, Water and Nutrient Relations. *In*: Physiological plant ecology II. P.S. Nobel, C.B. Osmond, H. Ziegler (eds). *Encyclopedia of Plant Physiology, Volume 12/B*. Springer, Berlin. Pp. 615–676.

Silvertown J., Franco M., Pisanty I., Mendoza A. **1993**. Comparative plant demography–relative importance of life-cycle components to the finite rate of increase in woody and herbaceous perennials. *Journal of Ecology* **81**(3):465–476.

Sullivan S., Rohde R. **2002**. On non-equilibrium in arid and semi-arid grazing systems. *Journal of Biogeography* **29**(12):1595–1618.

Svenning J.C., Skov F. **2004**. Limited filling of the potential range in European tree species. *Ecology Letters* **7**(7):565–573.

Swaine M.D., Lieberman D., Putz F.E. **1997**. The dynamics of tree populations in tropical forest: a review. *Journal of Tropical Ecology* **3**(4):359–366.

Thompson K., Grime J.P. **1983**. A comparative study of germination responses to diurnally-fluctuating temperatures. *Journal of Applied Ecology* **20**(1):141–156.

Vetaas O.R. **1992**. Micro-site effects of trees and shrubs in dry savannas. *Journal of Vegetation Science* **3**(3):337–344.

Washitani I., Masuda M. **1990**. A comparative study of the germination characteristics of seeds from a moist tall grassland community. *Functional Ecology* **4**(4):543–557.

Webb C.O. **2000**. Exploring the phylogenetic structure of ecological communities: An example for rain forest trees. *The American Naturalist* **156**(2):145–155.

West N.E. **1990**. Structure and function of microphytic soil crusts in wildland ecosystems of arid to semi-arid regions. *Advances in Ecological Research* **20**:180–223.

CHAPTER 3

Vegetation and biodiversity

3.1 Introduction

Knowledge of the biodiversity of vegetation present in semiarid, tropical and temperate forests is essential to understand the growth of trees associated in the vegetation. A few examples are cited herein.

The components of vegetation play an important role in the forest ecosystem. We mention here an example of vegetation in north-eastern Mexico, where we have studied the vegetation of the region. The present study was undertaken in the Tamaulipan thornscrub of Nuevo Leon, north-eastern region of Mexico. Various studies have been undertaken on the biodiversity of vegetation and its utilisation in Mexico.

A study was been made by González-Rodríguez et al. (2010) on the structural composition at three sites in Nuevo Leon, Mexico. During 2010, the plant diversity was estimated by the Shannon–Wiener index and the similarity between sites was calculated using the Jackard index. A total of 1741 individual plants belonging to 20 families were registered. Fabaceae had the highest number of species (10) followed by Euphorbiaceae (4) and Rhamnaceae (4), Rutaceae (3) and Cactaceae (2). The other 15 families were represented by only one species. The Shannon index showed that there were no statistical differences in biodiversity between sites; however, the Jackard index showed similarity among species between site one and site two. The most frequent species in the three sampling sites were: *Acacia rigidula* (255), followed by *Viguera stenoloba* (171), *Havardia pallens* (167), *Karwinskia humboldtiana* (132), *Forestiera angustifolia* and *Castela texana* (125). This group represented about 56% of the total. Less frequent species were: *Condalia spathulata* (7), *Ebenopsis ebano* (7), *Condalia hookeri* (6), *Wedelia acapulcensis* (5), *Cordia boissieri* (4), *Acacia farnesiana* and *Yucca treculeana* with (3), *Helietta parvifolia* (2) and *Croton torreyanus* (1).

In another study, carried out by Ramírez-Lozano et al. (2010) determined the relative abundance, relative dominance and relative frequency using dasometric parameters such as height and crown diameter. Plant diversity was estimated by the Shannon–Wiener index and similarity between sites was calculated using the Jackard index. A total of 13 710 individual plants belonging to 28 families were registered. Fabaceae had the highest number of species (10) followed by Fagaceae (4), Rutaceae (4), Euphorbiaceae (3), Olaceae (3), Cupressaceae

Autoecology and Ecophysiology of Woody Shrubs and Trees: Concepts and Applications, First Edition.
Edited by Ratikanta Maiti, Humbero Gonzalez Rodriguez and Natalya Sergeevna Ivanova.
© 2016 John Wiley & Sons, Ltd. Published 2016 by John Wiley & Sons, Ltd.

(3), Rhamnaceae (2) and Verbenaceae (2). Twenty families included only one species.

Maiti and Gonzalez-Rodriguez (2011) reported that *Agave lecheguilla* is a valuable fiber plant grown naturally in arid lands of Mexico for the extraction of fibre. The paper gives a short account of the plant, its distribution, methods of fibre extraction by the arid land farmers and its processing methods. It is a good source of economy for the arid land farmers.

Estrada-Castillón et al. (2012) published a book on flora and phytogeography of Cumbres at the Monterrey National Park, Nuevo Leon, Mexico. The authors describe that the north-eastern region of Mexico is characterised by climatic and landscape heterogeneity; its extensive plains, high mountains and scattered hills harbour an intricate and diverse mosaic of vegetation, characterised by a rich plant diversity and life forms. The heterogeneous physiography among different regions clearly show distinctive climatic zones, especially evident in the State of Nuevo Leon, where three physiographic provinces are recognised: Gran Llanura de Norteamerica (North American High Plains), Llanura Costera del Golfo Norte (North Coastal Gulf Plain) and Sierra Madre Oriental (INEGI, 2001). These have contrasting particularities of soils, vegetation types and plant diversity. The orthographic, edaphic and climatic factors of the physiographic provinces show close relationships between the vegetation-types, flora and plant endemism.

Domínguez Gómez et al. (2013) studied the plant structural diversity of the Tamaulipan thornscrub during the dry (summer) and wet (fall) seasons of 2009. The composition and structure of vegetation were studied at three sites

(S) in the State of Nuevo Leon, Mexico: Los Ramones (S1), China (S2) and Linares (S3). The main vegetation type was constituted by Tamaulipan thornscrub (TT). Ten plots were randomly established (10×10 m) at each site, where ecological indicators [abundance, dominance, frequency and importance value (IV)] were estimated. The species diversity was calculated by using the Shannon–Weiner Index. The similarity between sites was determined with the Jaccard Index. The total number of individuals during the dry and wet seasons was 1251 and 2457, respectively. In general, 57 species were found, of which 34 are common in both seasons. The main families were Fabaceae and Cactaceae. The genera with highest number of species were *Acacia* (5), *Croton* (3), *Echinocereus* (3) and *Opuntia* (2). The Shannon–Wiener Index showed statistical differences between sites and seasons. Sampling sites were determined as medium diversity, while the Jaccard index showed the highest similarity for Linares and the lowest for China in both seasons. The average cover for the three sites was 1061 m^2 and 1847 m^2 during the dry and wet seasons, respectively. S3 revealed the highest coverage value (1722.5 m^2). *Prosopis laevigata* showed the highest IV in both seasons in the S1 and S2 sites. However, in S3, *Lantana macropoda* and *Turnera diffusa* had the highest IV in both seasons. In general, the diversity of species between sites and seasons was homogeneous.

3.2 Climate

The climatic condition prevailing in a country contributes to the growth of vegetation. In this respect a study in semiarid region was carried out at the experimental station of Facultad de Ciencias Forestales, Universidad Autonoma de Nuevo Leon, located in the municipality of Linares (24° 47′N, 99° 32′W), at an elevation of 350 m. The climate is subtropical or semi-arid with a warm summer. The monthly mean air temperature varies from 14.7 °C in January to 23.7 °C in August, although during summer the temperature goes up to 45.7 °C. Average annual precipitation is around 805 mm with a bimodal distribution. The dominant type of vegetation is the Tamaulipan thornscrub or subtropical thornscrub woodland (INEGI, 2001). The dominant soil is deep, dark grey, lime-clay, vertisol with montmorrillonite, which shrinks and swells remarkably in response to a change in moisture content.

3.3 Hydrology

The growth and development of a plant is strongly affected by the availability of soil moisture, nutrients and solar radiation. Rainfall interception is a major component of the hydrological cycle that has not been extensively studied or modelled in plant species of semiarid and sub-tropical environments. Reily and Johnson (1982) studied the effects of altered hydrologic regime on tree growth along the Missouri River in North Dakota, following completion of a dam. Alterations in seasonal streamflow patterns, near elimination of over-bank flooding and apparent lowering of the water table during the early growing season following completion of the dam in 1953 and led to a significant decline in the postdam growth of *Ulmus americana*, *Fraxinus pennsylvanica*, *Acer negundo* and *Quercus macrocarpa*. Trees on terraces at the edge of the floodplain that received concentrated runoff from upland ravines (e.g. *Q. macrocarpa*) and those with deep root systems (e.g. *P. deltoides*) on low terraces close to the water table were least affected. The most pronounced change in tree growth was observed on high terraces that received little upland runoff (e.g. *U. americana*, *A. negundo*). Multiple regression analysis for *P. deltoides* growth demonstrated a distinct change from a correlation with spring streamflow in the predam period to correlation with rainfall parameters in the postdam period. Growth of *P. deltoides* and *Q. macrocarpa* on reference sites unaffected by damming of the Missouri River increased significantly in the postdam period.

Mitsch and Rust (1984) studied tree growth responses to flooding in a bottomland forest in the north-eastern. Water levels in the forest and streamflow data from the river were matched during a flood in the spring of 1979 to determine flooding duration from available stream discharge data collected in 1917–1978. Stream flow was sufficient to flood the forest for at least 10 days during 41 of the 61 years; the bottom land was not flooded at all in 11 of the years. Average annual tree ring growth (average ± standard deviation) for the period was 4.52 ± 1.89 mm/year for *Quercus bicolor*, 6.77 ± 2.3 mm/year for *Ulmus americana* and 6.26 ± 1.32 mm/year for *Fraxinus pennsylvanica*. The species revealed similar growth cycles in the 1940s and 1950s. Tree growth variables (radial growth and basal area growth) and flooding and streamflow variables were poorly correlated with significant relationships found only between annual growth of *Q. bicolor* and *U. americana* and average annual daily streamflow.

Putz and Chan (1986) investigated tree growth, dynamics and productivity in a mature mangrove forest in Malaysia. Growth of selected *Rhizophora apiculate* (Rhizophoraceae) trees was determined from 1920 to 1981 in a 0.16 ha plot of protected forest in the Matang Mangroves. Starting in 1950, the sample was increased to include monitoring the growth of all the trees more than 10 cm dbh (diameter at 1.3 m or above prop roots). Total above-ground dry weight (biomass) of the forest was estimated using stand tables and a regression equation of biomass on dbh calculated for destructively sampled *R. apiculate* trees from elsewhere in the Matang Mangroves. Net primary productivity (1950–1981) was calculated from estimated biomass increments and published litter-fall rates. *Rhizophora apiculate* has maintained its dominance of the plot since 1920 but *Bruguiera gymnorrhiza* (Rhizophoraceae) and several other more shade-tolerant species have increased in abundance. Mean mortality rate (1950–1981) for trees more than 10 cm dbh was 3.0% per year with a range of 1.3–5.4% per year. When trees fell over and hit other trees, the damaged trees usually died within 10 years. A major cause of mortality appeared to be sapwood-eating termites. It is suggested that *Rhizophora* spp. trees greater than 50 cm dbh and mangrove forests with total above-ground biomass exceeding 300 t/ha would develop in other areas outside the region affected by hurricanes if the forest was protected from human disturbance.

Xu et al. (2000) assessed the responses of surface hydrology and early loblolly pine (*Pinus taeda*) growth to soil disturbance and site preparation in a lower coastal plain wetland after harvesting of three 19-ha, 22 year old loblolly pine plantations in an Atlantic coastal wetland in the south-eastern USA. Overall surface hydrology and tree responses to the two harvest treatments and three site preparation levels (no preparation, bedding or mole-ploughing + bedding) were estimated by monitoring the water table dynamics and tree growth on a 20 × 20 m grid across the sites. The result revealed that surface soil disturbances affected the hydroperiod, by showing a large difference in water table elevation during the growing season between the wet-weather harvested and the dry-weather harvested sites. Bedding lowered the overall surface water table initially to a large extent, but this effect decreased rapidly during the first two years after stand establishment. Surface deformation, such as deep rutting or churning, appeared too.

Högberg et al. (2006) investigated tree growth and soil acidification in response to 30 years of experimental nitrogen loading on boreal forest relations among nitrogen load, soil acidification and forest growth was based on short-term (<15 years) experiments, or on surveys across gradients of N deposition that may also include variations in edaphic conditions and other pollutants. Tree growth initially showed positive response to all N treatments, but the longer term response was highly rate-dependent with no gain in N = 3. In contrast, the organic mor-layer (forest floor) in the N-treated plots had similar amounts per hectare of exchangeable base cations as in the tree growth, which was not correlated with the soil Ca/Al ratio (a suggested predictor of effects of soil acidity on tree growth). A boron deficiency occurred on N-treated plots, but was corrected at an early stage. Extractable NH_4^+ and NO_3^- were high in mor and mineral soils of ongoing N treatments, while NH_4^+ was elevated in the mor only in N3 plots. Ten years after termination of N addition in the N3 treatment, the pH had increased significantly in the mineral soil. There were also tendencies of higher soil base status and concentrations of base cations in the foliage. The results suggested the recovery of soil chemical properties, notably pH, may be quicker than predicted after removal of the N load. The long-term experiment demonstrated the fundamental importance of the rate of N application relative to the total amount of N applied, in particular with regard to tree growth and C sequestration. Hence, experiments adding high doses of N over short periods do not mimic the long-term effects of N deposition at lower rates.

Severe drought in moist tropical forests increases large carbon emissions by increasing forest flammability and tree mortality, thereby affecting tree growth. The frequency and severity of drought in the tropics may increase through stronger El Niño Southern Oscillation (ENSO) episodes, global warming and rainfall inhibition by land use change. Nepstad et al. (2004) discussed Amazon drought and its implications for forest flammability and tree growth. They presented a simple geographic information system in a soil water balance model, called RisQue (Risco de Queimada = Fire Risk) for the Amazon basin that they used to conduct an analysis of these patterns for 1996–2001. RisQue features a map of maximum plant-available soil water (PAW_{max}) developed using 1565 soil

texture profiles and empirical relationships between soil texture and critical soil water parameters. PAW is reduced by monthly evapotranspiration (ET) fields estimated using the Penman–Monteith equation and satellite-derived radiation inputs and recharged by monthly rain fields estimated from 266 meteorological stations. Modelled PAW to 10 m depth (PAW_{10m}) was similar to field measurements made in two Amazon forests. During the severe drought of 2001, PAW_{10m} fell to below 25% of PAW_{max} in 31% of the region's forests and fell below 50% PAW_{max} in half of the forests. Hence, approximately one-third of Amazon forests were susceptible to fire during the 2001 ENSO period. Field measurements also suggest that the ENSO drought of 2001 reduced carbon storage by approximately 0.2 Pg relative to years without severe soil moisture deficits. RisQue is sensitive to spin-up time, rooting depth and errors in ET estimates. Improvements in our ability to accurately model the soil moisture content of Amazon forests will depend upon a better understanding of forest rooting.

A study was made on rainfall interception parameters and to test the applicability of the Gash analytical model adjusted to four leguminous shrub species of Northeastern Mexico (*Pithecellobium pallens, P. ebano, Acacia rigidula* and *A. berlandieri*). The results showed that interception loss was statistically different among species and the differences were associated with plant biomass components and the structure of the shrub-top. Gash's analytical model for rainfall interception adequately predicted the total interception loss, since the maximum deviations between derived and modelled interception loss were 5%.

Therefore, the model is recommended to estimate rainfall interception for the plant species studied (Návar and Bryan, 1994).

Bibliography

Domínguez Gómez T.G., González Rodríguez H., Ramírez Lozano R.G., Estrada Castillón A.E., Cantú Silva I., Gómez Meza M.V., Villarreal Quintanilla J.A., Alvarado, M.S., Alanís Flores G. 2013. Diversidad estructural del matorral espinoso tamaulipeco durante las épocas seca y húmeda. *Revista Mexicana de Ciencias Forestales* **4**(17):106–123.

Estrada-Castillón E., Villarreal-Quintanilla J.A., Salinas-Rodríguez M.M., Jiménez-Pérez J., García-Aranda M.A. 2012. *Flora and Phytogeography of Cumbres de Monterrey National Park*, Nuevo Leon, Mexico.

González-Rodríguez H., Ramírez-Lozano R.G., Cantú-Silva I., Gómez-Meza M.V., Uvalle-Sauceda J.I. 2010. Composición y estructura de la vegetación en tres sitios del estado de Nuevo León, México. *Polibotánica* **29**:91–106.

Hanson P.J., Todd Jr. D.E., Amthor J.S. 2001. A six-year study of sapling and large-tree growth and mortality responses to natural and induced variability in precipitation and throughfall. *Tree Physiology* **21**(6):345–358.

Högberg P., Fan H., Quist M., Binkley D., Tamm C.O. 2006. Tree growth and soil acidification in response to 30 years of experimental nitrogen loading on boreal forest. *Global Change Biology* **12**(3):489–499.

INEGI 2001. *Cartas topográficas y edafología de Linares 1:50,000G14C59, Segunda Edición*. Instituto Nacional de Estadística y Geografía, Mexico City.

Maiti R.K., Gonzalez-Rodriguez H. 2011. Lechuguilla (*Agave lecheguilla*); an important commercial fiber plant and a source of income to the arid land farmers of Mexico. *International Journal of Bio-resource and Stress Management* **2**(1):104–110.

Mitsch W.J., Rust W.G. 1984. Tree growth responses to flooding in a bottomland forest in Northeastern Illinois. *Forest Science* **30**(2):499–510.

Návar J., Bryan R.B. 1994. Fitting the analytical model of rainfall interception of Gash to individual shrubs of semi-arid vegetation in northeastern Mexico. *Agricultural and Forest Meteorology* **68**(3/4):133–143.

Nepstad D., Lefebvre P., da Silva U.L., Tomasella J., Schlesinger P., Solórzano L., Moutinho P., Ray D., Benito J.G. 2004. Amazon drought and its implications for forest flammability and tree growth: a basin-wide analysis. *Global Change Biology* **10**(5):704–717.

Paulsen J., Weber U.M., Körner C. 2000. Tree growth near treeline: abrupt or gradual reduction with altitude? *Arctic, Antarctic and Alpine Research* **32**(1):14–20.

Putz F.E., Chan H.T. 1986. Tree growth, dynamics, and productivity in a mature mangrove forest in Malaysia. *Forest Ecology and Management* **17**(2/3):211–230.

Ramírez-Lozano R., Domínguez-Gómez T.G., González-Rodríguez H., Cantú-Silva I., Gómez Meza M.V., Sarquís-Ramírez J.I., Jurado E. 2013. Composición y diversidad de la vegetación en cuatro sitios del noreste de México. *Madera y Bosques* **19**(2):59–72.

Reily P.W., Johnson J. 1982. The effects of altered hydrologic regime on tree growth along the Missouri River in North Dakota. *Canadian Journal of Botany* **60**(11):2410–2423.

Xu Y.J., Burger J.A., Aust W.M., Patterson S.C. 2000. Responses of surface hydrology and early loblolly pine growth to soil disturbance and site preparation in a lower coastal plain wetland. *New Zealand Journal of Forestry Science* **30**(1/2):250–265.

CHAPTER 4

Case study: A trip to regions of biodiversity and rainforest in Riviera Maya

4.1 Introduction

We made a joyous and educational trip from 26 June to 6 July to Riviera Maya, Tulum, Sankan Reserve forest, Punta de Alan, Qintana Roo and Chichen Itza, Yucatan. These are southern regions of Mexico enjoying natural beauty, biodiversity of plants and wild animals, with remnants of the old Maya culture.

We got down from our two-hour flight from Monterrey to Cancun and started a long trip enjoying a great diversity of trees and shrubs in tropical rainforests on both sides of our road. We peeped from our travel van and watched keenly the beauty of rain forests for more than one hour on the way to Riviera Maya, the ancient region of Maya culture and arrived at our hotel.

Mexico is considered as a country of megadiversity for its richness in flora and fauna. Though we did not have an opportunity to take quantitative data we could visually observe the diversity in flora and fauna in the tropical rainforests of mangroves and other trees on both sides of the road from our vehicle. The tropical rainforests on both sides of the road showed a great diversity in shrubs and trees varying in height, crown and leaf canopy architecture and the leaves varying in size, shape and pigment intensity.

This huge diversity of woody species varied greatly in height and crown architecture, each species in its niche growing luxuriantly in harmony and capturing solar radiation for photosynthesis. Each grew in harmony with its neighborhood without any competition in the above-ground horizons and behaving the same in root system profile at different depths to absorb nutrients and water for its growth. Looking at the forest one can observe that species have attained different heights in search of solar radiation, thereby there is no competition among the species for getting solar radiation. Most of the species in semiarid Mexico have open canopy leaves with good efficiency in the capture of solar radiation for photosynthesis, generally highly branched, but occupying its own space without competition with its neighbour. Different species in the forest ecosystem varied in height thereby giving an opportunity for each species to grow and capture solar radiation to increase its height. In

Autoecology and Ecophysiology of Woody Shrubs and Trees: Concepts and Applications, First Edition.
Edited by Ratikanta Maiti, Humbero Gonzalez Rodriguez and Natalya Sergeevna Ivanova.
© 2016 John Wiley & Sons, Ltd. Published 2016 by John Wiley & Sons, Ltd.

this ecosystem the climbers climbed and occupied the upper part of crown, growing happily on the trees without affecting much the growth of its host tree. This mode of adaptation in a forest ecosystem occurs both in the rainforests and in the thornscrub of semiarid regions of Mexico.

The region, Riviera Maya, is situated stretching along Mexico's southern coastline on the bank of the turquoise waters of the Caribbean Maya Riviera; along 120 km of Caribbean soft sandy beaches with occasionally turbulent seas. This region of high biodiversity possesses green rainforests, ancient wells, several archaeological sites, natural undisturbed reserve forests, beautiful parks, pictures, native villages and many other tourist attractions. This region is rich in history and old traditions and experiences a fascinating adventure associated with the enchanting natural surroundings and charming birds. All these natural surroundings represent a fundamental part of life in the Maya civilisation. The forests and the ecosystem are well conserved depicting this natural life. Tourists can easily enjoy the beauty of the silent evergreen mangrove forests. The ecological and biological importance of this area is that it possesses two vitally important ecosystems: the mangrove swamps and the rainforest. The mangrove swamps are made up of groups of trees which, owing to their physiological characteristics, are able to survive and reach maturity on flooded land. These swamps are associated with coastal areas, streams, rivers and lagoons. Three types of Caribbean species of mangrove are found in the region: the red mangrove, black mangrove and white mangrove. Softwood mangrove areas are vital resources for their high productivity of

organic matter. These mangrove swamps serve as a breeding ground for a large diversity of fish, crustaceans and larger land-dwelling wild animals as well as marine species. The mangroves provide protection against storms and hurricanes.

The other outstanding ecosystem is the rainforest, which is characterised by the abundance and leafiness of tall trees, palms and exotic flowers. These act as shelter for many animals, such as deer, badgers, anteaters, grisons, squirrels, iguanas, eagles, paca and possums.

Mangrove plants are very tall, with thin stems, the leaf canopy is open, offering great opportunity in the capture of solar radiation for photosynthesis, leaves are lanceolate to broad and thick deep green with a shining waxy coating to prevent loss of water by transpiration. The stems are unbranched at the lower part but with a rudimentary lateral branch at the upper part of stems bearing leaves which help in the photosynthesis required for the elongation and growth of the stem. Solar radiation can penetrate vertically through species so that the leaves of the lateral branches receive light for photosynthesis and a part of solar radiation reaches the ground. Plants are closely spaced and, although having thin stems, neighbouring plants give support against bending, thereby maintaining a harmony with their neighbours. The topmost part of the plant is branched, giving an irregular contour. It seems the individuals have good neighbourly relations and share the solar radiation and give support against bending laterally. With the morning breeze, they swing as if dancing with neighbouring plants. The most important characteristic of mangrove plants in marshy lands is that the lower part of the stem of each plant is supported by several strong, special types of roots called rhizophores hanging from the branches and finally fixing their pegs in the ground Fleshy porous rhizophores arising from upper branches grow downwards, fix the tips in the soil and thereby support the upper branches. Numerous rhizophores are fixed on the ground soil so extensively that it is not possible to enter a forest of marshy mangroves, as if they are protecting their kingdom from trespassers. All the leaves starting from the upper branches to the lower ones receive light, owing to the horizontal orientation of leaves for photosynthesis and the continuous elongation and growth of the plants

It is very interesting to know the mechanism of adaptation of mangrove plants which grow luxuriantly very tall in highly saline soils. We did not observe any sign of responses to salinity stress in their growth habit. Roots help only to fix the plants, but do not absorb water. The huge number of fleshy rhizophores possess numerous rhizophores through which they absorb water vapour and undergo respiratory functions. Besides, the leaves have xerophytic characteristics, with thick leaves having a waxy shining coat which avoids the loss of water by transpiration. In addition, in our study it was observed that mangrove leaves possess compact palisade cells to avoid the loss of water through stomata.

Mangroves are able to filter out some salt at root level and some through its leaves; mangrove is also able to tolerate a much higher internal level of salinity. Its sap may be up to 10% more salty than seawater. They are also able to "breathe", absorbing oxygen through pore-like lenticels on their above-ground roots. This allows them to thrive in anaerobic soil, where there is a lack of oxygen. Their aerial roots which, while above-ground, spend part of their time submerged by high tide not only absorb oxygen, but are also able to transport it throughout the rest of the tree. Even though the mangrove can tolerate salinity, it also relies on freshwater to flush excess salt out of its system. Without freshwater flushing, the trees would die. Rain provides the freshwater needed for their survival.

I observed that light penetrates vertically through open canopy leaves and reaches the ground, thereby giving an opportunity for the leaves in the photosynthesis process. The flowers are large-sized and brilliant white in colour. It is observed that flowers start blooming at night (about 10 p.m.) up to the next day in the afternoon but are found to have fallen by the evening time. Thereby, the plant has to complete fertilisation during that period. The lower part of the stem does not bear lateral branches but the upper part bears lateral branches offering an opportunity for photosynthesis for plant growth. Near the seashores, the mangroves produce numerous highly branched plants. The neighbouring plants with their lateral branched rhizophores are fixed on the soil there by checking soil erosions against inundation of tidal waves.

The Blue Bay Grand Esmeralda has adopted the following strategy to conserve these ecosystems:

- They regulate the drainage of water and also conserve the water system, acting as a biological corridor for species of fauna.
- They use biodegradable chemical products in the different areas of the hotel, thereby avoiding damage to the environment.

- Their suppliers fulfil environmental regulations on the products they supply.
- They use a biological system at the sewage treatment plant and use the water from this in the watering system in order to make the most of this resource.
- Ecological walks for children are organised, thereby promoting the protection and importance of nature.
- In collaboration with a resident biologist they develop activities and methodologies to protect the environment.

We took a trip to Tulum, to the reserve forests of Sian Ka'an, finally reaching the sea beach of the Caribbean sea at Punta de Alan.

In order to reach the sea coast at Punta de Alan, we travelled in a tourist van on a corrugated and zigzag narrow stony road passing through the dense forests of the Reserve Biosphere of Sian Ka'an. This is rich in vegetation with a great diversity of flora and fauna. There exists large variability in crown and leaf canopy structure. Most of these trees have an open leaf canopy with highly branched thin stems, some with a closed canopy possess a stout main trunk with thick primary stems. Each species has a characteristic crown occupying its niche without disturbing its neighbours. Some are very tall, some medium in height and some short, thereby occupying all the available spaces and sharing the solar radiation for photosynthesis. Climbers grow over the crown, thereby competing with the host for sharing sunlight. Several species of palms are dominant near the roadsides. The palm leaves are used for thatching local houses and last for more than 10 years. No electricity supply is available in the region. Artificial gas light or solar light is used for local electricity. On the way our guide showed us two damaged Mayan houses, depicting a remnant of Maya culture. The guide showed us the gate of a Mayan house which was very low. Once one tourist entered but it was very difficult for him to come out. It took us more than 2.5 h to reach our destination at Punta de Alan, the sea beach on the Caribbean sea.

All of us were tired and very hungry. They served us a light snack and soft drinks. We got ready for a journey over the sea in a small boat. Each one of us had to wear a life jacket. Six persons of our family boarded a big boat with a driver who had more than 12 years of experience. The external Caribbean sea with bluish turbulent water welcomed us for a tour of 3 h. The driver told us the sea was less turbulent that day compared to another day. We were really lucky. More than 12 boats were full of passengers. We were thrilled and not afraid at all at the old age of 82. The driver was friendly and explained for us the beauty of the marine nature with its fauna and submerged flora; the marine algae which are the source of food for fish. He planned the tour so that we can see different fauna in the sea.

The sea seemed to us shallow because we could see blue green algae from the boat. My son Dr. Sandip and my grandson started taking photographs of the flying birds with a camera and then snapped the deep water algae and fish with an aquatic camera. It was a thrilling journey over the less turbulent sea, ups and downs with a splash of seawater on our face and clothes. The guide had plans to take us to different zones of marine animals and birds.

- Tortoise Zones: First he took us to zones of tortoise, very big in size with coloured specks on their bony cover, some more than 20 years old. They were reigning over in their kingdom floating over the sea and diving deep, raising their mouth up to breathe. It was really amazing journey in the tortoise region. The driver told us it was the time for catching their food, fish for which they dive up and down the sea, catching their prey.

- Paradise of fish fauna, coral and marine algae. The boat pilot asked my son to push his aquatic camera deep down to take photographs of varieties of fish fauna, coral and an up-growing bush of marine algae which we could see from our boat. Later, I observed it looked like a paradise, a kingdom of several varieties of fish of different colours and different sizes, red, black, some with beautiful ornamentation of white and black or red strips on the body scales swimming around and harbouring among algal flora and the coral on which they fed. They lived together in a symbiosis, forming an ecosystem of fauna and algal flora, with beautiful ornamentation. They were moving around in the bush of marine algae where they feed. With respect to the ecosystem algal flora it looks like a flower garden with algae of different colours, deep yellow, green, blue green, red, looking like globes with a bunch of ribbon-like filament, some looking like a bunch of red fruits. The algae of different colours attracted the fish as if inviting them to feed there. Both fish fauna and algal flora lived in a harmony. Probably they belonged to Pheophyta, Rodophyata, Cyanophyta containing yellow xanthophyll, carotenoids. Corals are a group of invertebrates are of different colours, red, white, black and yellow, looking like a garden under the sea. The ecosystem of algae looked like terrestrial forests, each algal population attaining a different height to absorb solar radiation reaching under the sea for their photosynthesis without affecting its neighbours. I had the intention to collect these algae for identification; but this is not allowed by the authorities. We should respect and conserve this valuable marine biodiversity. These algae containing different pigments had the capacity to undergo photosynthesis and produce different types of pigments such as fucoxanthin, fucoerythrin, phaeophytin, unlike carbohydrate, necessary for their growth. It seemed to me a zone of silence, peace like a paradise. We could enjoy the biodiversity of fish, coral and algae under the sea. Looking at this eye-catching marine ecosystem of fish fauna, coral and algal flora, I felt I was living in a paradise keeping me away from all sorts of worries from our modern world. We need to conserve this natural gift; we should conserve them carefully and maintain the marine biodiversity.

- Journey through mangroves and regions of crocodiles near sea beaches: territory of crocodiles. After visiting and enjoying the territory of tortoises and the biodiversity of underwater algal flora and fish fauna, the boat driver drove through silent lakes in the mangrove forests, the zones of crocodiles and birds. We tried to look for crocodiles but unfortunately we could not locate them during our available time. Probably they were taking a rest under some mangrove trees in between the rhizophores. On the other hand, we could see the ecosystem of mangroves bordering the sea beach. I could not identify the diversity of mangrove species as the boat was moving at high speed. I could observe how the mangroves were protecting the erosion of soils from the tidal waves. The plants were highly branched with thick lanceolate leaves with an open canopy for the capture of carbon through photosynthesis. The most interesting feature was I could observe the way the mangroves protect soil erosion by their strong highly horizontally ramifying rhizophores. These arose from the base of the tree, highly branched, growing horizontally and fixing these roots in the soil, looking like a roof of horizontal rhizophores supporting the tree with highly branched stems. During low tide the rhizophores absorbed water vapour from the sea necessary for photosynthesis and plant metabolism as well as oxygen for respiration. The plants were healthy in the highly saline aquatic environment and did not show symptoms of salinity stress. Being a botanist, I did not have any chance to observe the mechanism of adaptation of mangroves at the extreme saline environment. Though we did not have a chance to see crocodiles, the education on the mechanism of adaptation and protecting soil erosions gave me enough satisfaction. We could observe several birds sitting on the branches of the trees, some enjoying fish in their beaks recently captured from the seawater after diving. We could see a number of wild marine ducks, turkey and pelicans observing us while

we were driving across the sea. They did not feel disturbed by the nasty sounds of the boat because they are habituated to observe silently as the tourists drive over the sea.

4.2 Visit to dolphin territory, playground of dolphins

After visiting the zone of mangrove forests near the seashore, our boat driver took us to the territory of dolphins. The boat driver asked my son to get ready with his cameras, including the aquatic camera. We saw the huge dolphins, maybe more than 500 kg jumping over the seawater and diving down to catch their prey fishes moving around the algal and coral colonies. It was after noon, the time of food. With the aquatic camera my grandson took nice photographs of dolphins diving through tidal water, jumping up and down. They were playing like games and at the same time catching fish and engulfing them voraciously. Our driver was running behind them to give us an opportunity to see and take good photographs. I was really afraid and cautioned the driver not to dash and hurt them. A series of boats were behind us looking for dolphins. We were really causing damage to the dolphins' diversity in their territory. Though the authorities have asked guides and drivers not to disturb these precious creatures of the sea, some wanted to please the tourists by driving and running behind the dolphins and disturbing them as they fed. The dolphins were really hungry and keen to catch their prey and engulf them. This was the first time in my life I had the opportunity to see and enjoy the view of dolphins, thanks to my son who motivated us to take this tour. We human beings are selfish and disturb the territory of these precious marine creatures for our personal satisfaction. We were really happy and will never forget the games of dolphins jumping up and down, the gifts of nature.

4.3 Zones of pelicans and sea ducks

After an enjoyable visit to dolphin territory, the guide took us to a zone of pelicans with big coloured beaks sitting on the mango trees and looking at us without being disturbed by the roaring boat. This was a really enjoyable scene, the gift of nature.

4.4 Zones of sea dives and swimming

Finally the guide took us to a silent zone with shallow water and asked the tourists to jump down with a sea belt. It was a beautiful scene. Many of them floated on the seawater with their life jackets. My son, daughter in law, granddaughter and grandson got down and enjoyed swimming in the silent water for about 30 min. We did not get down due to our old age. Finally we arrived at the boarding zone. The guide asked us to take food served as a buffet.

4.5 Return journey through Sian Ka'an reserve forest

The guide asked us to board the van for the return journey to our hotel.

About 20 vans started this return journey, driving through deep forests over rugged roads, The condition of the roads was very bad, with potholes here and there, The driver and some tourists themselves drove the old vans, some of which stopped here and there, The guide had to drive to them and repair them. It was really a horrible journey. In my opinion the government should repair the roads for the comfortable and safe journey of tourists.

4.6 Food arrangement

The hotel authority made good arrangements for our breakfast, lunch and dinner in a common restaurant about 1 km from our hotel. They arranged transportation in a small train. Sometimes we used to walk along the road through the mangrove forests watching on both sides of the road the hanging rhizophores, palm trees, mangrove flowers, animals running away, very occasionally deers, lizards and so on.

At 9 p.m. at night we enjoyed Mexican traditional songs and dances in a big gathering. Many foreigners from different countries could really enjoy the traditional songs and dances typical of different states. All enjoyed the gathering and clapped.

4.7 Visit to Chichen Itza – Merida, Yucatan Peninsula

We booked a van for a trip to Chichen Itza, the historical pyramid of Maya culture located in Merida far south in the Yucatan Peninsula, Mexico. We started our journey early in the morning, again though tropical rainforest on both side of the road. I was enjoying the biodiversity of the rainforest and its ecosystem. Each species occupied its own niche and attained enough height to capture light for photosynthesis. It gave a wavy contour up and down on the horizon. The forest was dense consisting of shrubs, trees, climbers. As our vehicle was running very fast, it was difficult for me to identify the type of vegetation. I could observe that most of the tree species had an open leaf canopy with highly brached thin stems which favoured efficient capture of light and translocation of photosynthates in stems and other plant parts. They grew higher than those with closed canopy leaves. Some species possessed a closed leaf canopy which had a short stout stem with a large basal diameter and few primary branches with thicker stems. There were large variations in leaf morphology, size, shape, surface structure, intensity of leaf pigments.

On the basis of my analysis of the rainforest at both the Riviera Maya, the Sankaen Reserve Biosphere and that of the Yucatan, we could classify the plants in two distinct types.

1 Open leaf canopy: In this category, the woody species possessed an open leaf canopy giving an opportunity for the efficient capture of sunlight for photosynthesis. They possessed either lanceolate narrow, serrated leaves or broad big or small leaves. The most conspicuous characteristics of these species, was that they possessed highly branched stems with thinner stems.

2 Closed leaf canopy: These species possessed a very closed leaf canopy with overlapping lower leaves which received less light for photosynthesis.

Too many crowded leaves represented the wastage of solar energy and organic matter. These species possessed short thick basal stems which were bi- or tri-furcated to produce thick primary stems. These species possessed either narrow or broad big or small leaves as in the case of the first category. The leaves on the upper crown and lateral sides of the plant could only capture solar radiation for photosynthesis and the photosynthates were translocated to the basal stem and primary stems and other plant parts.

3 Intermediate category: With a decrease in leaf canopy the branching of stems was reduced and the basal stem diameter increased.

4.8 Visit to underground river, named "Cenote"

On the way to Chichen Itza, the guide was explaining us the history of the old Maya culture which was much more advanced scientifically than other cultures like Aztecas and Olmecas in Mexico. They were strong in Mathematics and Astronomy. They use symbols/scripts only to communicate with each other. As confirmed by modern astronomers, their predictions were perfect. He explained the marriage system. The boys mixed with the girls before marriage. The father of the bride had to give fruits, vegetables, food grains and other things as gifts to the girl at the time of marriage. During the marriage they danced and drink some kind of hard beverage.

He explained Mayas still live in their native village and maintain their traditional customs and dress. They hunt wild animals for their food. The land is fertile. They grow rice, sugarcane, vegetables and different types of fruits.

On the way, he took a bypass journey over rugged roads in order to take us to an underground river called: "Cenote". He explained that there are reported to be more than 7000 underground rivers which ultimately flow to the sea. One, two or three are open to the public. Utimately we arrived at the famous Cenote. There were more than 20 steps to go down to the underground river. Tourists who want to swim in the river have to take a bath in a bathroom first, to keep the river water hygienic. I did not go down but my son with his children went down to the river edge. I could watch a few tourists who were swimming happily. The water is bluish, clean and safe to drink. An the top there was a hut where the Mayas were selling their handicrafts and goods. Like us, they worship different gods for different purposes, showing a cultural similarity with the Hindu religion. My son purchased idols of gods which the Mayas still worship in their native villages. I asked my son to take a photograph with a Maya who was selling the handicrafts in a community shop shared by the village. They are fair in complex. The Mayan women are beautiful. We were told that, though Spanish married Mayas, thereby leading to a mixed culture, in the native village marriage is restricted only to native Mayas who speak their native language. I developed a respect for this culture.

Then the guide started the journey to our final destination, Chichen Itza. On the way he took us to a community restaurant for our lunch. Different types of food preparations were served as buffets. We had to wait about 45 min to get seats. We relished that food without complaint.

After lunch, the guide took us to the historic place of the great pyramid Chichen Itza. It was a very hot day with a scorching sun burning our forehead and arms. Both of us sat on a wooden bench under the shade of a big tree which had a closed canopy. Just in front of us the historical huge pyramid was standing as if inviting us to climb up. I had visited this place earlier and remembered all the places of interest. The pyramid has four sides with innumerable staircases to climb up. On our previous visit, we could climb up but the tourist department has now prohibited tourists from doing so. The tourists were asked by the guides to clap hands, which gave the musical sound of a bird from the topmost peak of the pyramid.

In 2007, NASA scientists visited the pyramid and tested that they could predict astronomical positions perfectly, far better than using modern science. Chichen Itza is now included as a miracle of the world for its architecture and its high astronomical interest. The guide was explaining to a group of tourists about the prehistory of this gigantic pyramid which depicts the early history of Maya culture. On the night of a particularly full moon, tourists from all over the place and foreigners gather on the ground in front of the pyramid. They sing songs and pray to the moon to come down the world. To a spectator it seems that the moon comes down from high in the sky and sits on the topmost floor of the pyramid and all the tourists clap hands and sing songs to please the moon. It is an amazing scene for all the tourists.

The pyramid is surrounded by typical ecosystems consisting of huge trees with a closed canopy and a huge basal trunk with a high basal diameter widely spaced. Under the shade of the trees there are small shops selling handicrafts typical of the Maya culture. Some sell trumpets which gives a sound like a typical animal or bird of the region. The surroundings are flooded with tourists from all over the world.

The sacred pyramid represents the glory of Maya culture and their advance in science. Though the Spanish demolished and destroyed several buildings and scattered idols of the gods and animals of the region, the pyramid remains intact. Probably they tried but failed to do any damage to this sacred building.

Being ignorant of the sanctity of the place, I want to mention only a few more things. Probably the rulers used to kill their victims by shooting them against pillars. Very near to this building we saw a place where pillars stood. On the left side, we could see a destroyed building which was the place used for games by the Mayas. All the guides took their tourists walking along zigzag paths and explaining the destruction of several buildings and sculptures of animals, birds and other creatures of the region. In one place they showed us piles of stones imitating maize cobs which represented symbols of fertility. Finally they took all the tourists to a famous sacred well. It has spiritual history. It was told that devotees jumped around the well, cut their heads, took out their hearts and throw them into the well to please the god. They felt themselves lucky to sacrifice their lives. Geologists and the tourist department have recovered many skulls, skeletons and huge ornaments from the well, some of which are preserved in the anthropological building in Mexico, with some others in the USA. The water of the well is highly contaminated.

The Spanish people abandoned this city owing to the non-availability of food grains, vegetables and fruits nearby but the pyramids remain here, representing the glory of the Maya culture.

Bibliography

Correa-Méndez F., Carrillo-Parra A., Rutiaga-Quiñones J.G., Márquez-Montesino F., González-Rodríguez H., Jurado-Ybarra E., Garza-Ocañas F. 2014. Contenido de humedad y sustancias inorgánicas en sub-productos maderables de pino para su uso en pélets y briquetas. *Revista Chapingo Serie Ciencias Forestales y del Ambiente* 20(1): 77–88.

Domínguez Gómez T.G., González Rodríguez H., Ramírez Lozano R.G., Estrada Castillón A.E., Cantú Silva I., Gómez Meza M.V., Villarreal Quintanilla J.A., Alvarado M.S., Alanís Flores G. 2013. Diversidad estructural del matorral espinoso tamaulipeco durante las épocas seca y húmeda. *Revista Mexicana de Ciencias Forestales* 4(17):106–123.

Estrada-Castillón E., Villarreal-Quintanilla J.A., Salinas-Rodríguez M.M., Jiménez-Pérez J., García-Aranda M.A. 2012. *Flora and Phytogeography of Cumbres de Monterrey National Park.* Nuevo Leon, Mexico.

González-Rodríguez H., Ramírez-Lozano R.G., Cantú-Silva I., Gómez-Meza M.V., Uvalle-Sauceda J.I. 2010. Composición y estructura de la vegetación en tres sitios del estado de Nuevo León, México. *Polibotánica* 29:91–106.

Maiti R.K., Gonzalez-Rodriguez H. 2011. Lechuguilla (*Agave lecheguilla*); an important commercial fiber plant and a source of income to the arid land farmers of Mexico.

International Journal of Bio-resource and Stress Management **2**(1):104–110.

Ramírez-Lozano R., Domínguez-Gómez T.G., González-Rodríguez H., Cantú-Silva I., Gómez Meza M.V., Sarquís-Ramírez J.I., Jurado E. **2013**. Composición y diversidad de la vegetación en cuatro sitios del noreste de México. *Madera y Bosques* **19**(2):59–72.

Silva C.I., González R.H. **2002**. Propiedades hidrológicas del dosel de bosques de pino-encino en el noreste de México. *CiENCiA UANL* **5**(1):72–77.

CHAPTER 5

Plant traits

5.1 Research advances in plant traits

Trees and shrubs occurring in a forest show a large variation in leaf morphology: in leaf shape, size, stem structure, branching pattern and several other morphological traits and also in their quantitative characters.

The main type of vegetation in northeastern Mexico, known as Tamaulipan thornscrub, is distinguished by a wide range of taxonomic groups exhibiting differences in growth patterns, leaf life spans, textures, growth dynamics and phenological development (Reid et al., 1990; McMurtry et al., 1996). This semiarid shrubland, that covers about 200 000 km^2 including southern Texas and northeastern Mexico, is characterised by an average annual precipitation that varies from 400 to 800 mm and a yearly potential evapotranspiration of about 2200 mm. Vegetation has been utilised as a forage source for domestic livestock and wildlife, fuelwood, timber for construction, medicine, agroforestry and reforestation practices in disturbed sites (Reid et al., 1990). The great diversity of native tree species in this region reflects the plasticity among shrub species derived from their development of effective mechanisms to cope with seasonal water stress.

In the following, we undertook a study focused on characterising the woody plant traits in Tamaulipan thornscrub. Bushy shrubs such as huizache (*Acacia farnesiana*), colima (*Zanthoxylum fagara*), coma (*Sideroxylon celastrinum*), chapote (*Diospyros texana*), panalero (*Forestiera angustifolia*), anacahuita (*Cordia boissieri*), brasil (*Condalia hookeri*) and granjeno (*Celtis pallida*) generally form part of the species found in this vegetation type. These species are used for varios domestic purposes such as fences, firewood and charcoal. Tamaulipan thornscrub plays an important role in the production of carbon in semiarid Mexico. Branches and wooden logs are buried in a trough and covered with soil to prevent the entry of air for respiration, but a hole is made in the centre to burn the underground wood for several days until they are converted to hard charcoal. This an important source of income for ranchers. The plants are characterised by different growth forms (life forms), the stems of the trees are woody and branched and the shrubs have woody stems but are highly branched.

5.1.1 Methodology

We characterised the different plant features of 15 native shrub species in Tamaulipan thornscrub. Data from 10 individuals

Autoecology and Ecophysiology of Woody Shrubs and Trees: Concepts and Applications, First Edition.
Edited by Ratikanta Maiti, Humbero Gonzalez Rodriguez and Natalya Sergeevna Ivanova.

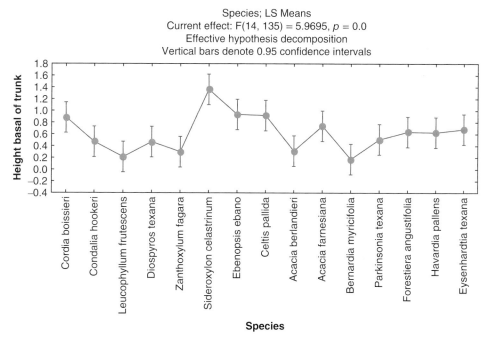

Figure 5.1 Basal height of main stem in different woody species, north-eastern Mexico.

of each species were taken from plants selected at random. Several variables were quantitated, such as basal height of main trunk, number of primary branches arising from main stem, main plant height, canopy cover, type of crown and canopy type.

There was a large variability in the basal trunk height (m) among the studied species, showing significant differences between them (Figure 5.1). *Sideroxylon celastrinum* had the highest basal height followed by *Ebenopsis ebano*, *Celtis pallida* and *Cordia boissieri*.

Similarly, total plant height (m) showed significant variability ($p = 0.05$), among which *Parkinsonia texana* had the highest plant height. Variation in plant height was also found among canopy type, open, semi-closed and closed canopy type ($p = 0.05$). Open canopy showed the greatest plant

height, followed by semiclosed and closed types. Canopy cover also show variation at $p = 0.05$, among which *P. texana* had the maximum canopy cover (9 m). Tree top canopy height showed variability at $p = 0.05$. Plant species also show variation in canopy type. Wood density (kg/m^3) showed significant variability between species at $p = 0.05$, among which *Bernar dia myricifolia* had the highest density, followed by *P. texana*, *Eysenhardtia texana* and *Ebenopsis ebano*. Other species had a low density. Wood density also showed variability between canopy types, showing highest density in open canopy followed by semiclosed and closed canopy (Figure 5.2).

We conducted a correlation analysis to identify relationships between the measured characteristics (see Tables 5.1–5.15).

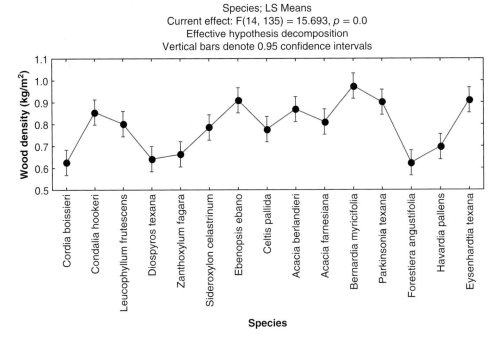

Figure 5.2 Wood density in different woody species, north-eastern Mexico.

Table 5.1 Correlation analysis between plant traits in *Cordia boissieri*.

Plant trait	1	2	3	4	5	6	7
1. Basal height of trunk (m)	–						
2. Diameter main stem (m)	−0.09	–					
3. Total height (m)	0.67	−0.37	–				
4. Canopy cover X on side (m)	0.55	−0.34	0.81	–			
5. Canopy cover Y on side (m)	0.49	−0.47	0.85	0.96	–		
6. Plant height (m)	−0.63	0.21	−0.57	−0.27	−0.37	–	
7. Wood density (kg/m³)	0.04	−0.13	−0.12	0.14	0.10	0.35	–

Table 5.2 Correlation analysis between plant traits in *Condalia hookeri*.

Plant trait	1	2	3	4	5	6	7
1. Basal height of trunk (m)	–						
2. Diameter main stem (m)	0.13	–					
3. Total height (m)	0.53	0.66	–				
4. Canopy cover X on side (m)	0.58	0.68	0.81	–			
5. Canopy cover Y on side (m)	0.37	0.69	0.79	0.90	–		
6. Plant height (m)	0.01	−0.33	0.05	0.00	−0.11	–	
7. Wood density (kg/m³)	−0.41	−0.63	−0.47	−0.38	−0.19	0.20	–

Table 5.3 Correlation analysis between plant traits in *Leucophyllum frutescens*.

Plant trait	1	2	3	4	5	6	7
1. Basal height of trunk (m)	–						
2. Diameter main stem (m)	0.20	–					
3. Total height (m)	−0.25	0.09	–				
4. Canopy cover X on side (m)	−0.04	−0.39	0.07	–			
5. Canopy cover Y on side (m)	−0.15	−0.63	0.31	0.85	–		
6. Plant height (m)						–	
7. Wood density (kg/m^3)	−0.54	−0.34	−0.39	−0.13	−0.03		–

Table 5.4 Correlation analysis between plant traits in *Diospyros texana*.

Plant trait	1	2	3	4	5	6	7
1. Basal height of trunk (m)	–						
2. Diameter main stem (m)	0.00	–					
3. Total height (m)	−0.21	−0.08	–				
4. Canopy cover X on side (m)	−0.25	−0.16	0.64	–			
5. Canopy cover Y on side (m)	−0.37	−0.50	0.21	0.31	–		
6. Plant height (m)						–	
7. Wood density (kg/m^3)	0.20	0.70	0.27	0.23	−0.25		–

Table 5.5 Correlation analysis between plant traits in *Zanthoxylum fagara*.

Plant trait	1	2	3	4	5	6	7
1. Basal height of trunk (m)	–						
2. Diameter main stem (m)	0.62	–					
3. Total height (m)	−0.59	−0.54	–				
4. Canopy cover X on side (m)	−0.30	−0.47	0.09	–			
5. Canopy cover Y on side (m)	−0.24	−0.46	0.31	0.42	–		
6. Plant height (m)						–	
7. Wood density (kg/m^3)	−0.38	−0.46	0.04	0.53	0.46		–

Table 5.6 Correlation analysis between plant traits in *Sideroxylon celastrinum*.

Plant trait	1	2	3	4	5	6	7
1. Basal height of trunk (m)	–						
2. Diameter main stem (m)	−0.30	–					
3. Total height (m)	−0.63	0.40	–				
4. Canopy cover X on side (m)	0.08	0.37	0.24	–			
5. Canopy cover Y on side (m)	0.15	−0.48	−0.38	−0.51	–		
6. Plant height (m)	−0.37	−0.33	0.11	−0.38	−0.14	–	
7. Wood density (kg/m^3)	−0.13	−0.05	−0.06	0.38	−0.09	−0.27	–

Table 5.7 Correlation analysis between plant traits in *Ebenopsis ebano*.

Plant trait	1	2	3	4	5	6	7
1. Basal height of trunk (m)	–						
2. Diameter main stem (m)	0.21	–					
3. Total height (m)	0.16	0.23	–				
4. Canopy cover X on side (m)	0.11	0.79	0.14	–			
5. Canopy cover Y on side (m)	−0.39	0.40	0.20	0.26	–		
6. Plant height (m)	0.04	0.19	0.16	0.12	0.10	–	
7. Wood density (kg/m³)	−0.42	−0.30	−0.21	−0.18	0.26	−0.51	–

Table 5.8 Correlation analysis between plant traits in *Celtis pallida*.

Plant trait	1	2	3	4	5	6	7
1. Basal height of trunk (m)	–						
2. Diameter main stem (m)	0.00	–					
3. Total height (m)	0.40	0.60	–				
4. Canopy cover X on side (m)	−0.12	0.52	0.63	–			
5. Canopy cover Y on side (m)	0.29	−0.04	0.44	0.42	–		
6. Plant height (m)	0.18	−0.96	−0.41	−0.42	0.21	–	
7. Wood density (kg/m³)	−0.19	0.32	−0.05	−0.29	0.07	−0.40	–

Table 5.9 Correlation analysis between plant traits in *Acacia berlandieri*.

Plant trait	1	2	3	4	5	6	7
1. Basal height of trunk (m)	–						
2. Diameter main stem (m)	0.41	–					
3. Total height (m)	−0.27	−0.67	–				
4. Canopy cover X on side (m)	0.13	0.42	−0.06	–			
5. Canopy cover Y on side (m)	0.13	0.35	0.13	0.31	–		
6. Plant height (m)	0.30	0.23	−0.20	−0.26	−0.44	–	
7. Wood density (kg/m³)	−0.79	−0.07	0.00	−0.20	0.15	−0.36	–

Table 5.10 Correlation analysis between plant traits in *Acacia farnesiana*.

Plant trait	1	2	3	4	5	6	7
1. Basal height of trunk (m)	–						
2. Diameter main stem (m)	−0.06	–					
3. Total height (m)	0.32	0.39	–				
4. Canopy cover X on side (m)	0.03	0.54	0.38	–			
5. Canopy cover Y on side (m)	−0.53	0.47	0.40	0.53	–		
6. Plant height (m)						–	
7. Wood density (kg/m³)	0.32	−0.23	0.05	−0.64	−0.42		–

Table 5.11 Correlation analysis between plant traits in *Bernardia myricifolia*.

Plant trait	1	2	3	4	5	6	7
1. Basal height of trunk (m)	–						
2. Diameter main stem (m)	0.53	–					
3. Total height (m)	−0.35	0.08	–				
4. Canopy cover X on side (m)	0.03	0.36	0.58	–			
5. Canopy cover Y on side (m)	0.16	0.66	0.31	0.37	–		
6. Plant height (m)						–	
7. Wood density (kg/m^3)	0.23	−0.42	−0.15	−0.11	−0.74		–

Table 5.12 Correlation analysis between plant traits in *Parkinsonia texana*

Plant trait	1	2	3	4	5	6	7
1. Basal height of trunk (m)	–						
2. Diameter main stem (m)	−0.24	–					
3. Total height (m)	−0.05	0.13	–				
4. Canopy cover X on side (m)	0.20	−0.60	0.18	–			
5. Canopy cover Y on side (m)	0.35	−0.66	0.17	0.90	–		
6. Plant height (m)						–	
7. Wood density (kg/m^3)	0.13	0.14	−0.20	−0.46	−0.19		–

Table 5.13 Correlation analysis between plant traits in *Forestiera angustifolia*.

Plant trait	1	2	3	4	5	6	7
1. Basal height of trunk (m)	–						
2. Diameter main stem (m)	0.51	–					
3. Total height (m)	0.58	0.31	–				
4. Canopy cover X on side (m)	0.30	0.49	0.42	–			
5. Canopy cover Y on side (m)	−0.17	−0.21	0.16	0.29	–		
6. Plant height (m)	−0.40	0.19	−0.71	0.05	−0.28	–	
7. Wood density (kg/m^3)	−0.49	−0.09	−0.65	−0.27	−0.23	0.56	–

Table 5.14 Correlation analysis between plant traits in *Havardia pallens*.

Plant trait	1	2	3	4	5	6	7
1. Basal height of trunk (m)	–						
2. Diameter main stem (m)	−0.22	–					
3. Total height (m)	−0.17	0.54	–				
4. Canopy cover X on side (m)	0.11	0.42	0.63	–			
5. Canopy cover Y on side (m)	−0.21	0.80	0.68	0.42	–		
6. Plant height (m)	0.16	−0.06	0.62	0.10	0.02	–	
7. Wood density (kg/m^3)	0.44	−0.63	−0.76	−0.46	−0.60	−0.36	–

Table 5.15 Correlation analysis between plant traits in *Eysenhardtia texana*.

Plant trait	1	2	3	4	5	6	7
1. Basal height of trunk (m)	−						
2. Diameter main stem (m)	−0.52	−					
3. Total height (m)	−0.18	0.77	−				
4. Canopy cover X on side (m)	−0.24	0.73	0.49	−			
5. Canopy cover Y on side (m)	−0.55	0.58	0.22	0.65	−		
6. Plant height (m)	−0.29	−0.54	−0.44	−0.74	−0.31	−	
7. Wood density (kg/m³)	0.52	−0.17	−0.04	−0.38	−0.41	−0.23	−

5.1.2 Results and Discussion

On the basis of the correlation analysis, we can distinguish three groups of tree species:

1 Tree species that have many statistically distinct and significant relationships between the measured characteristics.

2 Tree species that have a small quantity of statistically significant relationships between the measured characteristics.

3 Tree species that have no statistically significant relationships between the measured characteristics.

The first group includes *Cordia boissieri, Condalia hookeri, Eysenhardtia texana* and *Havardia pallens*. The statistically significant relationships are different for the different tree species. However, the significant relationship between total height and canopy cover (X, Y on side), between the canopy cover X on side and canopy cover Y on side were revealed for all tree species in this group. Two species of this group (*C. hookeri* and *E. texana*) showed a statistically significant relationship between diameter mainstem trunk and total height.

The second group includes the majority of tree species: *Leucophyllum frutescens, Diospyros texana, Ebenopsis ebano, Celtis pallida, Acacia berlandieri, A. farnesiana, Bernardia myricifolia, Parkinsonia texana* and *Forestiera angustifolia*. The statistically significant relationships are different for the different tree species.

The third group is the smallest and includes *Zanthoxylum fagara* and *Sideroxylon celastrinum*.

Thus, the most studied tree species has a small number of statistically significant relationships among measured characteristics; and this is different for different tree species.

In order to identify dependency between plant traits, we conducted a regression analysis. We made a regression analysis using a few variables in a few species, as mentioned below. We used the variables which revealed statistically significant relationships (Figures 5.3–5.14).

With respect to regression coefficients in *Cordia boissieri* it was observed that there existed a significant relation among canopy cover (X – Y on side; $R^2 = 0.91$). Canopy cover (X – Y on side) increased linearly with increasing total height (Figure 5.3).

For *Condalia hookeri* we got seven dependencies (Figure 5.4). All dependencies were linear. The strongest relationship was found between canopy cover X on side and canopy cover Y on side ($R^2 = 0.81$). Canopy cover (X – Y on side) increased

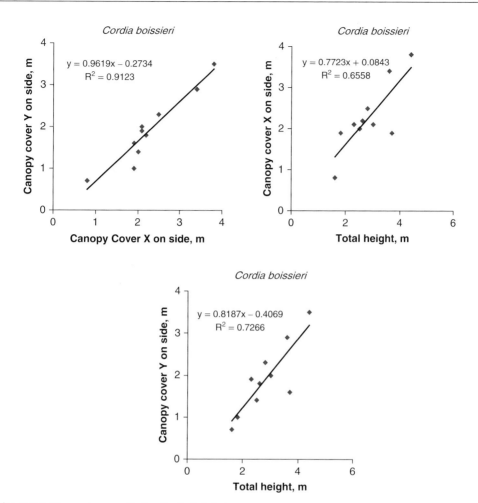

Figure 5.3 Regression results for *Cordia boissieri.*

linearly with increasing total height. This relationship was the same as for *Cordia boissieri.* Unlike *Cordia boissieri,* for *Condalia hookeri* we traced a direct linear dependence between diameter of main stem and canopy cover X and Y on side, total height and the inverse linear dependence between diameter of main stem and wood density.

Leucophyllum frutescens revealed only one relationship: a linear relationship between canopy cover X on side and canopy cover Y on side ($R^2 = 0.72$; Figure 5.5).

For *Diospyros texana* the regression analysis revealed a logarithmic relationship between diameter of main stem and wood density ($R^2 = 0.60$) and a linear relationship between total height and canopy cover X on side ($R^2 = 0.41$; Figure 5.6).

For *Ebenopsis ebano* the regression analysis revealed only one relationship: a linear relationship between diameter of main stem and canopy cover X on side ($R^2 = 0.62$; Figure 5.7).

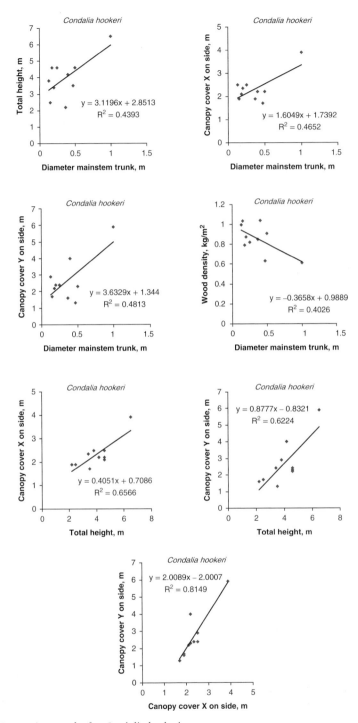

Figure 5.4 Regression results for *Condalia hookeri*.

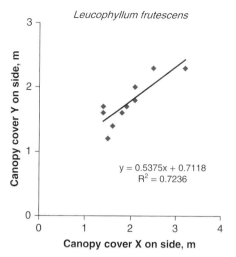

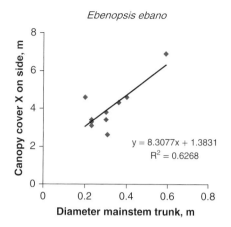

Figure 5.7 Regression results for *Ebenopsis ebano*.

Figure 5.5 Regression results for *Leucophyllum frutescens*.

For *Celtis pallida* the regression analysis revealed a logarithmic relationship between diameter of main stem and total height ($R^2 = 0.61$) and a linear relationship between total height and canopy cover X on side ($R^2 = 0.39$; Figure 5.8). We showed a similar finding for *Diospyros texana*.

For *Acacia berlandieri* the regression analysis revealed only an inverse linear

relationship between diameter of main stem and total height ($R^2 = 0.44$; Figure 5.9, left). For *A. farnesiana* it also revealed an inverse linear dependence between canopy cover X on side and wood density ($R^2 = 0.41$; Figure 5.9, right).

For *Bernardia myricifolia* the regression analysis revealed a linear relationship between diameter of main stem and canopy cover Y on side ($R^2 = 0.43$) and an inverse linear relationship between canopy cover Y on side and wood density

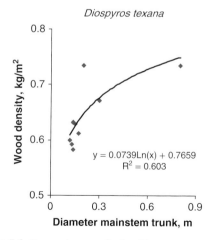

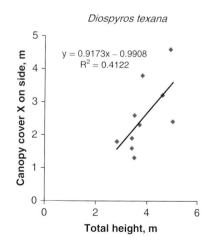

Figure 5.6 Regression results for *Diospyros texana*.

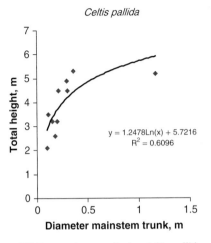

Figure 5.8 Regression results for *Celtis pallida*.

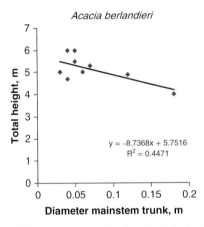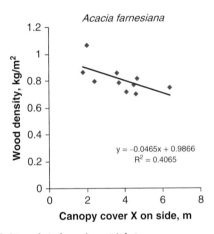

Figure 5.9 Regression results for *Acacia berlandieri* (left) and *A. farnesiana* (right).

($R^2 = 0.54$; Figure 5.10). Wood density decreased sharply with increasing canopy cover Y on side.

For *Parkinsonia texana* the regression revealed two inverse linear relationships between diameter of main stem and canopy cover X on side ($R^2 = 0.35$) and between diameter of main stem and canopy cover Y on side ($R^2 = 0.43$). It also showed a linear relationship between canopy cover X on side and canopy cover Y on side ($R^2 = 0.80$; Figure 5.11). Canopy cover Y on side decreased with increasing canopy cover X on side.

For *Forestiera angustifolia* the regression analysis revealed an inverse linear relationship between total height and wood density ($R^2 = 0.43$; Figure 5.12).

For *Havardia pallens* the regression analysis revealed a linear relationship between diameter of main stem and canopy cover Y on side ($R^2 = 0.64$) and an inverse

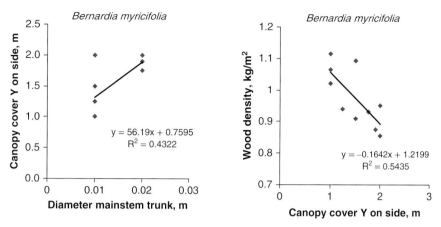

Figure 5.10 Regression results for *Bernardia myricifolia*.

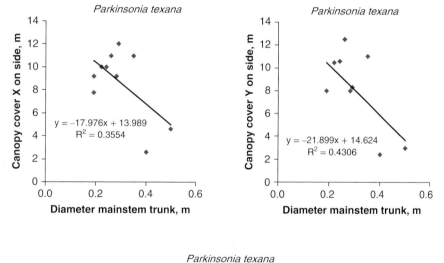

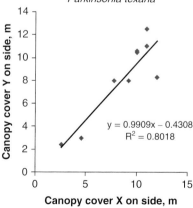

Figure 5.11 Regression results for *Parkinsonia texana*.

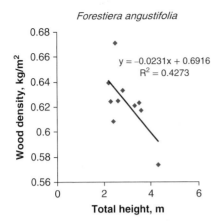

Figure 5.12   Regression results for *Forestiera angustifolia*.

logarithmic relationship between diameter of main stem and wood density ($R^2 = 0.43$; Figure 5.13).

For *Eysenhardtia texana* the regression analysis revealed two linear relationships between diameter of main stem and total height ($R^2 = 0.58$) and between diameter of main stem and canopy cover X on side ($R^2 = 0.53$; Figure 5.14).

Thus, different tree species have different relationships between the measured

parameters. Relationships are predominantly linear. The relationship between canopy cover X on side and canopy cover Y on side for *Cordia boissieri, Condalia hookeri, Parkinsonia texana* and *Leucophyllum frutescens* is expressed most strongly and it is shown by the best fit linear equation. Linear regression is also effective for describing the relationship between total height and canopy cover Y on side for *Cordia boissieri*.

In *Condalia hookeri* the main stem diameter showed a significant relation with total height ($R^2 = 0.439$) and diameter with canopy cover ($R^2 = 0.463$); but diameter showed a negative relationship with wood density ($R^2 = 0.402$). Total height showed a significant relationship with canopy cover ($R^2 = 0.656$).

Leucophyllum frutescens showed a significant relationship in canopy cover ($R^2 = 0.723$).

In *Diospyros texana*, diameter of main stem showed a significant relationship with wood density and total height with canopy cover ($R^2 = 0.422$).

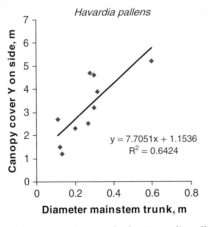

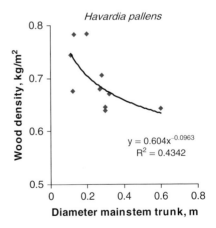

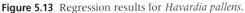

Figure 5.13 Regression results for *Havardia pallens*.

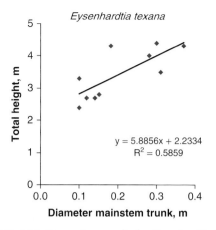 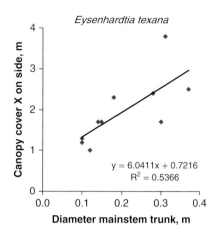

Figure 5.14 Regression results for *Eysenhardtia texana*.

In *Ebanopsis ebano*, basal diameter showed a significant relationship with canopy cover ($R^2 = 0.622$).

In *Celtis pallida*, basal diameter showed a significant relationship with total height ($R^2 = 0.606$), total height with canopy height ($R^2 = 0.391$).

In *Acacia berlandieri*, basal diameter showed a negative relationship with total height ($R^2 = 0.44$).

5.2 Branching pattern of trees

The branching patterns of trees show a large variability among trees depicting various architectural patterns. It is an interesting mechanism of tree species, for its capacity to capture solar energy is by leaves acting as a solar panel. Each species extends its branches arising from its main stem depending the available space with its neighbours. Foresters adopt various silvicultural practices to manage species varying in branching pattern. Various studies have been undertaken to describe and analyse the branching patterns of trees. A few scientists have developed mathematical models to analyse branching patterns. We discuss here a few published reports on the branching patterns of trees.

Leopold (1971) investigated the efficiency of branching patterns, the relation of average numbers and lengths of tree branches to branch size. The size of a branch is defined by branch order or its position in the hierarchy of tributaries. Similar to river drainage nets, a definite logarithmic relation exists between branch order and lengths and numbers. This relation can be quantified in tree branching systems and several random-walk models in both two and three dimensions. Besides, the most probable arrangement appears to minimise the total length of all stems in the branching system.

Steingraeber and Waller (1986) reported a non-stationarity of tree branching patterns and bifurcation ratios. Branching patterns in organisms have been analysed by using branch ordering and bifurcation ratio techniques. Data on four species revealed two specific patterns of non-stationarity that are directly related to morphological patterns of shoot development. Average bifurcation ratios appear to be inappropriate descriptions of tree branching patterns, since they are based solely on relative branch position and ignore the biologically important details of form and development.

Pickett and Kempf (1980) investigated the branching and leaf display of dominant forest shrubs and understorey trees in central New Jersey with an objective to determine: (a) whether branching differentiation occurs in shrubs which reach optimum development in different successional environments, (b) the contrast in branching of small trees between field and forest and (c) the nature of within-crown branching plasticity in a mature canopy tree. They observed that shrubs do not differ widely in gross branching structure (ratio of terminal to supporting branches) and proposed that branch angle, length and alteration of leaf orientation may be significant display characters. Small trees show a markedly variable response to open versus closed habitats, demonstrating the expected increase in branching ratio in open environments. Within, but not outside, the forest, earlier successional species were more variable in branching. A single canopy tree crown also demonstrated an alteration of leaf display components, including increased length and wider angle of branches, but not branch ratio in the shaded, lower crown. They suggest some of the morphological traits of shrub branching may be important in determining their leaf display. Finally, they discussed differences in shrub and tree habit, such as cloning and the presumably reduced costs of support in shrubs, which may explain the failure of shrubs to exploit the same component of branching strategy as trees.

Trees can be identified by the shape of their silhouette, such as V-shaped, columnar, pyramidal, round, oval. Trees exhibit a variety of forms depending on their branching patterns. The arrangement and position of the branches on a tree give the tree a definite shape. The three positions of branches can be: pointing upward from the trunk, pointing straight out from the truck, or pointing downward from the trunk.

There are three main forms:

1 Excurrent: the main stem goes straight for the entire height of the tree, with branches forming patterns, for example evergreens.

2 Decurrent: the main stem continues up about halfway, then splits into more than one main branch, for example fruit trees.

3 Columnar: the main trunk continues for the full height of the tree, with the branches forming only at the top, for example palm trees.

Heuret et al. (2003) studied the branching patterns and growth units of monocyclic or bicyclic annual shoots on the main axis of five year old red oaks in a plantation in south-western France. They described each growth unit as the production of axillary branches associated with each node in the form of a sequence. For a given category of growth units, homogeneous zones (i.e. zones in which composition in terms of type of axillary production did not change substantially) were identified on such sequences using a dedicated statistical model called a hidden semi-Markov chain. Branching patterns shown by the growth unit of monocyclic annual shoots and on the second growth unit of bicyclic annual shoots were very similar. Branches with a one-year delay in development tended to be polycyclic at the top of the growth unit and monocyclic lower down. The number of nodes shown by the branched zone of the growth unit of monocyclic annual shoots was stable, irrespective of the total number of nodes of the growth unit. In contrast, the second growth unit of bicyclic annual shoots exhibited a correlation between the number of nodes in the branching zone and the total number of nodes. These can be classified as monopodial, pseudomonopodial and sympodial.

Daniel Stolte (2013) studied the branch systems of trees in the University of Arizona. According to the author, this study represented the first empirical test of a theory that University of Arizona ecology professor Brian Enquist (1998) helped develop. That theory held that a tree's branching structure – specifically, the width and length of its branches – predicts how much carbon and water a tree exchanges with the environment in relation to its overall size, independent of species. This theory could be used to scale the size of plants to their function, such as amount of photosynthesis, water loss and respiration, especially in the light of climate change [Lisa Patrick Bentley (2013), a postdoctoral fellowship in Enquist's laboratory]. All of the tree species they studied had very similar branching patterns regardless of their difference in appearance. In addition, there were similar general ecological, biological and physical principles that resulted in a similar branching architecture across those species over the course of evolution.

According to Bentley (2013), there is a relationship between the size and shape of branches and how much water the trees lose through evaporation. She found the theory to be correct in that it allows for predictions about a tree's function, depending on its size, and that the theory's principles apply across species, despite their differences in appearance.

In conclusion it may be stated that both the crown architecture and branching patterns are the characteristics of a tree species which can be related to the adaptation of each species to a particular environment. The organisation of these features contribute to the productivity of a particular species in a particular environment. The intensity of these traits of each species vary in the understorey and open environment. Therefore, concerted research input needs to address this aspect, in order to assess the productivity of trees. Besides, there is a necessity to characterise the traits of each species in a particular forest ecosystem,

5.3   Tree crown architecture

The tree crown depicts the top part of the tree, which consists of branches that grow out from the main trunk and support the various leaves used for photosynthesis. The variability of crown architecture in different species of woody plants in a forest offers a beautiful landscape in any particular forest. It is of great interest to the landscape architect as well as to the foresters in their selection for a particular landscape. This is the most vital part of a tree species contributing to its productivity through the process of photosynthesis. This is highly dependent on the mode of branching arrangement, its size and leaf orientation in the branches. The branches form different shapes characteristic of a species. While all trees feature a crown, several types of crowns adorn different types of trees. Thus, tree crowns are adapted to fit their role. The variability in crown architecture among tree species in a forest ecosystem confers characteristics of great interest to aesthetic architects and siviculturists for the identification of a species.

The tree crowns are responsible for capturing solar radiation and thereby serving as a solar panel, not permitting the sun's rays to reach the ground. The crown of a tree refer to the upper layer of leaves. The classification of crown architecture of the trees and shrubs of the Tamaulipan thornscrub has three categories: open (<39% of the sky is obstructed by other crowns); partly closed (40–69% of the sky is obstructed by the crowns of other trees); closed (70–100%) of the sky is covered by other crowns).

The crown architecture of trees varies in form: globose, pyramidal, conical, elliptic, rectangular or irregular. The growth habit of the crown varies among these forms. In general, they are semierect, but branches may be vertical, erect, extended or open. These forms are influenced by the form and diameter of the crown and the height of the plant. Other variable is the distribution of branches: irregular, horizontal or ascending. The leaf apex varies, such as obtuse, pointed, round or apiculate.

Isebrands and Nelson (1982) described the crown architecture, branch morphology and distribution of leaves within the crown of short-rotation *Populus* "Tristis" related to biomass production in northern Wisconsin. The relationship of leaf area to above ground biomass productivity was also estimated for the same trees.

The first-order branches within the trees showed acrotony and were predominantly long shoots. No branching higher than third order was observed. Leaf size and specific leaf weight were greatest on the current terminal shoot and decreased from the upper portion of the crown to the base. Some 95% of the long shoots in the six year old trees were in the three uppermost vertical strata (5–8 m) and 95% of the short shoots were in the lowermost leaf-containing vertical strata (3–6 m). Long-shoot leaves had higher specific leaf weights than short-shoot leaves attached to branches on the same height growth increment. Leaf-area indices (LAI) were 7.6 and 8.8 $m^2\ m^{-2}$ for the five and six year old stands, respectively. Leaf area per tree was linearly related to the aboveground biomass of the tree. The linear regression line for the relationship between leaf area and diameter^2 × height (D^2H) for the six year old trees in the study was statistically different from that for the five year old trees. The results suggested this relationship may serve as a useful quantitative index of crown closure in poplar stands. The results also suggested some crown morphological criteria useful for the selection and breeding of improved poplar trees for short-rotation intensive culture.

Subsequently, O'Connell and Kelty (1994) analysed the crown architecture of understorey and open-grown white pine (*Pinus strobus* L.) saplings, using 15 understorey saplings and 15 open-grown saplings that were selected to have comparable heights (mean of 211 cm, range of 180–250 cm). Mean ages of understorey and open-grown trees were 25 and 8 years, respectively. Understorey trees showed a lower degree of apical control, shorter crown length and more horizontal branch angle, contributing to a broader crown shape than that of open-grown trees. Total leaf area was greater in open-grown saplings than in understorey saplings, but the ratio of whole-crown silhouette (projected) leaf area to total leaf area was significantly greater in understorey pine (0.154) than in open-grown pine (0.128), indicating that the crown and shoot structure of understorey trees exposed a greater percentage of leaf area to direct overhead light. Current-year production of understorey white pine gave significantly less than that of open-grown white pine, but a higher percentage of current-year production was allocated to foliage in shoots of understorey saplings. The overall change to a broader crown shape in understorey white pine was qualitatively similar, but much more limited than the changes that occurred in fir and spruce. This may prevent white pine from persisting in understorey shade as long as fir and spruce saplings.

Similarly, Ceulemans et al. (1990) studied crown architecture, including branching pattern, branch characteristics and orientation of proleptic and sylleptic branches in five poplar clones (*Populus deltoides*, *P. trichocarpa* and *P. trichocarpa* × *P. deltoids* hybrids), grown under intensive culture in the Pacific Northwest, USA. Branch characteristics measured were number, length, diameter, biomass, angle of origin and termination. The results suggested that genotype has a major influence on crown architecture in *Populus*. Clonal differences in branch characteristics and branching patterns were observed that showed striking differences in crown form and architecture. Branch angle and curvature showed significant difference between

clones and between height growth increments within clones. Branch length and diameter were significantly correlated in all clones. Sylleptic branches and the considerable leaf area they carry have important implications for whole tree light interception and thus play a critical role in the superior growth and productivity of certain hybrid poplar clones. The considerable variation in branch characteristics implies a strong justification for including them in selection and breeding programmes for *Populus*.

Gilmore and Seymour (1997) studied the crown architecture of *Abies balsamea* from four canopy positions. Data collected from four distinct canopy positions from each of 39 *A. balsamea* trees were used to construct models to describe the cumulative leaf area distribution within the crown and to predict the needle mass of individual branches, the average branch angle, branch diameter, branch length, crown radius per whorl and the average number of living branches per whorl. They tested the hypothesis that regression models were equal between canopy positions and that a model to predict branch needle mass was valid at the northern and southern extremes of the central climatic zone of Maine. Canopy position had an effect on the models constructed to predict needle mass, branch angle, branch diameter, branch length, crown radius and the number of living branches per whorl. However, compared with an expanded model that incorporated parameters calculated for each crown class, there was only a small loss in model precision when a general model constructed from data pooled from all crown classes was used to predict needle mass, branch angle and branch diameter. Regression equations unique to each crown class were needed to predict crown shape and leaf area distribution in the crown satisfactorily. The branch needle mass model, which was constructed from data collected at the southern extreme of the central climatic zone of Maine, consistently underestimated needle branch mass when applied to the northern extreme of the central climatic zone

Moorthy et al. (2011) made a field characterisation of olive (*Olea europaea* L.) tree crown architecture using terrestrial laser scanning (TLS) data. Intelligent laser ranging and imaging system (ILRIS-3D) data was obtained from individual tree crowns at olive (*Olea europaea*) plantations in Córdoba, Spain. From the observed 3D laser pulse returns, quantitative retrievals of tree crown structure and foliage assemblage were obtained. Best methodologies were developed to characterise diagnostic architectural parameters, such as tree height ($r^2 = 0.97$, rmse = 0.21 m), crown width ($r^2 = 0.97$, rmse = 0.13 m), crown height ($r^2 = 0.86$, rmse = 0.14 m), crown volume ($r^2 = 0.99$, rmse = 2.6 m^3) and plant area index (PAI) ($r^2 = 0.76$, rmse = 0.26 m^2 m^2). This research demonstrates that TLS systems can potentially be the new observational tool and benchmark for the precise characterisation of vegetation architecture for improved agricultural monitoring and management.

It may be concluded that tree crown architecture shows a large variation in shape, form, very few monopodial, mostly pseudomonopodial and few sympodial types. A combination of these classes of branching systems confers beauty to the landscape and the ecosystem. Very few species possess a typical architecture that an aesthetic architect may use for town planning. We selected species with a good architecture for planting in urban areas: *Cordia boissieri*, *Diospyros texana*, *Ebenopsis ebano*, *Celtis pallida*, *Parkinsonia texana* and *Havardia pallens* for planting in semiarid regions of Mexico. In general, most of the tree species possess open canopy leaves with medium narrow leaves, very few with broad leaves with ramifying branches, with open canopy leaves. Sympodial types with narrow leaves exist in between other types of crown architecture. Sympodial types show profuse branching from the base of the plant with spreading thinner branches around. There is a necessity to quantify the proportion of different plant species with medium broad leaves in a forest ecosystem for detecting the mode of co-existence and adaptation as well as for their capacity in the capture of solar radiation and sharing available spaces with neighbouring species.

In our visit to mangrove forests in tropical rainforest in Riviera Maya, a region of ancient Maya Culture, Mexico, mangrove species were observed to possess monopodial to pseudomonopodial branches arising from the main stem, dichotomously branched and possessing semi-broad thick leaves with a waxy coating to avoid water loss by transpiration. Mangroves possess open canopy leaves for the efficient capture of solar radiation for photosynthesis, necessary for vital functions. They produce enormous hanging rhizophorous roots with fleshy and porous holes for the absorption of moisture and oxygen for respiration. These rhizophores grow downwards eventually, fix themselves to the ground and serve as a support to the main plant, thereby acting as a barrier to stop tourists from entering mangrove forests. The inundated lower branches of rhizophores roots serve as a breeding ground for fishes, crabs and other sea animals. We also observed that the trees prevalent in rainforests on the way to Riviera Maya possess open canopy leaves with a high capacity to capture solar radiation. In our visit to arid lands of Arizona, Mezquites (*Prosopis* sp.) is the dominant gregarious tree that possesses pseudomonopodial branching and open canopy leaves. The plant possesses specific morpho-anatomical and physiological mechanism of resistance to drought in those arid lands.

5.4 Leaf traits

In a forest, different plant species grow together with a great diversity of leaf shapes. sizes and leaf surfaces characteristically varying in their intensity of leaf pigments and leaf anatomical traits related to their species' adaptation to the environment. Trees with leaves of different sizes growing together is probably a mechanism for co-existence in sharing the solar radiation for photosynthesis without any competition. In addition, tree species show a large diversity in attaining height, leaf canopy and crown architecture, thereby facilitating each species to obtain sunlight for its photosynthesis. Species with a closed canopy are probably not efficient in capturing solar radiation, as the lower leaves are not fully exposed to sunlight. Therefore, a combination of species with broad and narrow leaves help in a mutual coexistence for the capture of sunlight. Besides, several leaf anatomical traits are related to the adaptation of species to environmental stresses such as drought, resistance to insects and so on.

5.5 Variability in leaf canopy architecture

Leaves contribute greatly to plant growth and productivity for photosynthesis and nutrient contents. There exists a great diversity among plant species in growth form, leaf size and shape and canopy management. In addition, there exist some general relations across a wide range of species in leaf traits between species with respect to carbon fixation. The outer canopy leaves and its specific leaf area (SLA, leaf area per unit of mass) tends to be correlated with leaf nitrogen per unit dry mass, photosynthesis and dark respiration sites (Wright et al., 2004). The availability of nutrients in leaves is essential for efficient plant function. Besides, a large variation in traits among species favours nutrient conservation. The importance of leaf nutrient content in plant function is well documented with respect to nutrient conservation, life span, high leaf mass per area, low nutrient concentration and low photosynthetic capacity (Reich et al., 1997). Sufficient research activities have been undertaken on nutrient content and metabolism in leaves.

In our preliminary studies, we observed a large variability in the contents of macro- and micronutrients in native woody species of the semiarid regions of north-eastern Mexico. In our survey, both in the Tamaulipan thornscrub of semiarid Mexico and in our recent visits to rainforests of southern Mexico, we observed there exists a great variability in leaf canopy with respect to open canopy, semi-closed and closed canopy associated with variations in leaf size, shape and surface structure. Species can be classified on the basis of this important trait. The hypothesis is that tree species with open canopy have a high capacity for the capturing solar radiation for the photosynthetic process; thereby, high carbon sequestration is expected compared to those with closed canopies. We also observed that tree species with open canopy produced highly branched, thin stems compared to those with closed canopy which produced a thick basal stem area with thick primary branches. Those with open canopy had highly ramified thin stems. We went further to confirm our hypothesis by taking more quantitative data, yet to be analysed. In our recent study, we estimated the contents of macro- (Ca, K, Mg, N and P) and micronutrients (Cu, Fe, Mn and Zn) of 10 native shrubs and trees of semiarid Mexico. We observed a large variability in nutrient contents between species. We also observed a large variability in carbon content. We selected four species which contained high carbon concentration, about 50%. Thus the mean value of C content ranged from 37 to 50%. The species with carbon content close to 50% were: *Leucophyllum frutescens*, *Forestiera angustifolia*, *Bumelia celestrina* and *Acacia berlandieri*. These species may be recommended for reforestation purpose in areas contaminated with a high carbon dioxide load. Interestingly, we observed that these four species possess an open canopy architecture, revealing that the shrub species have a high capacity to capture solar radiation for the efficient production of photosynthates and high carbon fixation. Further studies are in process on the quantification of plant traits and its relation to plant productivity. This could be a promising line of research in the field of forest science.

To further test our hypothesis based on our results and understanding of the relevant literature, we concluded that our hypothesis should be that photosynthetic efficiency and carbon fixation are related to leaf canopy architecture. In order that we fully address this question, differences in the ecology of individual species and different kinds of plant communities that may occur must be evaluated. For this reason, we propose three types of plant communities. These may be native multi-species natural or seminatural vegetation, monospecific plantations (monoculture) of a given species and planted mixtures of some form of sets of species. This division is expected to encompass all potential variability with respect to factors such as allelopathic effects and individual habitat species preferences as well as biomass gain and canopy architecture resulting from genetic variability, not only within species but also the performance of individuals with respect to habitat differentiation.

For the majority of these species, surveys from field measurements or sample plots will provide almost certainly an inadequate number for species level analysis as it is most common in tropical and subtropical plant communities to have a very low dominance of individual species on a per hectare basis, with consequently a very large number of them. Therefore, different species will have to be clustered (aggregated in ecological groups for the purposes of the analysis). This grouping will be related to two kinds of variables. The first will be canopy architecture while the second will be biomass growth rate and more specifically diameter growth rate. The second above mentioned characteristic is related to both photosynthetic efficiency and carbon sequestration because of its ability to promote the growth of woody perennials as a fundamental growth index, as has been proven worldwide by a variety of studies. It also provides a cheap and reliable index in relation to other measures like canopy biomass gain because it has been shown to be correlated well with these and provides a simple and viable alternative to destructive sampling techniques. A priori and a posteriori grouping of species are both perfectly acceptable methodologies to cluster species and with respect to this, non-continuous variables are perfectly acceptable for a given trait such as type of canopy architecture or mean leaf size in maturity.

The sampling scheme and grouping methodology described above will also take into consideration the significant differences in the studied variables with respect to the stage in the life cycle of a given species. Studies have shown this to differentiate and it is interesting to conduct separate analyses of this to determine more specific differences, if the finalised datasets are adequate for the purposes of statistical analysis.

Ideally, a wide range of habitat disturbance classes will also be measured. These will include former agricultural land use that is now abandoned, among other kinds of habitats. This is because certain studies of this kind used data from undisturbed vegetation only and the statistical relationships proved weak, while others that included disturbed vegetation results were more statistically significant, possible because of greater differentiation between fields due to the greater spectrum of habitat conditions. Studies which include both disturbed and undisturbed ecosystems have this. For the Tamaulipan thornscrub ecosystem in the north-eastern region of Mexico, the desire to address such issues is very relevant because of the endangered status of the ecosystem warranting immediate action to be taken towards devising schemes for sustainable management, conservation and restoration of this habitat; such relationships need to be known and quantified with a very high level of precision. This type of research will also assist relevant work elsewhere by enriching the necessary knowledge base. Care must be taken in conducting such measurements in the field as the relative health of individual specimens of a given species may vary significantly and care must be taken to measure only healthy or relatively healthy individuals by using some form of canopy health status score.

Inter-institutional collaboration will be essential in addressing this hypothesis, not only for collecting data but also for issues related to arriving at the most parsimonious working model for testing the scientific hypothesis (Maiti et al., 2014).

5.6 Variability in leaf traits of 13 native woody species in semiarid regions of North-eastern Mexico

Ratikanta Maiti[1], Humberto Gonzalez-Rodriguez[1] and Theodore Karfakis[2]

[1] *Facultad de Ciencias Forestàles, Universidad Autonoma de Nuevo Leon, Linares, Mexico*
[2] *Terra Sylvestris Non-Governmental Organisation, Kalamos Lefkados, Greece*

The present study was undertaken with the goal of analysing the morphology and variability of leaf length, width, petiole length, total leaf length, fresh and dry weight of individual leaves of 14 native shrub species in northeastern Mexico. The native species *Cordia boissieri*, *Condalia hookeri*, *Sargentia greggii*, *Diospyros texana*, *Zanthoxylum fagara*, *Sideroxylon celastrinum*, *Karwinskia humboldtiana*, *Celtis pallida*, *Guaiacum angustifolium*, *Prosopis laevigata*, *Celtis laevigata*, *Parkinsonia texana*, *Forestiera angustifolia* and *Havardia pallens* were chosen due to their ecological and economic importance to the rural population and the large variability in morphological characteristics between them. Descriptive statistical analyses showed that there was large variability in these leaf traits between different species. Principal component analysis (PCA) revealed that it was possible to produce two axes that could explain more than 83% of the observed variation and could therefore be used in the future for separating tree species in ecological guilds and to study species- and vegetation community-level responses to perturbations and individual performance in the field or under experimental conditions. This suggests that similar advances are possible for other species both in the region of study and elsewhere.

5.6.1 Introduction

Leaves play a vital role in the growth and development of a plant through the process of carbon assimilation, exchange of gases through stomatas and loss of excess water through transpiration. They vary largely in size, form, shape, surface structure, thickness and so on. Enormous research activities have been directed towards leaf traits and their role in plant metabolism.

The rates of leaf emergence and expansion influence leaf growth and its duration.

In this respect, a study was undertaken by Sun et al. (2006) on the effect of leaf traits on leaf emergence phenology, timing of leaf emergence, leaf expansion rate, duration of leaf emergence and expansion, leaf mass per area in temperate woody species in eastern Chinese *Quercus fabri* forests. Regression analysis across species revealed the relationship between leaf phenology and the leaf traits mentioned. Species with small leaves emerged earlier than the species with large leaves. Leaf expansion rate was positively correlated with leaf area and timing of leaf emergence but no significant relationship was found between leaf size and leaf emergence period. In this respect, Campanella and Bertller (2009) assessed leafing patterns (rate, timing and duration of leafing and leaf chemistry) of four evergreen shrubs in the Patagonian Monte, Argentina. They observed two species, *Chuquiraga* which produced new leaves concentrated in massive short leafing (5–48 days) and *Larrea* whose leaves emerged gradually (128–258 days). It is concluded that the difference in leafing pattern could provide evidence of ecological differentiation among co-existing species.

The anatomical and morphological leaf traits of *Quercus ilex*, *Phillyrea latifolia* and *Cistus incanus* were studied. The specific leaf weight (SLW) and total leaf thickness, leaf density index and leaf inclination changed according to leaf age. Leaf folding may be related to the less xeromorphic leaf structure of *C. incanus* (Gratani and Bonbelli, 2000).

Gratani and Varone (2004) studied leaf morphology, water relations, leaf lifespan and gas exchange in *Erica arborea* L., co-occuring in the Mediterranean maquis.

The variation in these traits are dependent on the prevalence of drought and recovery which may be related to a narrow root system. This could be related to a variation in root tissues.

Calagari et al. (2006) studied morphological variation in leaf traits of *Populus euphratica* Oliv. in natural populations. In this study, leaf length, maximum leaf width, leaf area, distance between middle of maximum leaf width and leaf blade, distance between maximum of leaf width and midrib and the ratio of petiole length to leaf length are good discriminating criteria for classifying various populations.

Markesteijn et al. (2007) reported the influence of light on leaf trait variation in 43 tropical dry forest tree species. They studied leaf trait variation and the relation between trait plasticity and light demand, maximum adult stature and ontogenetic changes in crown exposure of the species. Leaf trait variation was mainly related to differences between species and to a minor extent to differences in light availability. Traits related to the palisade layer, thickness of the outer cell wall and N(area) and P(area) showed the greatest plasticity, suggesting their importance for leaf function in different light environments. Short-lived species showed the highest trait plasticity. Overall plasticity was modest and rarely associated with juvenile light requirements, adult stature or ontogenetic changes in crown exposure. Dry forest tree species possessed a lower light-related plasticity than wet forest species, probably because wet forests cast a deeper shade. In dry forests, light availability may be less limiting and low water availability may limit leaf trait plasticity in response to irradiance.

Gianoli et al. (2009) studied the distribution and abundance of vines along the light gradient in a southern temperate rain forest. Canopy openness was significantly different in the three light environments. Their study in a temperate rain forest raised the question that the widespread notion of vines may be a consequence of the abundance of some lianas in disturbed tropical forest sites.

Moreno et al. (2010) studied changes in the traits of shrub canopies across an aridity gradient in northern Patagonia, Argentina, where precipitation was low (125–150 mm) and the mean annual temperature varied from 8 °C to 13.5 °C. They characterised the presence of spiny leaves, leaf pubescences, thorny stems and photosynthetic stems. Richness and diversity of shrub species and morphotypes were positively correlated with aridity. Increased diversification in the traits, species and morphotypes in shrub canopies with increasing aridity support the hypothesis that the variability in aridity provides an ecological differentiation between shrub species, facilitating their coexistence.

Gotsch et al. (2010) studied the variation in leaf traits and water relations in 12 evergreen and semideciduous woody species in seasonal wet and dry forest of Costa Rica. Over two years they analysed leaf nitrogen (N), leaf carbon (C), specific leaf area (SLA), cuticle thickness, leaf thickness, leaf lifespan (LLIS), leaf water content and canopy openness, among other traits. The species showed large variations between these traits, but season, forest and their interactions had shown a great influence on specific leaf trait variations. Leaf traits that contributed most to variation across sites were leaf carbon, leaf water potential, leaf thickness and SLA. Traits that contributed most to variation across seasons were leaf toughness, leaf water potential and leaf water content.

Soil factors are driving forces in influencing spatial distribution and functional traits of plant species. Four morphological and 19 chemical leaf traits (macronutrients, trace elements and N signatures) were analysed in 17 woody plant species. In this study (Domínguez et al., 2012), leaf mass per area (LMA) and leaf dry matter content (LDMC) were significantly related to many leaf nutrient concentrations, but only when using abundance-weighted values at community level. Among-traits links were much weaker for the cross-species analysis. Nitrogen isotopic signatures were useful for understanding different resource-use strategies. Community-weighted LMA and LDMC were negatively related to light availability, contrary to what was expected.

Spatial variation in filters imposed by the abiotic environment causes variation in functional traits within and between dependent plant species. This is abundantly clear for plant species along elevational gradients, where parallel abiotic selection pressures give rise to predictable variation in leaf phenotypes among ecosystems. Three key leaf functional traits are associated with axes of variation in both resource competition and stress tolerance: leaf mass:area ratio (LMA), leaf nitrogen content per unit mass (N_{mass}) and N content per unit area (N_{area}). The results indicate that environmental filtering both selects locally adapted genotypes within plant species and constrains species to elevational ranges based on their range of potential leaf trait values (Read et al., 2014).

Rios et al. (2014) studied divergence and phylogenetic variation of ecophysiological traits in lianas and trees. The mean phylogenetic distance was 1.2 times greater among liana species than among tree species. They found support for the expected pattern of greater species divergence in lianas, but did not find consistent patterns regarding ecophysiological trait evolution and divergence.

Leaf traits govern several plant functions related to the growth and development of the plant. Although there are positive intraspecific relationships between leaf area per unit dry weight (SLA) and between concentrations of P and N, but these relations are affected by soil concentration. It was reported that physiological activity in foliage was co-limited by both P and N availability in the woodland meadow but there were positive relationships between SLA and nitrogen concentration in the woodland meadow and in the bog. There was a positive relationship between SLA and P concentration but no relationship between SLA and N concentration increasing with dcreasing SLA. Total variation in foliar structural and chemical chracteristics was similar in all sites (Niinemets and Kull, 2003). Poorter and Bongers (2006) compared the leaf traits and plant performance of 53 co-occurring tree species in a semi-evergreen tropical moist forest community. The species showed large variations between all leaf traits such as life span, specific leaf area, nitrogen, assimilation rate, respiration rate, stomatal conductance and photosynthetic water use efficiency. Photosynthetic traits were strongly related with leaf traits; and specific leaf area predicted mass-based rates of assimilation and respiration. Leaf life span predicted many other leaf characteristics. Leaf traits were closely correlated with the growth, survival and light requirement of the species. Leaf investment strategies varied on a carbon gain in short-term against long-term leaf persistence. Thereby, it was linked to a variation in whole-plant growth and survival. Leaf traits were good predictors of plant performance. High growth in gaps was promoted by cheap, short-lived and physiologically active leaves (Poorter and Bonger, 2006).

Variation of morphological and chemical traits of perennial grasses in arid ecosystems are largely influenced by the relative abundance of shrubs. Tiller height, blade length and blade area of *Poa ligularis* increased significantly with increasing relative shrub cover, while the concentration of soluble phenolics in the leaf blades decreased with increasing shrub cover. Nitrogen concentration and specific blade area decreased with increasing shrub cover (Moreno and Bertlier, 2012).

In order to determine suitability for afforestation, different parameters, such as relative crop growth rate (CGR), net assimilation rate, specific leaf area (SLA) and leaf area index were calculated. CGR was better correlated with NAR. At early stage CGR and diameter could be used as selection criteria for selection for afforestation (Lamers et al., 2006).

A decline in herbage use reduced individual leaf mass, specific leaf area and shoot digestibility but increased leaf C and dry matter content (Louault et al., 2005).

A study was carried out at the experimental station of Facultad de Ciencias Forestales, Universidad Autonoma de Nuevo Leon, located in the municipality of Linares (24°47′N, 99°32′W) at an elevation of 350 m. The climate is subtropical or semiarid with warm summer, monthly mean air temperature vary from 14.7 °C in January to 23 °C in August, although during summer the temperature goes up to 40 °C. Average annual precipitation is around 805 mm with a bimodal distribution. The dominant type of vegetation is the Tamaulipan thornscrub or subtropical thornscrub woodland (SPP-INEGI, 1986). The dominant soil is deep, dark grey, lime-clay, vertisol with montmorrillonite, which shrinks and swells remarkably in response to changes in moisture content.

The leaf traits of the following native plants of north-eastern Mexico were included in this study (Table 5.16).

Twenty mature leaves of each species were selected at random and measured for leaf length (cm), leaf breadth (cm), petiole length (cm), dry weight (mg). At first, two exploratory analyses were performed. The first was descriptive statistics and the second was a series of normality tests for each separate variable using the Anderson–Darling normality test. Plots were also done to check on the presence and properties of outliers in the datasets. Next, PCA analysis was done by standardising the variables and using a correlation matrix for the analyses. All analyses where performed in Minitab V 17 software.

Fourteen trees and woody native species used for different purposes in the region were selected for this research. Fully expanded leaves from terminal branches were selected through a stratified and random sampling from each of the 14 species. Leaves were taken to the laboratory immediately and measurements were taken. Trees were selected at random from the

Table 5.16 Native plants of northeastern Mexico included in the study.

Species	Family	Growth habit
Cordia boissieri A. DC.	Boraginaceae	Shrub
Condalia hookeri M. C. Johnst.	Rhamnaceae	Tree
Diospyros texana Scheele.	Ebenaceae	Tree
Sideroxylon celastrinum (Kunth) Pennington	Sapotaceae	Tree
Karwinskia humboldtiana (R. & S.) Zucc.	Rhamnaceae	Treea
Zanthoxylum fagara (L.) Sarg.	Rutaceae	Shrub
Celtis pallida Torr.	Ulmaceae	Shrub
Prosopis laevigata (Humb. &Bonpl. Ex Willd.) M. C. Johnston	Mimosaceae	Tree
Forestiera angustifolia Torr.	Oleaceae	Shrub
Guaiacum angustifolium (Engelm.) Gray	Zygophyllaceae	Shrub
Sargentia greggii S. Watson	Rutaceae	Shrub
Celtis laevigata Willd.	Ulmaceae	Tree
Havardia pallens (Benth.) Britton & Rose	Fabaceae	Shrub

Table 5.17 Statistical means of all respective leaf traits for the 14 woody plant species selected.

Species	Leaf length (cm)	Leaf width (cm)	Petiole length (cm)	Total leaf legth (cm)	Leaf area (cm^2)	Leaf dry weight (mg)	Specific leaf area (cm^2 mg^{-1})	Leaf fresh weight (mg)
Cordia boissieri	6.26	3.26	1.28	7.35	14.94	0.18	77.39	0.34
Condalia hookeri	3.88	2.06	0.37	4.25	5.07	0.02	186.21	0.07
Sargentia greggii	9.37	4.40	0.69	10.06	29.58	0.38	82.02	0.78
Diospyros texana	3.94	2.03	0.25	4.19	6.19	0.05	134.59	0.1
Zanthoxylum fagara	6.70	2.39	0.22	6.92	6.86	0.06	116.61	0.14
Sideroxylon celastrinum	3.18	1.48	0.31	3.49	3.03	0.03	127.20	0.09
Karwinskia humboldtiana	5.12	2.31	0.6	5.73	9.80	0.05	215.35	0.12
Celtis pallida	3.54	2.06	0.35	3.89	5.84	0.04	128.62	0.12
Guaiacum angustifolium	3.71	2.68	0.77	4.48	3.75	0.03	120.08	0.06
Prosopis laevigata	7.36	7.56	2.69	10.05	13.78	0.16	86.17	0.33
Celtis laevigata	7.53	3.72	1.31	8.84	19.38	0.19	100.72	0.34
Parkinsonia texana	2.05	2.86	0.74	2.79	2.73	0.02	127.22	0.05
Forestiera angustifolia	3.68	0.59	0.1	3.78	1.75	0.01	152.31	0.02
Havardia pallens	3.33	4.18	1.34	4.67	5.44	0.03	141.47	0.08

Tamaulipan thornscrub, Linares, Nuevo Leon, Mexico. All species are native to arid and semiarid zones in Mexico and adjacent USA territories.

Large variations in leaf traits between the species were observed that were undoubtedly due to inherent species-specific differences (Table 5.17). Leaf length on average varied from 2.0 to 9.37 cm. Highest leaf length was observed in *Sargentia greggii* and *Celtis laevigata* (9.37 and 7.53 cm, respectively) while

Prosopis laevigata and *Zanthoxylum fagara* also showed high values (7.36 and 6.7 cm, respectively). Leaf width ranged from 2.0 to 4.5 cm. *P. laevigata* and *S. greggii* had the maximum width observed (7.56 and 4.4 cm, respectively). Petiole length ranges from 0.1 to 2.7 cm. Maximum petiole length was found in *P. laevigata* (2.69 cm). Minimum petiole length was observed in *Forestiera angustifolia* (0.1 cm) and *Zanthoxylum fagara* (0.2 cm). Total leaf length varied from 2.7 to 10.0 cm. Greatest leaf length was observed in *S. greggii* (10 cm) and *P. laevigata* (10 cm). Minimum leaf length was observed in *Parkinsonia texana* (2.79 cm) and *Sideroxylon celastrinum* (3.49 cm). The maximum leaf area was observed in *Sargentia greggii* (29.58 cm^2) and *Celtis laevigata* (19.38 cm^2). Minimum leaf area was in *Forestiera angustifolia* (1.75 cm^2) and *Sideroxylon celastrinum* (3.03 cm^2). Leaf highest dry weight was observed in *Sargentia greggii* (0.38 mg), *Celtis laevigata* (0.19 mg) and *Prosopis laevigata* (0.16 mg). Minimum dry weight was observed in *Forestiera angustifolia* (0.01 mg) and *Condalia hookeri* (0.02 mg). Specific leaf area (SLA) varied from 77.39 cm^2 mg^{-1} (*Cordia boissieri*) to 215.35 cm^2 mg^{-1} (*Karwinskia humboldtiana*). Leaf fresh weight varied on average from 0.02 mg (*Forestiera angustifolia*) to 0.78 mg (*Sargentia greggii*). Similar observations to those of this study have been made elsewhere for a similar vegetation community. It was found that the height, blade length and blade area increased significantly with shrub cover. The large variations in leaf traits among these ecologically important woody species of the Tamaulipan thornscrub are therefore very likely to contribute to species diversification, co-existence and adaptation to the semiarid condition, as has been reported elsewhere.

5.6.2   Principal component analysis

Principal component analysis (PCA; Table 5.18 and Figure 5.15) revealed that the first two PCA axes accounted for the largest proportion of observed variation; and hence we can keep these for examining further hypotheses. The outlier plot for the PCA analysis reasserted the significant presence of outliers (Figure 5.16). This may simply be because of very high interspecific differences. These results indicate that it is possible to address questions regarding leaf ecophysiology for individual species for specific scientific hypotheses by using the two PCA axes. This very importantly includes areas like leaf physiological response to perturbations such as atmospheric pollution and animal browsing, among other constraints. These results also show that specific variables describing leaf physiology are highly important in determining differences and thereby responses to environmental stresses. It will also be possible to describe interspecific differences in terms of leaf ecophysiology using the two PCA axes in the future for the species used in this study. Finally it

Table 5.18 Eigenvalues and respective probabilities of the principal component analysis.

Eigenvalue	Proportion	Cumulative
5.5534	0.694	0.694
1.1566	0.145	0.839
0.7961	0.1	0.938
0.2913	0.036	0.975

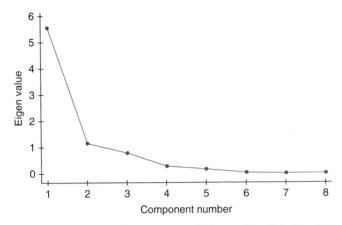

Figure 5.15 Screen plot of the principal components relative to the original variables.

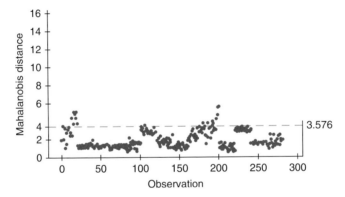

Figure 5.16 Outlier plot of the principal components of the analysis.

Table 5.19 Principal component axes coefficients for the analysis.

Variable	PC1	PC2	PC3	PC4
Fresh weight (mg)	0.398	0.248	0.038	−0.340
Leaf length (cm)	0.385	0.200	−0.196	0.564
Leaf width (cm)	0.313	−0.541	−0.075	−0.324
Petiole length (cm)	0.259	−0.694	−0.072	0.050
Total length (cm)	0.402	−0.029	−0.188	0.495
Leaf area (cm^2)	0.395	0.250	−0.160	−0.261
Dry weight (mg)	0.399	0.244	0.073	−0.331
SLA (cm^2 mg^{-1})	−0.226	0.047	−0.940	−0.189

is possible to use the first two PCA axes found significant (Table 5.19) for future examinations of hypotheses regarding ecosystem functional responses to perturbations or examining forest dynamics in communities composed of these or similar species. These results are in agreement with other similar studies looking at large woody shrubs and tree species in other parts of the world.

5.6.3   Conclusions

There was an obviously very large inter-specific diversity among the species studied. It was possible, however, to address the issue of finding a methodology for examining differences in leaf morphology between the different species through the new variables (axes) produced by principal component analysis. It was also possible to limit the inherently large variability using these and to produce variables that were considerably more amenable to statistical analysis. More species need to be added through additional sampling to the continuum and redo the PCA analyses in the future with these included in order to see if this relationship holds for other species and for a broader range of leaf morphological characteristics. Following this final analysis, it will be possible to address more diverse issues in the endangered Tamaulipan scrub ecosystem, like response to anthropogenic perturbations and climate change. Until that time, using the species in this study as indicator species, it is possible to address the above ecologically important issues with an acceptable level of accuracy for issues requiring both temporary and permanent sampling programs.

Bibliography

Aiba S.-I., Kohyama T. **1997**. Crown architecture and life-history traits of 14 tree species in a warm-temperate rain forest: significance of spatial heterogeneity. *Journal of Ecology* **85**:611–624.

Bentley L.P. **2013**. *Tree achitecture*. http://phys.org/news/2013-08-tree-architecture-reveals-secrets-forest.html#jCp

Calagari M., Modirrahmati, A.R., Asadi F. **2006**. Morphological variation in in leaf traits of *Populus euphratica* Oliv. natural populations. *International Journal of Agriculture and Biology* **8**(6):754–758.

Campanella M.V., Bertller M.B. **2009**. Leafing patterns and leaf traits of four evergreen shrubs in the Patagonian Monte, Argentina. *Acta Oceologica* **35**(6):831–837.

Ceulemans R., Stettler R.F., Hinckley T.M., Isebrands J.G., Heilman P.E. **1990**. Crown architecture of *Populus* clones as determined by branch orientation and branch characteristics. *Tree Physiology* **7**(1/4):157–167.

Domínguez M.T., Aponte C., Pérez-Ramos I.M., García L.V., Villar R., Marañón T. **2012**. Relationships between leaf morphological traits, nutrient concentrations and isotopic signatures for Mediterranean woody plant species and communities. *Plant and Soil* **357**(1/2):407–424.

Enquist B. **1998**. *The branch systems of trees*. http://phys.org/news/2013-08-tree-architecture-reveals-secrets-forest.html#jCp

Gianoli E., Saldaña A., Jiménez-Castillo M., Valladares F. **2009**. Distribution and abundance of vines along the light in a southern temperate rain forest. *Journal of Vegetation Science* **21**(1):66–73.

Gilmore D.W., Seymour R.S. **1997**. Crown architecture of *Abies balsamea* from four canopy positions. *Tree Physiology* **17**(2): 71–80.

Gotsch S.G., Powers J.S., Lerdau, M.T. **2010**. Leaf traits and water relation of 12 evergreen species in Costa Rican wet and dry forests: patterns of intra-specific variation across forests and seasons. *Plant Ecology* **211**(1):133–146.

Gratani L., Bonbelli A. **2000**. Correlation between leaf age and other leaf traits in three Mediterranean maquis shrub species: *Quercus ilex, Phillyrea latifolia* and *Cistus incanus*. *Environmental and Experimental Botany* **43**(2): 141–153.

Gratani L., Varone L. **2004**. Leaf key traits of *Erica arborea* L. co-occuring in the Mediterranean maquis. *Flora – Morphology, Distribution, Functional Ecology of Plants* **199**(1): 58–89.

Heuret P., Guédon Y., Guérard N., Barthémy D. **2003**. Analysing branching pattern in plantations of young red oak trees (*Quercus rubra* L., Fagaceae). *Annals of Botany* **91**(4):479–492.

Isebrands J.G., Nelson N.D. **1982**. Crown architecture of short-rotation, intensively cultured *Populus* II. Branch morphology and distribution of leaves within the crown of *Populus* 'Tristis' as related to biomass production. *Canadian Journal of Forest Research* **12**(4):853–864.

Lamers J.P.A., Khamzina A., Worbes M. **2006**. The analysis of physiological and morphological attributes of ten tree species for early determination of their suitability to afforest degraded landscapes in the Aral Sea Basin of Uzbekistan. *Forest Ecology and Management* **221**:249–239.

Leopold L.B. **1971**. Trees and streams: The efficiency of branching patterns. *Journal of Theoretical Biology* **31**(2):339–354.

Louault F., Pillar V.D., Aufrere J., Garnier E., Soussana J.-F. **2005**. Plant traits and functional types in response to reduced disturbance in a semi-natural grassland. *Journal of Vegetation Science* **16**:151–160.

Maiti R., González-Rodríguez H., Cantú-Silva I. et al. **2014**. Editorial. *IJBSM* **5**:6.

Markesteijn L., Poorte, L., Bongers. F. **2007**. Light-dependent leaf trait variation in 43 tropical dry forest tree species. *American Journal of Botany* **9**(4):515–525.

McMurtry C.R., Barnes P.W., Nelson J.A., Archer S.R. **1996**. *Physiological responses of woody vegetation to irrigation in a Texas subtropical savanna. La Copita Research Area: 1996 Consolidated Progress Report*. Texas Agricultural Experiment Station–Corpus Christi, Texas A&M University System, College Station, TX, pp. 33–37.

Moorthy I., Miller J.R., Jimenez Berni J.A., Zarco-Tejada P., Hu B., Chen J. **2011**. Field characterization of olive (*Olea europaea* L.) tree crown architecture using terrestrial laser scanning data. *Agricultural and Forest Meteorology* **151**(2):204–214.

Moreno I., Bertlier M.B. **2012**. Variation of morphological and chemical traits of perennial grasses in arid ecosystems. Are these patterns influenced by the by the relative abundance of shrubs? *Acta Oecologica* **41**:39–45.

Moreno L., Bertiller M.B., Carrera A.L. **2010**. Changes in traits of shrub canopies across an aridity gradient in northern Patagonia, Argentina. *Basic and Applied Ecology* **11**(8):693–701.

Návar J., Bryan R.B. **1994**. Fitting the analytical model of rainfall interception of Gash to individual shrubs of semi-arid vegetation in northeastern Mexico. *Agricultural and Forest Meteorology* **68**(3/4):133–143.

Návar J., Méndez E., Nájera A., Graciano J., Dale V., Parresol B. **2004**. Biomass equations for shrub species of Tamaulipan thornscrub of North-eastern Mexico. *Journal of Arid Environments* **59**(4):657–674.

Niinemets U., Kull K. **2003**. Leaf structure vs. nutrient relationships vary with soil conditions in temperate shrubs and trees. *Acta Oecologica* **24**:209–219.

O'Connell B.M., Kelty M.J. **1994**. Crown architecture of understory and open-grown white pine (*Pinus strobus* L.) saplings. *Tree Physiology* **14**(1):89–102.

Pickett S.T.A., Kempf J.S. **1980**. Branching patterns in forest shrubs and understory trees in relation to habitat. *New Phytologist* **86**(2):219–228.

Poorter L., Bongers F. **2006**. Leaf traits are good predictors of plant performance across 53 rain forest species. *Ecology* **8**(7):1733–1743.

Read Q.D., Moorhead L.C., Swenson N.G., Bailey J.K., Sanders N.J. **2014**. Convergent effects of elevation on functional leaf traits within and among species. *Functional Ecology. Special Issue: Climate change and species range shifts* **28**(1):37–45.

Reich P.B., Grigal D.F., Aber J.D., Gower S.T. **1997**. Nitrogen mineralization and productivity in 50 hardwood and conifer stands on diverse soils. *Ecology* **78**(2):335–347.

Reid N., Marroquín J., Beyer M.P. **1990**. Utilization of shrubs and trees for browse, fuel-wood and timber in the Tamaulipan thornscrub, northeastern Mexico. *Forest Ecology and Management* **36**:61–79.

Rios R.S., Salgado-Luarte C., Gianoli E. **2014**. Species divergence and phylogenetic variation of ecophysiological traits in lianas and trees. *PLoS ONE* **9**(6):e9987.

SPP-INEGI. **1986**. *Síntesis Geografía del Estado del Nuevo León*. Instituto Nacional de Geografía Estadística e Información, México, D.F.

Steingraeber D.A., Waller D.M. **1986**. Non-stationarity of tree branching patterns and bifurcation ratios. *Proceedings of the Royal Society of London. Series B, Biological Sciences* **228**:187–194.

Stolte D.S. **2003**. One tree's architecture reveals secrets of a forest, study finds. (http://phys.org/news/2013-08-tree-architecture-reveals-secrets-forest.html).

Sun S., Jim D., Li R. **2006**. Leaf emergence in relation to leaf traits in temperate woody species in East-Chinese *Quercus fabri* forests. *Acta Oecologica* **30**(2):212–222.

Wright I.J., Reich P.B., Westoby M., et al. **2004**. The worldwide leaf economics spectrum. *Nature* **428**:821–827.

CHAPTER 6

Leaf traits

6.1 Introduction

Foliar features are subjected to continuous selection processes due to the close relationship with the environment where they grow and develop. The study of these attributes conducted by Medrano-Sanchéz (2014) found patterns in foliar morphology were a function of their ability to adjust to environmental conditions through adaptive differentiation. In this study, the morphological leaf characteristics of the following shrub native species of the Tamaulipan thornscrub were analyzed: *Leucophyllum frutescens*, *Acacia rigidula*, *Sideroxylon celastrinum*, *Acacia berlandieri*, *Cordia boissieri*, *Celtis pallida*, *Forestiera angustifolia*, *Amyris madrensis*, *Bernardia myricifolia* and *Lantana macropoda*. The specific objectives were: (a) to know the variation of these foliar traits between the species and (b) to determine if there is a relation among these foliar features. Ten morphological leaf characteristics were studied (leaf fresh weight, petiole length, leaf length, total leaf length, leaf dry weight, leaf area, dry matter, specific leaf area, leaf water content, leaf width and leaf length:width ratio) in 50 leaves belonging to five individuals per species. Results showed that there was an ample and significant spectrum of variation among the studied leaf characteristics. The morphological variation found among foliar traits was highly significant among shrub species. Besides, highly significant relationships were observed among the studied leaf characteristics. The morphological leaf variation was related to phylogenetic diversity and to local variations (Figure 6.1).

Spearman's correlation coefficient (above diagonal line) and probability value (below diagonal line) of different leaf traits of different woody species, northeastern Mexico, is depicted in Table 6.1.

6.2 Leaf anatomy

Leaf anatomical traits play an important role in the taxonomic delimitation of a species and its adaptation to environments, such as arid or semiarid conditions. Apart from variability in leaf shape, size, margin and dentation among plant species, the compact or loose palisade layer also varies. The density of trichomes on the leaf surface has been related to insect tolerance and drought resistance in crops like sorghum, cotton and so on. A few studies have been directed in these aspects

---

Autoecology and Ecophysiology of Woody Shrubs and Trees: Concepts and Applications, First Edition.
Edited by Ratikanta Maiti, Humbero Gonzalez Rodriguez and Natalya Sergeevna Ivanova.
© 2016 John Wiley & Sons, Ltd. Published 2016 by John Wiley & Sons, Ltd.

Figure 6.1 Variability in leaf morphology among a few native species of north-eastern Mexico. 1 *Amyris madrensis*, 2 *Acacia rigidula*, 3 *Forestiera angustifolia*, 4 *Celtis pallida*, 5 *Sideroxylon celastrinum*, 6 *Acacia berlandieri*, 7 *Cordia boissieri*, 8 *Leucophyllum frutescens*, 9 *Lantana macropoda*, 10 *Bernardia myricifolia*. (See insert for color representation.)

in woody tree species, as mentioned below.

6.3 Taxonomy

Kristic et al. (2002) studied the variability of the leaf anatomical chracteristics of *Solanum nigrum* from different habitats across Europe, centering on Yugoslavia. The most widespread weed species from the genus *Solanum* is *S. nigrum* L. The

influence of ecological factors on a plant organism and resulting plant adaptations are most evident in leaf morphology and anatomy. Therefore, the anatomical structure of leaves and leaf epidermal tissue of *S. nigrum* was analysed and compared among plants that originated from different habitats, in order to determine leaf structural adaptations. It was assessed that *S. nigrum* lamina had a mesomorphic structure with some xero-heliomorphic adaptations. The differences in number of hairs, stomata

Table 6.1 Spearman's correlation coefficient (above diagonal line) and probability value (below diagonal line) of different leaf traits ($n = 500$).

Leaf trait	V1	V2	V3	V4	V5	V6	V7	V8	V9	V10
V1. Fresh weight		0.877	0.932	0.785	0.897	0.959	0.969	0.988	−0.581	0.110
V2. Length	0.001		0.821	0.723	0.982	0.884	0.801	0.897	−0.287	0.277
V3. Breadth	0.001	0.001		0.857	0.871	0.956	0.911	0.919	−0.730	0.227
V4. Ptiole length	0.001	0.001	0.001		0.827	0.842	0.797	0.763	−0.655	0.216
V5.Total length	0.001	0.001	0.001	0.001		0.921	0.839	0.908	−0.389	0.290
V6. Leaf area	0.001	0.001	0.001	0.001	0.001		0.933	0.948	−0.596	0.299
V7. Dry weight	0.001	0.001	0.001	0.001	0.001	0.001		0.925	−0.645	0.012
V8. Water content	0.001	0.001	0.001	0.001	0.001	0.001	0.001		−0.542	0.181
V9. Leaf form	0.001	0.001	0.001	0.001	0.001	0.001	0.001	0.001		0.006
V10. Specific leaf area	0.014	0.001	0.001	0.001	0.001	0.001	0.793	0.001	0.895	

number, thickness of lamina, palisade and spongy tissue, as well as the size of mesophyll cells were noticed. The highest values for most parameters was observed for plants from cultivated soil. The largest variations of the examined characters were found for leaves from ruderal habitats, where environmental conditions were most variable (Kristic et al., 2002).

A study was made on the leaf anatomy of 15 species of vascular plants occurring in the coastal zones of Falcon State, Venezuela, to evaluate the potential adaptive value of leaf anatomical features to a saline environment. Transverse sections and macerates of leaf material preserved in FAA were prepared for microscopic analysis. Results showed that the development of water-storing tissue in the mesophyll and/or epidermal cells was the main characteristic associated with the saline habitat in these species. Other characteristics of potential adaptive value were: presence of trichomes, stomata protected by papillae, crystals in mesophyll cells, secretory tissues and Kranz anatomy (García et al., 2008).

A study was made on the morphological and anatomical characteristics of cones, leaves and branchlets of *Pseudotsuga* in Mexico (Reyes Hernández et al., 2005). They studied the vegetative and reproductive organs of 293 trees obtained from 19 localities in different geographic regions. Analysis of variance showed a significant difference in morpho-anatomical characters, sufficient to separate the different species.

Arambarri et al. (2011) studied the foliar anatomy of shrubs in a semiarid region in Argentina. The study revealed that some of the characters related to adaptation were stellate trichomes in *Capparicordis tweediana* and *Ruprechtia triflora*, cystolitic trichomes in *Celtis* spp., multicellular peltate scales in *Zanthoxylum coco*, the presence of papillose epidermis (e.g. *Schinopsis lorentzii*), cyclocytic stomata in *Bulnesia sarmientoi*, *Maytenus vitis-idaea*, *Moya spinosa* and *Schinopsis* spp., idioblastic crystalifer epidermal cells (*Scutia buxifolia*), crystaliferous epidermis (*Maytenus vitis-idaea*), multilayered epidermis (*Jodina rhombifolia*), presence of hypodermis in *Castela coccinea*, *Maytenus vitis-idaea*, *Prosopis ruscifolia* and *Ziziphus mistol*, bicollateral vascular bundles in *Lycium cestroides*, presence of crystal sand in *Calycophyllum multiflorum* and *Lycium cestroides* and an absolute absence of crystals in the family Capparaceae.

The genera and species of Piperaceae show a considerable structural diversity of leaves and especially stems. Souza et al. (2004) made a comparative morphological and anatomical study of the leaves and stems of three common Brazilian species of this family (*Peperomia dahlstedtii* C.DC., *Ottonia martiana* Miq. and *Piper diospyrifolium* Kunth), the vegetative organs of which had previously been little studied. *P. dahlstedtii* was an epiphyte and had a herbaceous stem with whorled leaves phyllotaxis and a polystelic structure, a multiseriate adaxial leaf epidermis and calcium oxalate monocrystals in parenchyma and collenchyma petiole cells. *O. martiana* and *P. diospyrifolium* showed strong similarities, both being terrestrial plants, with alternate phyllotaxis, stele with medullary bundles and dorsiventral leaves with an epidermis and subepidermic layer. In *O. martiana* the stomatal complex was staurocytic and presented silica crystal sand in the parenchymal petiole and midrib cells. In *P. diospyrifolium* the stomatal complex was tetracytic and there were calcium oxalate raphide crystals in the parenchyma of the petiole and midrib cells. On the other hand, the three species showed some structural likenesses, in that all had hypostomatic and dorsiventral leaves, oily cells in petiole and mesophyll, secretory trichomes and an endodermis with Casparian strips.

Chernetskyy (2012) studied the role of morpho-anatomical traits in the taxonomic determination of Kalanchidae analysing the morphological and anatomical structure of the leaves of 35 species of the genus *Kalanchoë* Adans. (Crassulaceae DC.) from the taxonomic aspect of the subfamily Kalanchoideae Berg. Based on his own studies and literature analyses of the flower morphology, embryology, karyology, vascular anatomy of stems and molecular genetics, the author concluded that the most appropriate taxonomic system of the subfamily Kalanchoideae assumed the existence of one genus *Kalanchoë* divided into three sections: *Bryophyllum* (Salisb.) Boit. and Mann., *Eukalanchoë* Boit. and Mann. and *Kitchingia* (Bak.) Boit. and Mann. Distinguishing three separate genera *Bryophyllum* Salisb., *Kalanchoë* Adans. and *Kitchingia* Bak., as had been the case throughout the history of the subfamily Kalanchoideae, was hardly possible due to existence of "intermediate" species.

6.4 Adaption to environments

A study was made to evaluate leaf anatomical traits that contribute to a better understanding of plant–environment relationships and to the development of technologies for the sustainable use and conservation of Chacoan forests in Argentina. The density of epidermal cells, stomata and trichomes was determined. Type of mesophyll and the type and distribution of vascular and esclerenchymatic tissues were also analysed. Most trees in the Chacoan Oriental District had hypostomatic leaves with dorsiventral mesophyll, a high density of epidermal cells ($4000–7000\,\mathrm{mm^{-2}}$)

and an intermediate density of stomata ($300–500\,mm^{-2}$; Arambarri et al., 2012).

Mulroy and Rundel (1977) studied adaptations to desert environments in the warm deserts of the south-western USA and north-western Mexico. Two groups of ephemeral annual plants were recognised. One group, the winter annuals, consisted of species that germinated and completed their life cycles during the winter and spring months; the other group, the summer annuals, included species that germinated and completed their life cycles during the summer and early fall months (Went, 1948; Shreve, 1951). Normally, the seasonal occurrences of these two groups of species are highly predictable and are determined by specific temperature and moisture combinations required for germination (Juhren et al., 1956).

Water appears to be the major environmental constraint in a Mediterranean climate and global change effects are likely to provide more frequent and longer drought periods. Water shortage has significant demographic effects on ecosystem composition, with some plant species thriving in arid environments because of a combination of several anatomical and physiological adaptations. A study was made on the anatomical basis of resistance in plant species from a typical thermo-Mediterranean ecosystem. Observations confirmed the presence of several adaptive properties for the macchia ecosystem. Most of the species examined showed the presence in their internal leaf tissues of ergastic substances, mainly tannins and calcium oxalate, which have defensive functions, crucial in the adaptive resistance of plants to water stress. Nearly all the species presented adaptations for

protection against photodamage possibly induced from strong UV-B solar irradiance in the summer. The more significant anatomical features are the trichomes, covering the abaxial surfaces of leaves. Such structures are able to regulate the water budget of the plants, both by influencing the diffusion boundary layer of the leaf surface and by regulating the leaf optical parameters and, hence, the leaf temperature. In many species, when trichomes or wax layers reduce radiation absorbance, two or three layers of palisade parenchyma are present, presumably to provide a better efficiency in utilising light for photosynthesis. In almost every plant examined, stomata are sunken or well protected. Rotondi et al. (2003) determined the annual courses of the leaf water potential, net photosynthesis and transpiration rates of the four dominant species of the population (*Juniperus phoenicea* L., *Pistacia lentiscus* L., *Phillyrea angustifolia* L. and *Chamaerops humilis* L.) and the ecosystem.

A study was undertaken on the morpho-anatomical characteristics of *Celtis ehrenbergiana*. It was observed that the shade leaf structure was bifacial, the epidermis had wavy-sinuous anticlinal cell walls and the mesophyll was formed by palisade, spongy parenchyma and angular-lacunar collenchyma. The sun leaf type had a thick leaf-limb, leathery and dark green. The sun leaf structure was equifacial, the epidermis had straight anticlinal cell walls and the mesophyll was formed by homogeneous palisade parenchyma and angular-massive collenchyma. There were very few or no stomata on the adaxial surface of a shade type leaf, but they were numerous on the sun type leaf. *Celtis ehrenbergiana* exhibited phenotypic plasticity.

This means it has advantages over other species less adaptable and might survive climate changes (Nughes et al.. 2013).

Ephemeral annuals, in decided contrast to the perennial desert flora, are commonly considered to exhibit no striking adaptations to the desert climate but rather are thought to be mesophytic in nature and escape un-favourable conditions of soil water stress and high insolation by rapid completion of their life cycles during the brief periods when temperature and moisture regimes are favourable. However, studies indicated that both winter and summer annuals commonly showed specific physiological and morphological adaptations to desert environments (Mulroy and Rundel, 1977).

The genetic potential of different plant species in different environmental conditions differs in relation to different physiological, biochemical and anatomical characteristics. Of these varying attributes, the leaf anatomical characteristics play the most important role for establishing a cultivar in varied environmental conditions. Noman et al. (2006) studied the inter-cultivar genetic potential of *Hibiscus* in relation to leaf anatomical characteristics. To fulfil the study requirements, *H. rosa-sinensis* and its six cultivars (which were well adapted to their specific natural habitats) were collected from different locations in Faisalabad District, Pakistan, that had great environmental changes around the year. Results showed significant variability among the cultivars' anatomical characteristics. Cultivars Lemon shiffon and Wilder's white emerged more promising that the others by possessing a greater epidermal thickness, increased epidermal cell area, high cortical cell area and increased stomatal density than the other cultivars. On the other hand, cultivars Cooperi alba, Mrs. George Davis and Frank green possessed the least cortex cell area, the lowest xylem region thickness and minimum phloem region thickness, respectively. Overall, it can be concluded that anatomical genetic potential has provided cultivars Lemon chiffon and Wilder's white with an enormous capability to grow well under variable environments.

In the context of literature surveys, it is well documented that leaf anatomical traits play an important role in taxonomy and the adaptation of tree species to environmental stresses such as drought, cold and high temperature. The density of trichomes, cuticular thickness, compact palisade and the presence of tannins, phenolics and other exudates are related to resistance to water loss. On the basis of these traits, the species can be categorised for their adaptation to different environments of abiotic and biotic stress. There is a great necessity to direct research inputs in this direction.

6.5   Leaf surface anatomy

A preliminary study was undertaken on the dermal anatomy of a few native species of north-eastern Mexico. We present here the dermal structure of a few plant species that grow and develop in the north-eastern region of Mexico. Simple techniques have been developed by R.K. Maiti to study the epidermal tissue of leaves.

Technique 1. Small pieces of thermocol are slowly dissolved in a small amount of xylene in a petri dish and brought to the constituency of honey. Then, the solution is applied with the help of the little finger to both the upper and lower surface of the leaves of each species in the region between the midrib and margin. Then they are left on the table to dry. Once dried, a piece of transparent tape is applied and pressed on the region with a finger. Finally the tape is taken off with an impression of the leaf and pressed onto a slide in the same direction. Now it is permanent and ready to observe under the microscope.

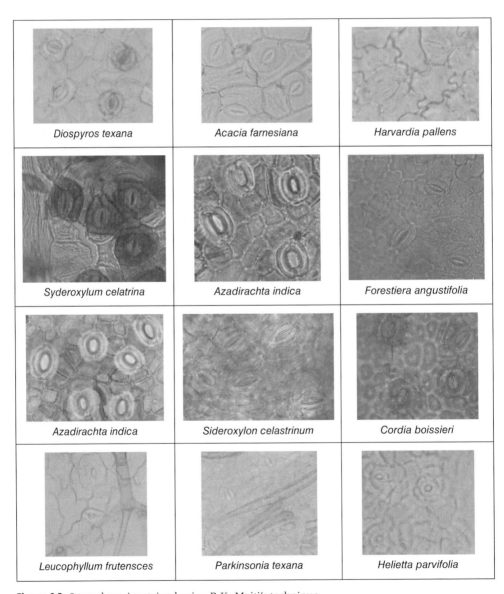

Figure 6.2 Several species stained using R.K. Maiti's technique.

Technique 2. In this technique, a small portion of the leaf lamina of each species is put in a mixture of 10% chromic acid:10% nitric acid in a test tube and the mouth of the test tube is plugged with cotton. Then, several test tubes containing several species are kept boiling in a beaker in a water bath for 15 min. This clears the leaf lamina. Then a portion of the leaf lamina is transferred to a slide and covered with a cover slip with a few drops of glycerine and a stain and observed under a microscope. Several stained species are shown in Figure 6.2.

There is a large variability in the type and size of stomata, an epidermal cell structure which can be utilised in the taxonomic delimitation of a species and as an adaptation to abiotic and biotic stresses. Therefore, there is a great necessity to study the leaf surface structure of different species, to help in the taxonomic delimitation of a species and as an adaptation of the species to xeric environments.

Bibliography

Arambarri A., Monti C., Bayón N., Hernández M., Novoa M.C., Colares M. **2012**. Ecoanatomía foliar de arbustos y árboles del Distrito Chaqueño Oriental de la Argentina. *Bonplandia* **21**(1):5–26.

Arambarri A.M., Novoa M.C., Bayón N.D., Hernández M.P., Colares M.N., Monti C. **2011**. Anatomía foliar de arbustos y árboles medicinales de la región chaqueña semiárida de la Argentina. *Dominguezia* **27**(1):5–24.

Chernetskyy M.A. **2012**. The role of mrphoanatomical traits of the leaves in the taxonomy of Kalanchoideae Berg. Subfamily (Crassulaceae DC.). *Modern Phytomorphology* **1**:15–18.

García M., Jáuregui D., Medina E. **2008**. Adaptaciones anatómicas foliares en especies de Angiospermas que crecen en lazona costera del Estado de Falcón (Venezuela). *Acta Botánica Venezuelica* **31**(1):291–306.

Juhren M.F., Went F.W., Phillips E. **1956**. Ecology of desert plants. IV. Combined field and laboratory work on germination of annuals in the Joshua Tree National Monument, California. *Ecology* **37**:318–330.

Kristic T.N., Merkulov L.S., Boza P.P. **2002**. The variability of leaf anatomical characteristics of *Solanum nigrum* L. (Solanales, Solanaceae) from different habitats. *Proceedings for Natural Sciences* **102**:59–70.

Mulroy T.W., Rundel P.W. **1977**. Annual Plants: Adaptations to desert environments. *BioScience* **27**(2):109–114.

Noman A., Ali A., Hameed M., Mehmood T., Iftikhar T. **2006**. Comparison of leaf anatomical characteristics of *Hibiscus rosa-sinesis* grown in Faslabad region. *Pakistan Journal of Botany* **46**(1):199–206.

Nughes L., Colares M., Hernández M., Arambarri A. **2013**. Morfo-anatomía de las hojas de *Celtis ehrenbergiana* (Celtidaceae) desarrolladas bajo condiciones naturales de sol y sombra. *Bonplandia* **22**(2):159–170.

Reyes Hernández V.J., Vargas Hernández J.J., López Upton J., Vaquera Huerta H. **2005**. Variación morfológica y anatómica en poblaciones Mexicanas de *Pseudotsuga* (Pinaceae). *Acta Botanica Mexicana* **70**:47–67.

Rotondi A., Rossi F., Asunis C., Cesaraccio C. **2003**. Leaf xeromorphic adaptations of some plants of a coastal Mediterranean macchia ecosystem. *Journal of Mediterranean Ecology* **4**(3/4):25–35.

Shreve F. **1951**. *Vegetation of the Sonoran Desert*. Carnegie Institution of Washington Publication 591, Washington, D.C.

Souza L.A., Moscheta I.S., Oliveira J.H.G. **2004**. Comparative morphology and anatomy of the leaf and stem of *Peperomia dahlstedtii* C.DC., *Ottonia martiana* MIQ and *Piper diospyrifolium* Kunth (Piperaceae). *Gayana Botánica* **61**(1):6–17.

Went F.W. **1948**. Ecology of desert plants. I. Observations on germination in the Joshua Tree National Monument, California. *Ecology* **29**:242–253.

Zhuang L., Chen, Y.-N., Li W.-H., Wang Z.-K. **2011**. Anatomical and morphological characteristics of *Populus euphratica* in the lower reaches of Tarim River under extreme drought environment. *Journal of Arid Land* **3**(4):261–267.

CHAPTER 7

Wood characteristics

7.1 Introduction

The apical growth of a tree is initiated by the apical meristem. Besides the basics of the roots, the main stem (trunk), and the leaves and branches, there are growing points at the tips of the stems and roots, called apical meristems. These growing points, through cell division, are responsible for the tree's vertical growth. In addition, another meristem called the vascular cambium or lateral meristem is present in between the primary xylem and phloem referred to simply as the cambium. This tissue is responsible for practically all of the horizontal growth on a tree. The cambium consists of two types of cells, fusiform and ray cells that, by cell division, form new bark outward, and also new wood inward. Fusiform cells divide longitudinally to give rise to vessels and fibre cells and the ray gives rise to rays. The seasonal growing activity of the cambium is responsible for the formation of growth rings seen in cut wood: the cambium is most active in the spring (this wood is sometimes referred to as spring wood or earlywood), with growth slowing in the summer (called summer wood or late wood), and completely ceasing in the winter. These differences in growing cycles from year to year form annual rings, which are accurate indicator of a tree's age. Softwoods: latewood tends to be darker, denser and has smaller diameter tracheids; earlywood is lighter, softer and has larger diameter tracheids. Hardwoods (ring-porous): latewood has smaller and less frequent pores; earlywood has larger, more numerous pores.

7.2 Different types of wood

Wood is a porous and fibrous structural tissue found in the stems and roots of trees and other woody plants. It is composed of cellulose fibres embedded in a matrix of lignin which resists compression. Wood is the secondary xylem in the stems of trees and in the roots of trees or shrubs. In a living tree, it gives a support function, enabling woody plants to grow large or to stand up by themselves. It helps in the transfer of water and nutrients to the leaves and other growing tissues.

Hardwoods are produced by angiosperms. Many species are deciduous. Those of temperate regions lose their leaves every autumn as temperatures fall and are dormant in the winter but those of tropical regions may shed their leaves in response to seasonal or sporadic periods of drought. Hardwoods possess a more

Autoecology and Ecophysiology of Woody Shrubs and Trees: Concepts and Applications, First Edition.
Edited by Ratikanta Maiti, Humbero Gonzalez Rodriguez and Natalya Sergeevna Ivanova.
© 2016 John Wiley & Sons, Ltd. Published 2016 by John Wiley & Sons, Ltd.

complex structure than softwoods. The main characteristics separating "hardwoods" from softwoods is the presence of pores, or vessels. The vessels may show considerable variation in size, shape of perforation plates (simple, scalariform, reticulate, foraminate) and structure of cell wall, such as spiral thickenings. The wood from these trees is generally harder than that of softwoods. Hardwood from deciduous species, such as oak, normally shows annual growth rings, but these may be absent in some tropical hardwoods. Next, we describe the macroscopic characteristics of a transverse section of wood of a few native species of north-east Mexico.

7.3   Wood density of a few woody species

We measured the wood density of a few tree species from the north-eastern region of Mexico. Table 7.1 shows the wood density of 11 woody species, demonstrating the variability in wood density among the studied species.

7.4   Wood anatomy and wood fibres

Ratikanta Maiti, Artemillo Carrillo Parra and Humberto Gonzalez Rodriguez

Facultad de Ciencias Forestales, Universidad Autonoma de Nuevo Leon, Linares, Mexico

Wood anatomical traits of timbers are related to the determination of timber quality, which is of great importance to the wood industry and forest science. Wood is formed by the secondary growth of cambium which produces secondary xylem vessels, wood fibres and wood parenchyma. Among these tissues, the wood fibre cells contribute greatly to the strength and quality of wood, depending on the degree of lignification in the cell walls. Timber with highly lignified fibre cells is expected to contribute strength and to produce a useful hardwood for furniture manufacturers, while softwoods (having thin-walled fibre cells and a high amount of wood parenchyma) are suitable for soft furniture, fencing or paper pulp. The amount of lignification and length of fine cells vary in different environments. Therefore, there is a necessity to seperate the hardwood species from the others more suited for the fabrication of papers and other purposes.

All vascular plants possess a specialised tissue called xylem, which produces xylem vessels. These give mechanical support

Table 7.1 Woody density of 11 tree species from north-eastern Mexico.

Species	Density (g/cm³)	Standard deviation
Diospyros texana Scheele	0.641	0.054
Sideroxylon celastrinum (Kunth) T.D. Penn	0.785	0.077
Parkinsonia texana (A. Gray) S. Watson.	0.9	0.104
Prosopis laevigata (Humb. and Bonpl. Ex. Willd) M.C. Johnst.	0.95	0.077
Celtis pallida Torr.	0.776	0.064
Zanthoxylum fagara (L.) Sarg.	0.66	0.043
Condalia hookeri M. Johnst.	0.85	0.143
Foriestiera angustifolia Torr.	0.633	0.032
Celtis laevigata Willd.	0.717	0.034
Karwinskia humboldtiana (Schult.) Zucc.	0.884	0.08
Havardia pallens (Benth.) Britton & Rose.	0.706	0.06

and transport water, mineral nutrients and phytohormone signals. It may be possible to predict the quality of wood of a particular species on the basis of wood anatomy.

7.4.1   Structure, function and development

Unlike the earliest plants that evolved from primitive water-living ancestors and did not possess conductive tissue, the stems and roots of modern plants possess highly specialised conductive tissue, xylem and phloem, for the transport of water and nutrients from the roots to the stems and food from the leaves to the growing organs of the plant. Phloem transports photosynthetic products and plant growth hormones (phytohormones) mainly from the leaves to the rest of the plant. Xylem transports water, mineral nutrients and phytohormones from the roots to the leaves and other organs.

7.4.2   Xylem structure and variability

Xylem tissue possesses different cell types which vary across the different plant organs.

Xylem cell types. The structure of xylem is determined by the size, shape and lignification of xylem cell types, in shape and cell wall thickness.

Two types of xylem cells are found in plants, primary xylem formed from primordial tissue and secondary xylem formed by cambium. The primary wall in xylem is made of cellulose microfibrils, thereby allowing the wall to stretch and expand. The secondary wall is deposited on the inner side of the primary wall during the process of cell elongation and expansion. The cellulose microfibrils in the secondary wall are arranged in a regular fashion with alternating layers at fixed angles. The cellulose microfibrils are arranged in the secondary wall in three strata: S1, S2 and S3. Besides, xylem tissue contains parenchyma cells which store water, mineral nutrients and carbohydrate. They are used for most storage functions and metabolism. Many xylem parenchyma cells possess secondary lignified walls.

Xylem possesses sclerenchyma cells which provide mechanical support defences and water transport. The conducting cells of xylem are called tracheal elements. There are two types of tracheal elements, tracheids and vessel elements. Vessel elements are connected end to end through perforations in their end walls to form a vessel, characteristic of angiosperms. Tracheids are connected through large, circular bordered pits concentrated at the tapered ends, a characteristic of conifers. Primary xylem occurs in separate vascular bundles. The xylem tissue in a young stem and roots are arranged in separate primary vascular bundles with phloem tissue. The primary xylem of stems generally contains early differentiating protoxylem present on the inner side of xylem, with differentiating metaxylem on the outer side. In dicotyledons, secondary xylem is derived by cambium and increases during the life span of a plant. In dicotyledonous plants, cambium is present between the primary xylem and phloem. Along its outer side, cambium produces secondary phloem, thereby pushing the primary phloem further outwards. Along its inner side, cambium produces secondary xylem. The cambium present between primary phloem and xylem alternately forms interfascicular cambium ready to form secondary xylem inwards towards the pith and secondary phloem outwards, thereby pushing the primary phloem further outwards. The production of secondary xylem-producing wood is largely dependent on the activity of cambial cells influenced by seasonal environment. During spring, owing to a favourable season, it produces thick wood, but during the winter the cambial activity is reduced, forming thin wood, thereby producing a distinct annual ring between spring and winter wood. The age of a plant can be provisionally determined by counting the number of rings in a transverse section of the stem of a big tree.

7.4.3   Evolution of secondary xylem

The evolutionary pattern of secondary xylem is well documented. The evolutionary pattern of a plant can be determined on the basis of tracheal or xylem vessels. Primitive plants possess tracheids which are long, pointed and possess bordered pits. With advances in the evolutionary process, the length of tracheids was gradually reduced. Plants possessing long, thin xylem vessels with an inclined end are considered primitive. They also possess scalariform pits. With evolutionary advances, xylem vessels reduce in length and increase in diameter, pits change from scalariform to oval and their arrangement changes from opposing to alternating, the vessel diameter increases and the end walls are straight. A highly advanced plant possesses short truncated vessels with straight end walls and round alternating pits.

Wood is a hard, fibrous structural tissue present in the stems and roots of woody plants. Its main use is for furniture and in buildings. It has been used for thousands of years for both of these, for fire wood and as a construction material. With respect to its structure, wood is the secondary xylem derived by an intercalary meristem consisting of cambium in the stems of trees. Cambium consists of two types of cells: (a) fusiform cells giving fibre cells and xylem vessels and (b) the ray initials giving ray cells which form medullary cells. Fibre cells are composed of cellulose fibrils embedded in a matrix of lignin which resists compression. It gives mechanical strength to the plant and also helps to transfer water and nutrients to the leaves and other growing tissues.

Next, we discuss a few research advances in wood anatomical and physical characteristics.

Wood is composed mainly of three types of cells, such as wood fibre cells (sclerenchyma), vessels and wood parenchyma derived by secondary cambium. Fibre cells with lignified cell walls offer strength to the wood. The microscopic structure of wood includes annual rings and rays. It produces characteristic grain patterns in different species of trees. In transverse or cross-sections, the annual rings appear like concentric bands, with rays extending outward. The central core of wood represents the first year of growth since the pith is no longer present. A smaller series of concentric rings (knot) is a lateral branch embedded in the main trunk.

All the tissue inside the cambium layer to the center of the tree is xylem or wood. All the tissue outside the cambium layer (including the phloem and cork layers) forms the bark. The term phellem includes the corky bark and the tissue outside the phloem is called the phellogen. The wood of a tree trunk is mostly dead xylem tissue. The darker, central region is called heartwood. The cells in this region do not or no longer conduct water. They appear darker owing to the accumulation of resins, gums and tannins. The lighter, younger region of wood closer to the cambium is called sapwood. Although they are dead, the cells in this region conduct water and minerals from the soil. Xylem cells are alive when they are produced by the meristematic cambium, then they lose their cell contents and become hollow.

Root pressure does not adequately explain the rise of water in plant stems. In fact, the pressure required to force water up inside tall stems would greatly exceed the force of root pressure. In addition, root pressure does not operate when soil moisture is low; and even when soil moisture is high it is too weak to force water up a tall plant. Water molecules are actually pulled up from the leaves through minute tubular cells in the xylem tissue.

Wood anatomical and structural features observed and measured in tree-rings have proved to be useful in dendrochronology. They have added understanding and new insights to processes going on in trees with structural features that have shown linkages to environmental parameters, not given by other factors. The review carried out by Wimmer (2002) emphasises work done primarily on continuous and non-continuous wood anatomical features measured in dated tree-rings, reflecting internal and external conditions and processes. This study also includes new results from a study conducted in the East-Ore Mountains, Germany, where several anatomical features in the rings of trees grown under severe stresses were measured. Dated tree-rings show how environmental changes have caused modifications or adaptations of structural features. The measurement of many structural features in tree-rings remains tedious, although fast scanning devices have now been made available for some features such as cell size or microfibril angle. Overall, wood anatomy indicates that the growth and development of trees are dynamic processes. All these aspects, which are commonly illustrated in two and three dimensions, have in reality a fourth dimension – time.

Increasing concentrations of ions flowing through the xylem of plants produce rapid, substantial and reversible decreases in hydraulic resistance (Zwieniecki et al., 2001). Changes in hydraulic resistance in response to solution ion concentration, pH and non-polar solvents are consistent with this process being mediated by hydrogels. The effect is localised to intervessel bordered pits, suggesting that microchannels in the pit membranes are altered by the swelling and de-swelling of pectins, which are known as hydrogels. The existence of an ion-mediated response breaks the long-held paradigm of the xylem as a system of inert pipes and suggests a mechanism by which plants may regulate their internal flow regime. One of the properties of polysaccharide hydrogels is to swell or shrink due to imbibition. When pectins swell, pores in the membranes are pressed, slowing the water flow to a trickle. But when pectins shrink, the pores can open wide and water flushes across the xylem membrane toward thirsty leaves above. This remarkable control of water movement may allow the plant to respond to drought conditions.

According to Eric Meier (2015) in sharp contrast to the simple anatomy of softwoods, the hardwoods of the world exhibit a dazzling array of endgrain patterns and intricate motifs; and it is in this complexity that the challenge (and joy) of wood identification really comes alive. The decrease in the maximum effective opening diameter with an increase in the thickness of the cross section is greater for the heartwood than for the sapwood (Stamm, 1972).

Dendrochronological studies dealing with roots, stems and branches are very rare or often take the form of short notes. The difficulties of detecting rings and of quantifying radial growth in roots have already been described for various species. In oak the anatomical root structure differs from that of stemwood. The roots are radial-porous or diffuse-porous and there is often no clear distinction between individual rings. Our study used visual and radiographic techniques to examine the radial increment in roots of sessile oak (*Quercus petraea* L.) and compared this with radial growth in branches and along the stem. Coarse roots were cut from four 30 to 34 year old trees that had been uprooted mechanically and disks were taken at different distances from the stem-root base. Ring widths were measured in the stem at height 0.3 m, at breast height (1.3 m), beneath the crown, in branches of the crown, and in roots every 20 cm. The ring widths were cross-dated and the heterogeneity of growth within a root and within the root system were analysed. Asymmetric growth frequently occurred in roots so that ovals, I-beam and T-beam shapes were developed. With the method used in our study the annual growth layers close to the central cylinder could be distinguished,

as could those beneath the bark. Pointer years were detected in all sections of the tree and permitted correction of ring widths in roots. Root system, stem and branch showed a basic similarity in their radial sequence of ring width. The annual biomass increment was weaker and more variable with several more consecutive changes in the roots than in the stems. The root/shoot ratio reached a minimum rather early, beginning at the cambial age of 20 years (Drexhage et al., 1999).

Vestured pits were found in the vessels of 25 New Zealand species belonging to six dicotyledonous families. There were large variations in the extent and type of vesturing. Vesture was also observed in the fibre pits of some species (Meylan and Butterfield, 1974).

The tension wood of *Laetia procera* Poepp. (Flourtiaceae), a non-tropical forest species, contains a special secondary wall structure, with alternate arrangement of thick and thin layers. The species possesses a typical secondary wall structure with S1 + S2 + S3. Using UV microspectrophotometry it was observed that, in the thick secondary wall the cellulose microfibril angle is very low (very close to the fibre axis) and the cellulose microfibrils are well organised; but in a thin layer the cellulose microfibrils are less organised and oriented with a large angle in the axis of the cell. Thick layers are highly lignified (Ruelle et al., 2007).

Prosopis pallida is one of the most economically and ecologically important tree species in the arid and semi-arid lands of the American continent. Sections of *P. pallida* were used to describe its wood anatomy and to determine whether annual rings were visible or not. Results showed that *P. pallida* has well-differentiated annual growth rings and is therefore suitable for dendrochronological studies. Tree-ring chronologies correlate well with precipitation events related to El Niño Southern Oscillation phases. A master chronology for the northern area of Peru was built with these data, and some physiological derivations from the anatomy of *P. pallida* wood are discussed (López et al., 2005).

Structural heartwood characteristics for *Prosopis laevigata* (Humb. and Bonpl. *ex* Willd.) M.C. Johnst., including a histometrical evaluation, were studied by light microscopy coupled with a digitised image analysis system. The growth ring boundaries of the semi-ring-porous or diffuse porous wood are often marked by a marginal parenchyma band. Average fibre length is 975 µm and the fibres are thick-walled with a single cell wall thickness of 13 µm on average. The average diameter of the vessels which are arranged in non-specific patterns differs significantly between earlywood (116 µm) and latewood (44 µm). The chemical distribution of lignin and phenolic deposits in the tissue was investigated by means of scanning UV microspectrophotometry. Thereby, in heartwood tissue the deposition of extractives was detected in vessels, pit canals, parenchyma cells, fibre lumina and partly also in the S2 layers of the fibres. Monosaccharides were qualitatively and quantitatively determined by borate complex anion exchange chromatography. Holocellulose content was between 61.5 and 64.7% and Klason lignin content between 29.8 and 31.4%. Subsequent extraction of the soluble compounds was performed with petrol/ether, acetone/water and methanol/water by accelerated solvent extraction. Total extractives content in heartwood ranges between 14 to 16% on a dry weight basis. Major compounds in acetone/water extracts were identified as (–)-epicatechin, (+)-catechin and taxifolin and quantitatively determined by liquid chromatography (Carrillo et al., 2008).

A comparative study has been made on macroscopic and microscopic anatomical characteristics of five species of the family Rosaceae, *Crateagus mexicana*, *Pyrus cummunis*, *P. malus*, *Prunus americana* and *P. domestica*. The wood species showed similar macro and microscopic characteristics (Pérez Olvera et al., 2008).

Anatomical features of wood have a great variation among species as a result of genetic and environmental factors. The anatomical heartwood characteristics of *Prosopis laevigata* species from two areas with differences in temperature and rain precipitation on north-east Mexico were compared. Fibre length (µm), diameter of vessels (µm) and the area of the vessels (μm^2) were measured using light microscopy coupled with a digitised-image analysis system. The differences were statistically analyzed with analysis of variance. Statistical differences were found between fibre lengths ($p < 0.0001$) and vessel diameters ($p < 0.001$) from the two localities. The locality Linares, Nuevo Leon, Mexico, with higher precipitation and lower temperature, showed higher fibre length and higher vessel diameter than China, Nuevo Leon. Hard environmental conditions, where low precipitation values and high temperatures prevail, condition the *P. laevigata* trees to reduce the risk of losing water (Carrillo-Parra et al., 2013).

Interlocked grains record any change in the orientation of axial elements. In this report, vessel and fibre orientations in *Acacia mangium* Willd were compared macroscopically and microscopically to analyse the interlocked grain. A method to print the cylindrical surface of a dry wood disk after bark exfoliation was devised to evaluate the stem axis and circumferential grain fluctuation and revealed circumferential heterogeneity in the vessel orientation. Fibre orientation manifested on some radial splits also was heterogeneous. A 3 mm thick transverse plate was used to estimate vessel orientation with soft X-ray photography, which enabled a wider-ranging evaluation than microscopy. Serial tangential thin plates and sections were used to measure fibre orientation angle with reflecting and polarised light microscopy, respectively, and fast Fourier transform. Both vessel and fibre orientations had a similar radial tendency and distinct inversion of the grain. However, the vessel orientation had a larger amplitude of change than fibre orientation (Ogata et al., 2003).

In order to study the relationship between the altitudinal distribution of *Quercus laceyi*, Poulos et al. (2007) studied the relative water content, water potentials, stomatal conductance and chlorophyll fluorescence. *Q. laceyi* was drought tolerant, while *Q. sideroxyla* was drought avoidant. Leaf spectral reflectance increased with time in response to decreases in leaf photosynthetic pigment concentrations in latter weeks of the drought. The results suggest a close association between the altitudinal distributions of these species and their adaptation to water stress (Poulos et al., 2007).

A study was made on the anatomy and ultra structure of three species of wood of *Prosopis* growing in a heterogeneous dry forest in Chaqueño Park. The species studied were: *P. vinalillo*, *P. alba* and *P. nigra*. The results show that all three species are very similar and consistent with the structural features of the subfamily Mimosoideae. However, the vessels/mm^2 was quite variable between species and between individuals of the same species. In the cases of inter-vessel, vessel-ray and vessel-parenchyma pits it is important to describe shape, type, distribution and ornamentation. Samples observed under scanning electron microscope displayed ornaments in pits and striations on the vessel walls. These striations were shown to be characteristics of three *Prosopis* species studied (Bolzón de Muniz et al., 2010).

7.4.4   Wood fibres

Wood fibres (sclerenchyma) have a complex ultrastructure and are composed of several cell wall layers. These cell wall layers vary in their contents of cellulose, hemicelluloses and lignin. The chemistry and ultrastructure of wood components determine the properties of this lignocellulosic fibre in both annual and perennial plants; and they also determine the longitudinal growth stress generated during cell wall maturation as the result of the biosynthetic and biochemical processes during cell wall formation. There are various types of reaction wood, revealing extreme cases of macromolecular and ultrastructural organisation. Approximately half of the angiosperms species produce tension wood where the secondary wall is partially replaced by a so-called "gelatinous layer" from which lignin is absent and made up of axially oriented cellulosic microfibrils.

The major components of plant cell walls are cellulose, hemicelluloses and lignin. A cellulose chain consists of about 10 000 glucose units. In general, 30–40 parallel cellulose chains associate to form microfibrils containing crystalline and amorphous parts. The orientation of microfibrils in the various cell wall layers determines the strength properties of individual fibres and solid wood (CEMARE 2015).

Fibre cell length, diameter and cell wall thickness are related to fibre quality and show variations in different positions and growth rate (Bhat et al., 1990; Helniska-Raczkowska and Fabisiak, 1991; Honjo et al., 2005; Tsuchiya and Furukawa, 2009).

The radial variation in length, cell wall thickness, total diameter and lumen diameter were observed in a study in Venezuela. The results revealed a gradual increase of the first two parameters with increasing distance from the pith to the periphery of the trunk and an inverse relation between fibre diameter and lumen diameter (Velásquez et al., 2014).

The fibre cell wall is composed of cellulose impregnated in a matrix of lignin. Barnett and Bonham (2004) studied cellulose microfibril angles in the cell wall of wood fibres. The term microfibril angle (MFA) in wood science is defined as the angle between the direction of the helical windings of cellulose microfibrils in the secondary cell wall of fibres and tracheids and the long axis of cell. The cellulose microfibrils are oriented in the S2 layer of the cell walls that forms the greatest proportion of the wall thickness and is responsible for the physical properties of wood. The authors made a review on the organisation of the cellulose component of the secondary wall of fibres and tracheids and the various methods that have been used for the measurement of MFA. They observed a large variation of MFA within the trees observed between juvenile (or core) wood and mature (or outer) wood. These differences in MFA have an effect on the properties of wood, with respect to its stiffness. The large MFA in juvenile wood contributes to low stiffness and gives the sapling the flexibility necessary to survive high winds without breaking. This fact has an increasing importance in forestry for short rotation cropping of fast-growing species. They are presently grown mainly for pulp and pressure has built up for increased timber production and improvedtimber quality by reducing the juvenile wood MFA. The mechanism for the orientation of microfibril deposition is still a matter of debate, but the application of molecular techniques likely to enable modification of this process is not yet explored (Barnett and Bonham, 2004).

van Leeuwen et al. (2011) made an assessment of standing wood and fibre quality using ground and airborne laser scanning. Physical and chemical characteristics of wood show variations with both tree and site characteristics. But these characteristics are largely dependent at tree level on crown development, stem shape and taper, branch size and branch location, knot size, type and placement, age. All of these influence the wood properties; but stocking density, moisture, nutrient availability, climate, competition, disturbance and stand age have also been confirmed as key determinants of wood quality. The authors identified a number of key wood quality attributes (i.e. basic wood density, cell perimeter, cell coarseness, fibre length, microfibril angle) and established links between these properties and forest structure and site attributes. This technique is recommended to predict wood quality in standing timber (van Leeuwen et al., 2011).

The properties of wood and wood-based materials are strongly dependent on the properties of its fibres, that is the cell wall properties. The ability to characterise these in order to increase our understanding of structure–property relationships is thus highly important. Eder et al. (2013) presented a brief overview of the state of the art in experimental techniques to characterise the mechanical properties of wood at both the level of the single cell and that of the cell wall. Challenges, opportunities, drawbacks and limitations of single fibre tensile tests and nanoindentation are discussed with respect to the wood material properties.

7.4.5   Physical and mechanical properties

Thornscrubs are a vegetation type from North-eastern Mexico consisting of 60–80 tree and shrub species that are used for a wide range of construction, decorative and energy purposes. Basic research is lacking to establish additional uses and thereby increase their value in the timber industry. Carrillo et al. (2011) studied basic density, modulus of elasticity, modulus of rupture and their relationship with their properties.

The quality size of the charcoal of branches was acceptable according to France quality. The quality of charcoal from branches can be improved by controlling

air intakes to prevent an increment of temperature (Bustamante-García et al., 2013).

The natural durability of wood was determined according to the European Norm 350-1. Highly significant differences ($p < 0.001$) were found between the durability of woody species. Species with lower mass loss after exposure to *Coniophora puteana* were *Ebenopsis ebano* (6.3%), *Condalia hookeri* (8.6%) and *Cordia boissieri* (11.8%). *E. ebano* (7.1%), *Condalia hookeri* (8.2%) and *Cordia boissieri* (11.5%) showed lower mass losses after exposure to *Trametes versicolor*. According to the international standard, all three woody species were classified as very durable and durable species (Carrillo et al., 2013).

7.4.6   Guide to studying wood anatomy

Before the identification of any timber species, it is necessary to be familiar with the general characteristics and macroscopic studies on wood structure. Depending on xylem structure, the sections are generally transverse, radial or tangential.

A transverse section is cut perpendicular to the length of the trunk. In this plain, one can observe the growth rings and their characteristics, the ring breadths, the percentage of early and late wood and the type of transition between them. Rays are largely observed as lines which cross the growth ring at a right angle. Other microscopic elements are type of pore, grouping and arrangement of pores, size of pores, size of rays, type of parenchyma, texture, type of transition between soft- and heartwood.

A radial section is perpendicular to the rings; and a tangential section is parallel to the length of the trunk, or perpendicular to the rays or tangential to the growth rings.

7.4.6.1 Anatomical description

Cortex: phloem tissue, generated by cambium, two layers clearly visible.

External cortex: phloematic tissue for protection against atmosphere.

Internal cortex: phloematic tissue, functional layer for products of photosynthesis.

Cambium: secondary meristem, with cell division producing cortex outside and xylem inside.

Wood: gives support and transport water and nutrients through xylem vessels.

Soft wood: contains living cells, stores reserve substances, susceptible to insects and other organisms.

Heartwood: present in the centre of the wood formed by internal layers; does not contain living cells, the reserve materials are transformed into phenolic compounds; generally dark, stores various classes of products, oils, gums, resins, tannins, aromatic substances and pigments impregnated in the cell walls.

Pith: central part of stem formed mainly by parenchymatous tissue, susceptible to attack by biological organisms.

Rays: composed of similar types of cells which are extended radially in the wood crossing through growth rings.

Growth rings: in a transverse section of wood, one can see circles of concentric rings, each ring representing the quantity of wood produced by cambium under favourable conditions in spring; in winter the cambial activity is reduced, producing less secondary wood and thereby distinguishing clearly winter from spring wood.

Table 7.2 Differences between coniferous and broad-leaved woods.

Coniferous	Broad-leaved
Without pores	With pores
Homogeneous structure	Heterogeneous structure
Growth rings well defined	Growth rings little distinguished
Rays less defined	Rays defined
Soft wood and heartwood generally less marked	Soft and heartwood well marked
Two types of parenchyma	

Table 7.2 shows the differences between coniferous and broad-leaved woods.

The categories of parenchyma cells are described in the following list.

A. Apotracheal parenchyma: parenchyma present independent of pores or vessels. This is further subclassified:
  • Apotracheal diffuse: when individual cells are found in dispersed forms in contact with pores.
  • Apotracheal diffuse or grouped: when the parenchyma cells are grouped forming tangential lines.

B. Parenchyma paratracheal: when the parenchyma surrounds the pores in a circle, more or less oval. These cells are thin or thick:
  • Parenchyma aliform: when the parenchyma cells surrounds completely but also extends laterally like wings.
  • Parenchyma aliform confluent: aliform parenchyma but whose lateral extensions permit the union of pores.

C. Parenchyma in bands: parenchyma forms bands, lines, or concentric lines which may be in contact with pores:

- Bands or lines thin.
- Bands broad.
- Scalariform: the bands of parenchyma are stretched so that rays are arranged horizontally.
- Marginal: the bands of marginal parenchyma is found near the growth rings.
- Reticulate: bands whose breadth is similar to that of rays, forming a reticulate appearance.

7.4.7 General description of wood anatomy

The wood anatomy of a plant may follow this pattern:

- *Transverse section*. Pores (vessels) regular or irregular in form, size, irregularly distributed termed as diffuse porous, or arranged in rings called ring porous. Isolated or in groups, multiples. Parenchyma type vescicentric, aliform, aliform confluent, apotracheal parenchyma in bands, scalariform, terminal or marginal. Sclerenchyma profuse or scant. Medullary rays distinct, wavy or straight.
- *Tangential section*. Rays homogenous (similar size), or hetrogenous (dissimilar size), shape, long or short, profuse or few. Vessels long or short, pits round, oval, alternate or opposite.
- *Radial section*. Ray cells, multilayered, stratified or non-stratified.

7.4.7.1 Description of the wood anatomy of a few woody species of North-east Mexico

Microscopic characteristics (transverse section):

1 *Acacia amentacea* (see Figure 7.1)

Pores are rings to diffuse porous, not uniform in size, shape, majority of pores solitary, very few are in multiple of two, oval to almost rectangular, mostly big in size. Axial parenchyma and paratracheal perenchyma aliform, confluent. Apotracheal parenchyma aggregated in tangential broad bands. Narrow marginal parenchyma observed. Medular rays distinct, little wavy, separating radial bands of pores. Sclerenchyma not abundant. Fibre cells long with broad lumen, thin walled. Vessels small, truncated with slightly inclined end walls with broad perforation plate, pits round, numerous, alternate in arrangement. Wood tissue compact, seem to be associated with thick-walled fibre cells and numerous pores (vessels).

2 *Acacia berlandieri* (see Figure 7.2)

Pores diffuse porous, mostly solitary, a few in groups of two or three, numerous, contain gummy substance. Not uniform in size. Pores oval in shape, mostly large, some are very small. Axial parenchyma confluent. Apotracheal parenchyma in the form of broad band, scalariform, marginal parenchyma is visible. Medulary rays thin to broad, traverse through wood tissue. Rays short, more or less spindle-shaped, mostly two or three cells in breadth, a few uniseriate, heterogeneous, cells oval in shape. Vessels truncated, broad, short, more or less with straight perforation plates, pits elongated scalariform, alternating in arrangement. Medullary ray cells stratified, multilayered, stratified. Vessels broad, very short, broad,

Figure 7.1 Transverse section of *Acacia amentacea*. (See insert for color representation.)

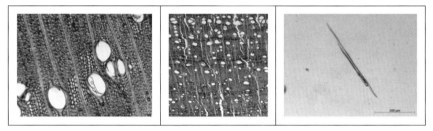

Figure 7.2 Transverse section of *Acacia berlandieri*. (See insert for color representation.)

Figure 7.3 Transverse section of *Acacia farnesiana*. (See insert for color representation.)

truncated with straight perforation plate, pits oval in shape, alternating in arrangement, evolutionarily more advanced. Wood tissue is compact with thick-walled fibre cells, profuse pores, seems to be hard.

3 *Acacia farnesiana* (see Figure 7.3)

Pores diffuse porous, ovoidal in shape, scanty, mostly solitary, big and small-sized. Paratracheal parenchyma aliform confluentric and confluent, surrounded by many cells, apotracheal parenchyma in broad band, scalariform, medullary rays medium broad and thinly traverse through the wood tissue. Rays spindle-shaped short, 2-4 celled broad, heterogeous, small celled. Rays multiseriate, broad, stratified fibre cells long, pointed, broad lumened with thin wall. Suitable for paper pulp. Vessels broad, short, truncated with straight perforation plate,

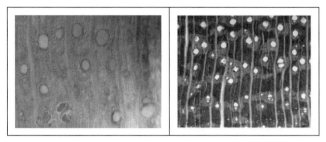

Figure 7.4 Transverse section of *Acacia shaffneri*. (See insert for color representation.)

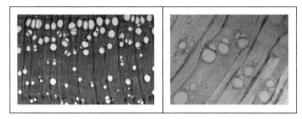

Figure 7.5 Transverse section of *Acacia wrightii*. (See insert for color representation.)

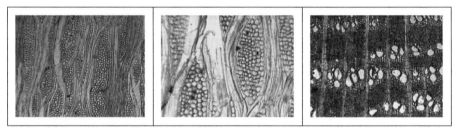

Figure 7.6 Transverse section of a stem of *Cordia boissieri* showing the organization of primary and secondary xylem. (See insert for color representation.)

pits elongated, alternate, evolutionarily advanced. Wood tissue compact, hard.

4 *Acacia shaffneri* (see Figure 7.4)

Pores diffuse, mostly solitary, a few of two to three cells, oval in shape, profuse in numbers. Paratracheal parenchyma aliform confluent, apotracheal parenchyma in broad bands, scalariform, medulary rays, medium thick traverse through wood tissue. Vessels contain gummy materials.

5 *Acacia wrightii* (see Figure 7.5)

Pores diffuse porous, numerous, small-sized, oval, some are very small. Wood tissue highly compact revealing the wood is very hard. Paratracheal parenchyma aliform confluent, apotracheal in bands, scalariform. Rays narrow, long, 2-4 celled, heterogeous, ray cells non-stratified.

6 *Cordia boissieri* (see Figure 7.6)

Pores arranged in rings, diffuse porous, arranged in groups of 3–5 cells,

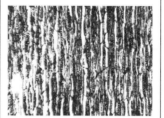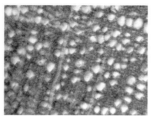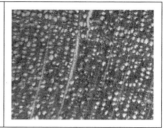

Figure 7.7 Transverse section of *Helietta parviflora*. (See insert for color representation.)

ovoidal but compressed. Parenchyma vascicentric aliform. Apotracheal parenchyma in bands. Wood contains a large amount of soft parenchymatous tissue, therefore the wood is soft. Rays spindle-shaped moderately long, maximum 3–6 cells in breadth, tapering, heterogeneous, compressed rays spindle-shaped, very broad, mutilayered, composed of highly compact small round cells, seems to be homogenous. Owing to the profuse quantity of ray and parenchymatous cells, the wood is soft. Rays are multiseriate, non-stratified. Vessels cylindric, broad or medium in length, pits large, alternate in arrangement.

7 *Helietta parviflora* (see Figure 7.7)

Pores, numerous, pores mostly isolated, a few in radial groups of three or four cells, pores are oval in shape, diffuse porous. Wood tissue compact. Paratracheal parenchyma vascicentric, apotracheal parenchyma diffuse. Owing to compact tissue, the wood seems to be very hard. Pores partially ring porous, in a ring of a few cells, mostly in multiple of three or four cells, large-sized, somewhat ovoidal. Pratracheal parenchyma vascicentric. Apotracheal parenchyma in

bands, somewhat scalariform. Rays uniseriate to biseriate, heterogeneous, short, cells round. Vessels cylindrical, long with oblique perforation plate, pits round, alternating. Vessels broad, long, cylindrical, pits oval, large-sized, alternating in arrangement.

8 *Condalia hookeri* (see Figure 7.8)

Pores seem to be diffuse porous arranged in clusters of several pores. Paratracheal parenchyma vascicentric, apotyracheal in bands, wood tissue highly compact with profuse sclerenchyma, thereby imparting very hard wood. Rays are small, mostly bi- or tri-seriate, heterogeneous. Wood contains numerous exudates, gums. Wood tissue is loose, probably offering soft wood.

9 *Diospyros palmeri* (see Figure 7.9)

Pores, oval or compressed, diffuse porous. Parenchyma paratracheal vascicentric. Apotracheal parenchyma diffuse, terminal parenchyma present, fibres profuse, thereby imparting very hard wood. Rays numerous, mostly uniseriate, a few biseriate composed of ovoidal cells, heterogeneous.

10 *Zanthoxylum fagara* (see Figure 7.10)

Pores ring, porous, ovoid, not many, mostly solitary, a few in radial groups

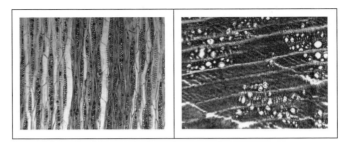

Figure 7.8 Transverse section of *Condalia hookeri*. (See insert for color representation.)

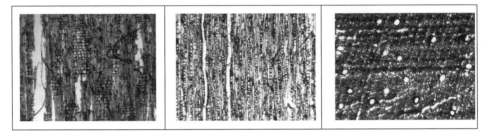

Figure 7.9 Transverse section of *Diospyros palmeri*. (See insert for color representation.)

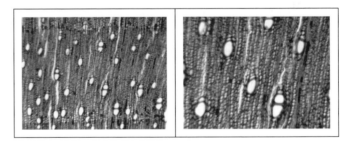

Figure 7.10 Transverse section of *Zanthoxylum fagara*. (See insert for color representation.)

of two or three cells, ovoidal in shape. Paratracheal parenchyma, vascicentric. Apotracheal parenchyma scalariform. Terminal parenchyma present near the annual ring. Apotracheal parenchyma aggregated. Wood is composed of mostly soft parenchymatous tissue, thereby imparting soft wood. Rays short, mostly uniseriate or bi- to triseriate, heterogenous.

11 *Karwinskia humboldtiana* (see Figure 7.11)

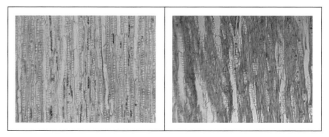

Figure 7.11 Transverse section of *Karwinskia humboldtiana*. (See insert for color representation.)

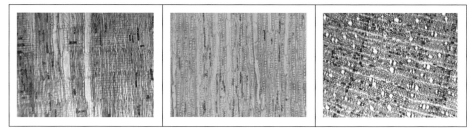

Figure 7.12 Transverse section of *Diospyros texana*. (See insert for color representation.)

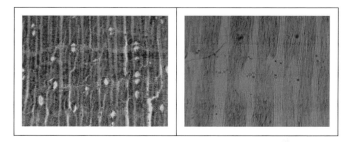

Figure 7.13 Transverse section of *Celtis pallida*. (See insert for color representation.)

12 *Diospyros texana* (see Figure 7.12)

Pores diffuse porous, numerous, round, solitary, a few in groups of two or three. Paratracheal parenchyma vascicentric. Apotracheal parenchyma scalariform in bands. Rays numerous, uniseriate, seem to be homogeneous and mostly rectangular cells. Vessels medium long with slightly inclined perforation plate.

13 *Celtis pallida* (see Figure 7.13)

Pores scarce, ring porous, mostly solitary, a few in a radial ring of two cells. Paratracheal parenchyma vascicentric. Apotracheal parenchyma

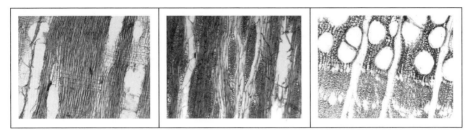

Figure 7.14 Transverse section of *Celtis laevigata*. (See insert for color representation.)

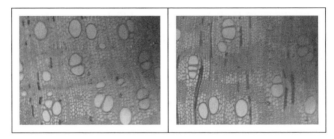

Figure 7.15 Transverse section of *Caesalpinia mexicana*. (See insert for color representation.)

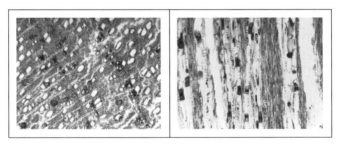

Figure 7.16 Transverse section of *Eysenthardtia polystachya*. (See insert for color representation.)

scalariform. Wood tissue compact, imparting hardness to the wood.

14 *Celtis laevigata* (see Figure 7.14)

Pores solitary, a few of two cells, ring porous. Paratracheal parenchyma vascicentric. Apotracheal parenchyma in the form of bands. Marginal parenchyma few. The presence of profuse parenchyma imparts softness to the wood, not suitable for furniture. Rays broad mostly multiseriate, heterogeneous. Vessels medium long

with slightly inclined perforation plate.

15 *Caesalpinia mexicana* (see Figure 7.15)

Pores diffuse porous, large-sized, mostly solitary, a few in groups of two. Parenchyma aliform to confluent. Apotracheal parenchyma aggregated. Wood is composed of numerous soft tissues, thereby imparting softness to the wood. Rays uniseriate, homogenous.

16 *Eysenthardtia polystachya* (see Figure 7.16)

7.5 Characterisation of wood fibres of shrubs and tree species of the Tamaulipan thornscrub, north-eastern Mexico

7.5.1 Background

The shrubs and trees of the Tamaulipan thornscrub in the semiarid regions of north-eastern Mexico are of great economic importance for various uses such as timber for furniture, fences, posts, firewood, sources of forage for wild grazing animals (by possessing macro- and micronutrients required by these animals), herbs, medicine and reforestation (Reid et al., 1990; Stienen et al., 1990). The utilisation of wood for different purposes depends on the structure, physical and chemical characteristics of wood elements, mainly fibre cells. Studies have been undertaken on these aspects by researchers. Wood is a hard, fibrous structural tissue present in the stems and roots of woody species. Its main use is for furniture and building construction, besides fences, paper pulp and other products. With respect to its structure, wood is the secondary xylem derived by intercalary meristem consisting of the cambium in the stems of trees. Cambium consists of two types of cells: (a) fusiform cells giving fibre cells and xylem vessels and (b) ray initials giving ray cells forming medullary cells. Fibre cells are composed of cellulose fibrils embedded in a matrix made of lignin, which resists compression. It gives mechanical strength to the plant and also helps to transfer water and nutrients to the leaves and other growing tissues.

Wood fibres have a complex ultrastructure and are composed of several cell wall layers. These cell wall layers vary in their contents of cellulose, hemicelluloses and lignin. The chemistry and ultrastructure of wood components determines the properties of this lignocellulosic fibre of perennial plants, it also determines longitudinal growth stress generated during cell wall maturation as a result of the biosynthetic and biochemical processes during cell wall formation. There exist various types of reaction wood revealing extreme cases of macromolecular and ultrastructural organisation. Approximately half of the angiosperms species produce tension wood where the secondary wall is partially replaced by a so-called "gelatinous layer" where lignin is absent and the layer is made of axially oriented cellulosic microfibrils. The final fibre properties contribute greatly to the quality of pulp and paper as well as timber and its products.

7.5.2 Methodology

We adopted two techniques for macerating the wood for fibre studies:

Technique 1. A small portion of wood was kept in a test tube in a mixture of 10% chromic acid:10% nitric acid, then the test tube mouth was plugged with cotton and the test tube kept in an incubator for 24 h. Then, the macerated wood tissue was washed carefully, stained with safranin and then mounted in a slide with glycerine for micro-observation.

Technique 2. R.K. Maiti developed a simple technique for the maceration of wood for fibre characterisation. One disk 0.1 m thick was taken from two primary branch of a tree from each species. A few small pieces of wood from each species were dipped in test tubes of concentrated nitric acid, which were then plugged with cotton. Then, the test tubes were kept in a boiling water bath until the wood pieces started disintegrating. Then, the acid was decanted slowly and the macerated wood elements were washed several times with distilled water. Then, the macerated fibre cells were stained with safranin (1%) and observed under a microscope; photographs were taken with a digital camera fixed to the microscope. Fifty fibre cells of each species were measured using an ocular and a stage micrometer.

We took measurements in terms of length, breadth and cell wall of 50 fibre cells with the help of an ocular and a stage micrometer.

7.5.3 Results

The following depicts the morphology of wood fibre cells in the woody species of the Tamaulipan thornscrub. Wood fibre cells were observed at 10 × and 40×. Descriptions:

1 *Acacia berlandieri*
   - The lumen is a little broad, the cell wall thin but little lignified, the apex is pointed or round, the fibre cells a little long and broad.
2 *Acacia farnesiana*
   - The lumen is broad, cell wall thick, the apex is pointed, fibre cells uniform, some non-uniform.
3 *Acacia rigidula*
   - The lumen is broad, cell wall thin, the apex is pointed.
4 *Acacia shaffneri*
   - The lumen is a little broad, cell wall thin, the apex is pointed.
5 *Acacia wrightii*
   - The lumen is thin, cell wall is very thin, the apex is round.
6 *Bernardia myricifolia*
   - The lumen is a little broad, cell wall is thin, the apex is pointed, fibre cells thin and a little long.
7 *Caesalpinia mexicana*
   - The lumen is broad, cell wall very thin, the apex is pointed, fibre cells a little broad and a little long, uniform.
8 *Capsicum annuum*
   - The lumen is a little broad, cell wall thin, but a little lignified, the fibres long, the apex is pointed.
9 *Celtis laevigata*
   - The lumen is a little broad, cell wall is thin, but a little lignified, the apex is round, a few pointed.
10 *Celtis pallida*
   - The lumen is thin, cell wall thin, the apex is pointed and round.
11 *Condalia hookeri*
   - The lumen is thin, cell wall is thin, the apex is pointed.
12 *Cordia boissieri*
   - The lumen is a little broad, cell wall is thin, the apex is pointed and round.
13 *Croton terreyanus*
   - The lumen is thin, cell wall is thin but a little lignified, the apex is pointed or round.
14 *Diospyros palmeri*
   - The lumen is a little broad, cell wall is thin, the apex is pointed.
15 *Ehretia anacua*
   - The lumen is a little broad, cell wall is thin, the apex is pointed
16 *Eysenhardtia polystachya*
   - The lumen is a little broad, cell wall is thin, the apex is pointed, the fibre cells are uniform.

17 *Forestiera angustifolia*
- The lumen is thick, cell wall is thin, the apex is pointed, fibres are long.

18 *Fraxinus greggii*
- The lumen is thin, cell wall is thin, but a little lignified, the apex is pointed.

19 *Gochnatia hypoleuca*
- The lumen is thin, cell wall thick, the apex is pointed or round.

20 *Helietta parvifolia*
- The lumen is thin, cell wall is thin, the apex is pointed.

21 *Karwinskia humboldtiana*
- The lumen is thin, cell wall is thin, the apex is pointed.

22 *Lantana macropoda*
- The lumen is a little broad, cell wall is thin but lignified, the apex is pointed or round, the fibre cells are small, large and thin.

23 *Leucophyllum frutescens*
- The lumen is thin, the apex is pointed or round, the fibre cells are small and thin.

24 *Morus celtidifolia*
- The lumen is a little broad, the apex is pointed, the fibre cells are thin, uniform or non-uniform.

25 *Parkinsonia aculeate*
- The lumen is a little broad, cell wall is thin, but a little lignified, the apex is pointed.

26 *Prosopis laevigata*
- The lumen is a little broad, cell wall is thin, the apex is pointed or round.

27 *Quercus polymorpha*
- The lumen is very thin, cell wall is thin, the apex is pointed, the fibre cells are a little long and thin, the majority non-uniform.

28 *Salix lasiolepis*
- The lumen is broad, the cell wall is thin, the apex is pointed or round.

29 *Zantoxylum fagara*
- The lumen is thin, cell wall is thin but lignified, the apex is pointed or round.

30 *Ziziphus obstusifolia*
- The lumen is thin, cell wall is thin, the apex is pointed.

The species show a large variation in fibre cell length, breadth and wall thickness as shown in Table 7.3.

Microphotographs of wood fibre cells of a few species, seen at 10× and 40×, are shown in Figures 7.17–7.25.

7.5.4 Discussion

The main objective of this study was to investigate the variability in the morphology, size, shape and dimensions of fibres of 14 woody species. No attempts have been made to the physical quality and ultrastructure of the fibre cells. Many studies have been directed on the growth, ultrastructure and orientation of microfibrils in fibre cell walls. In the present study, large variations were observed in the size, shape of fibre cells among woody trees and one shrub. On the basis of this we classify the species on the morphological characteristics of the fibre cells of the species studied.

Cell wall thin: *Acacia berlandieri, Bernardia myricifolia, Helietta parvifolia, Leucophyllum frutescens, Quercus virginiana* and *Tribulus terrestris*. The woods of these species may be recommended for soft furniture, fences and paper pulp.

Cell wall thin, lumen broad: *Acacia berlandieri, Caesalpinia mexicana, Eysenhardtia polystachya, Forestiera angustifolia,*

Table 7.3 Mean and standard deviation (SD) of fibre cell length, breadth and wall thickness of different plant species of Tamaulipan thornscrub, north-eastern Mexico.

Plant species	Fibre cell length (μm)		Breadth (μm)		Wall thickness (μm)		Length to breadth ratio	
	Mean	SD	Mean	SD	Mean	SD	Mean	SD
Acacia berlandieri	464.01	159.70	23.52	7.40	2.98	1.42	19.73	21.58
A. farnesiana	598.58	255.24	16.66	8.22	2.94	1.10	35.93	31.05
A. rigidula	581.53	186.73	15.28	7.18	2.79	0.85	38.06	26.01
A. shaffneri	501.17	248.84	16.85	9.90	2.94	1.48	29.74	25.14
A. wrightii	438.06	214.92	15.68	6.56	2.89	0.95	27.94	32.76
Bernardia myricifolia	450.59	105.04	22.73	8.72	3.13	1.48	19.82	12.05
Caesalpinia mexicana	464.71	110.31	20.18	8.49	3.28	1.75	23.03	12.99
Capsicum annuum	437.26	93.12	18.62	10.33	2.84	1.03	23.48	9.01
Celtis laevigata	657.58	246.60	17.64	12.04	3.08	1.46	37.28	20.48
C. pallida	471.97	162.82	14.11	6.9	3.08	1.08	33.45	23.60
Condalia hookeri	398.47	161.13	17.64	10.09	3.23	1.52	22.59	15.97
Cordia boissieri	518.03	155.09	16.46	7.25	3.72	1.8	31.47	21.39
Croton terreyanus	453.53	101.78	14.73	7.67	3.23	1.15	30.79	13.27
Diospyros palmeri	486.47	152.47	18.03	8.48	3.08	1.08	26.98	17.98
Ehretia anacua	644.66	270.67	11.56	3.80	4.45	3.11	55.77	71.23
Eysenhardtia polystachya	477.65	96.88	19.99	8.62	2.84	1.14	23.89	11.24
Forestiera angustifolia	488.43	72.21	23.91	5.99	4.80	0.85	20.43	12.06
Fraxinus greggii	535.67	150.90	17.64	13.42	2.74	1.17	30.37	11.24
Gochnatia hypoleuca	563.30	176.16	17.24	5.79	2.94	1.21	32.67	30.42
Helietta parvifolia	647.78	209.75	15.48	4.88	2.94	1.31	41.85	42.98
Karwinskia humboldtiana	507.33	219.48	15.68	10.47	3.08	1.19	32.36	20.96
Lantana macropoda	391.60	86.18	14.30	7.97	2.59	0.76	27.38	10.81
Leucophyllum frutescens	438.45	94.49	16.66	7.47	3.08	1.19	26.32	12.65
Morus celtidifolia	454.72	130.86	15.68	6.26	2.74	0.80	29.02	20.90
Parkinsonia aculeata	473.33	110.03	31.64	8.55	6.86	0.98	14.96	12.87
Eysenhardtia polystachya	477.65	96.88	19.99	8.62	2.84	1.14	23.89	11.24
Forestiera angustifolia	488.43	72.21	23.91	5.99	4.80	0.85	20.43	12.06
Fraxinus greggii	535.67	150.90	17.64	13.42	2.74	1.17	30.37	11.24
Gochnatia hypoleuca	563.30	176.16	17.24	5.79	2.94	1.21	32.67	30.42
Helietta parvifolia	647.78	209.75	15.48	4.88	2.94	1.32	41.85	42.98
Karwinskia humboldtiana	507.33	219.48	15.68	10.47	3.08	1.19	32.36	20.96
Lantana macropoda	391.60	86.18	14.30	7.97	2.59	0.76	27.38	10.81
Leucophyllum frutescens	438.45	94.49	16.66	7.47	3.08	1.19	26.32	12.65
Morus celtidifolia	454.72	130.86	15.68	6.26	2.74	0.80	29.03	20.90
Parkinsonia aculeata	473.33	110.03	31.64	8.55	6.86	0.98	14.96	12.87
Prosopis laevigata	412.97	190.85	10.78	2.96	2.84	0.90	38.31	64.48
Quercus polymorpha	709.39	190.17	13.72	6.26	2.59	0.58	51.70	30.38
Salix lasiolepis	455.11	150.25	16.07	6.78	3.43	1.30	28.32	22.16
Zantoxylum fagara	454.13	155.80	16.07	8.57	3.23	1.15	28.26	18.18
Ziziphus obtusifolia	591.09	150.72	20.58	7.20	2.98	1.33	28.72	20.93

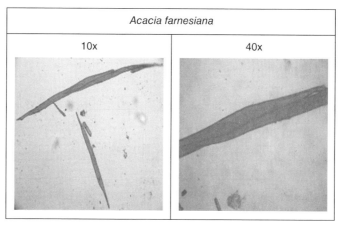

Figure 7.17 Microphotographs of wood fibre cells of *Acacia farnesiana*, seen at 10× and 40×. (See insert for color representation.)

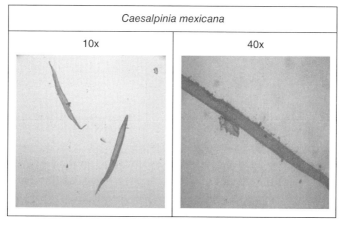

Figure 7.18 Microphotographs of wood fibre cells of *Caesalpinia mexicana*, seen at 10× and 40×. (See insert for color representation.)

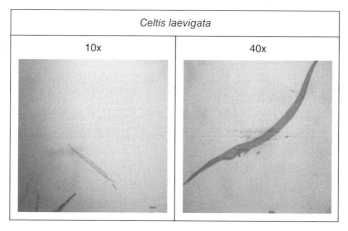

Figure 7.19 Microphotographs of wood fibre cells of *Celtis laevigata*, seen at 10× and 40×. (See insert for color representation.)

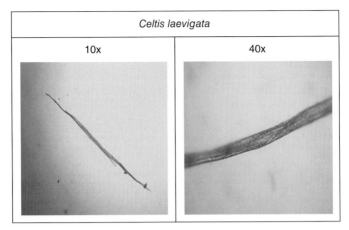

Figure 7.20 Microphotographs of wood fibre cells of *Celtis pallida,* seen at 10× and 40×. (See insert for color representation.)

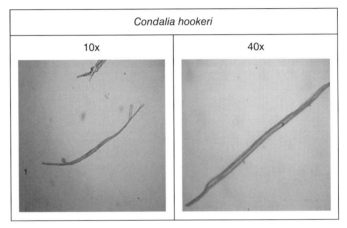

Figure 7.21 Microphotographs of wood fibre cells of *Condalia hookeri,* seen at 10× and 40×. (See insert for color representation.)

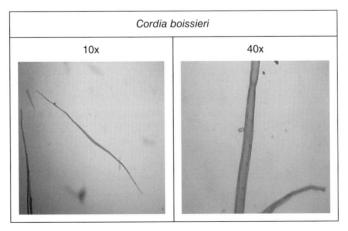

Figure 7.22 Microphotographs of wood fibre cells of *Cordia boissieri,* seen at 10× and 40×. (See insert for color representation.)

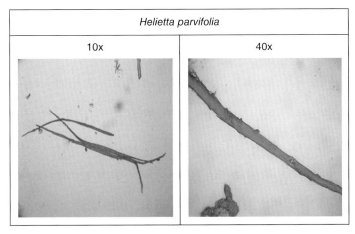

Figure 7.23 Microphotographs of wood fibre cells of *Helietta parvifolia,* seen at 10× and 40×. (See insert for color representation.)

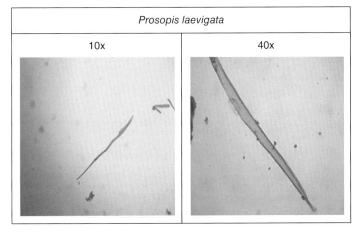

Figure 7.24 Microphotographs of wood fibre cells of *Prosopis laevigata,* seen at 10× and 40×. (See insert for color representation.)

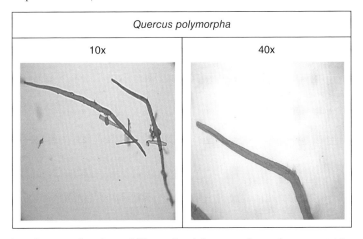

Figure 7.25 Microphotographs of wood fibre cells of *Quercus polymorpha,* seen at 10× and 40×. (See insert for color representation.)

Lantana macropoda and *Morus celtidifolia*. The wood of these species may be recommended for the preparation of paper pulp and good quality paper.

Cell wall thick, lignified, lumen narrow: *Fouquiera splendens* and *Salvia officinalis*. These species may be recommended for furniture and construction.

In the present study there were large variations in fibre cell length between different wood species on the basis of which we can tentatively classify the different species.

Fibre cell medium long: *Fouquiera splendens* (563 µm), *Forestiera angustifolia* (488 µm), *Eysenhardtia polystachya* (478 µm), *Acacia farnesiana* (473 µm), *Acacia berlandieri* (464 µm). It is expected that the wood of these species may offer moderate wood strength in its products.

Fibre cell length: *Quercus virginiana* (709 µm), *Helietta parvifolia* (648 µm), *Tribulus terrestris* (591 µm), *Acacia farnesiana* (473 µm), *Morus celtidifolia* (455 µm), *Salvia officinalis* (454 µm). It is expected that the timbers of these species may offer a greater wood strength in its products and papers.

Fibre cells short (less than 450 µm). *Lantana macropoda* (392 µm), *Capsicum annuum* (437 µm), *Leucophyllum frutescens* (438 µm). It is expected these timbers may offer poor wood strength in its products.

The present study was limited only to one locality in order to study the variability in fibre cell morphology and fibre cell length, which can vary in different localities, environments and positions in the trees which have been documented by various authors.

We did not attempt to study the relationship between fibre cell length and timber quality, which has been reported by various authors.

In addition to fibre cell length, it is well known that a high L/B ratio contributes to the strength of bast fibres. In the present study, we selected species having a high L/B ratio: *Ehretia anacua* (56), *Quercus polymorpha* (51), *Acacia rigidula* (38), *Celtis laevigata* (37), *Acacia farnesiana* (36), *Prosopis laevigata* (38) and *Celtis pallida* (33). The wood of these species may be recommended for better paper pulp and furniture.

7.5.5 Conclusions and research needs

Our study indicates that there exists a large variability in wood anatomical traits which can related to species identification and the quality determination of the species. There is also a large variability in the morphology, length and wall lignification of the fibre cells in each wood species. The intensity of lignification contributes to the strength and high quality of timber for furniture. Soft wood containing a high amount of parenchymatous tissue and thin walled fibre cells are used for the fabrication of soft furniture and fences. Woods having fibre cells with broad lumen and thin wall could be suitable for the manufacture of paper, as documented in the literature. Therefore, there is a great necessity to evaluate the wood anatomical structures of trees in a forest and to classify them for their suitability for various uses on the basis of wood anatomical structure. The selected wood of a particular species could be tested for its physical and chemical properties in a wood technology laboratory for confirmation.

Bibliography

Barnett J.R., Bonham V.A. **2004**. Cellulose microfibril angle in the cell wall of wood fibres. *Biological Reviews of the Cambridge Philosophical Society* **79**(2):461–72.

Bhat K.M., Bhat K.V., Dhamodaran T.K. **1990**. Wood diameter and fibre length of *Eucalyptus grandis* growth in Kerala, India. *Wood and Fiber Science* **22**(1):54–61.

Bolzón de Muniz G.I., Nisgoski S., Lonelí-Ramírez G. **2010**. Anatomía y ultraestructura de la madera de tres especies de *Prosopis* (Leguminosae – Mimosoidae) del Praque Chaqueño seco, Argentina. *Madera y Bosques* **16**(4):21–38.

Bustamante-García V., Carrillo-Parra A., González-Rodríguez H., Ramírez-Lozano R.G., Corral-Rivas J.J., Garza-Ocañas F. **2013**. Evaluation of chemical production process from forest residues of *Quercus sideroxy* Hunb, and Bonpl in a Brazilian beehive kiln. *Industrial Crops and Products* **42**(203):160–174.

Carrillo A., Foroughbach R., Bustamante V., Wehenkel C., González H. **2013**. Natural durability of wood of ten native species from northeastern Mexico. *Forest Science and Practice* **15**(2):160–166.

Carrillo A., Garza M., Núñez M. de J., Garza F., Foroughbakhch R., Sandoval S. **2011**. Physical and mechanical wood properties of 14 timber species from Northeastern Mexico. *Annals of Forest Science* **68**:675–679.

Carrillo A., Mayer I., Koch G., Hapla F. **2008**. Wood anatomical charactereacterists and chemical composition of *Prosopis laevigata* grown in the Northeast of Mexico. *IAWA Journal* **29**(1):25–34.

Carrillo-Parra A., Foroughbachk-Pournavab R., Bustamante-Gracia V., Sandoval-Torres S., Garza-Ocañas F., Moreno-Limón S. **2013**. Differences of wood elements of *Prosopis laevigata* from two areas of norteastern Mexico. *American Journal of Plant Sciences* **4**:56–60.

CEMARE **2015**. *CEMARE Workplan*, http://www.forestry.gov.uk/pdf/CEMARE_Workplan.pdf/$file/CEMARE_Workplan.pdf, last accessed 1 October 2015.

Drexhage M., Huber F., Colin F. **1999**. Comparison of radial increment and volume growth in stems and roots of of *Quercus petraea*. *Plant and Soil* **217**(1/2):101–110.

Eder M., Arnould O., Dunlop J.W.C., Hornatowska J., Salmén L. **2013**. Experimental micromechanical characterisation of wood cell walls. *Wood Science and Technology* **45**(3): 461–472.

Fahn A. **1990**. *Plant Anatomy*. Pergamon Press, New York.

Fujiwara S., Yangk K.C. **2000**. The relationship between cell length and ring width and circunferencial growth rate in Canadian species. *IAWA Journal* **21**(3):335–345.

Helniska-Raczkowska L., Fabisiak E. **1991**. Radial variation and growth rate in the length of axial elements of sessile oak wood. *IAWA Journal* **12**(3):257–262.

Honjo K., Furukawa I., Sahri M.H. **2005**. Radial variation of fibre length increment in *Acacia mangium*. *IAWA Journal* **26**(3):339–352.

López S.C., Sabaté S., Gracia C.A., Rodríguez R. **2005**. Wood anatomy, description of anual rings and responses to ENSO events of *Prosopis pallida* H.B.K., a wide-spread woody plant of arid and semi-arid lands of Latin America. *Journal of Arid Environments* **61**:541–554.

Meier E. **2015** *Wood Database*. http://www.wood-database.com/wood-articles/hardwood-anatomy/, last accessed 1 October 2015.

Meylan B.A., Butterfield B.G. **1974**. Occurrence of vestrured pit in the vessels and fibres of New Zealand woods. *New Zealand Journal of Botany* **12**(1):3–18.

Ogata Y., Fujita M., Nobuchi T., Sahri M.H. **2003**. Macrospopic and anatomical investigation of interlocked grain in Acacia mangium. *IAWA Journal* **24**(1):13–26.

Pérez Olvera C. de la P., Mendoza Aguirre A., Caja Romero J., Pacheco L. **2008**. Anatomía de la madera de cinco especies de la familia Rosaceae. *Madera y Bosque* **14**(1):81–105.

Poulos H.M. Goodale U.M., Berlyn G.P. **2007**. Drought response of two Mexican oak species, *Quercus laceyi* and *Q. sideroxyla* (Fagaceae), in relation to elevational position. *American Journal of Botany* **94**(5):809–818.

Reid N., Marroquín J., Beyer M.P. **1990**. Utilization of shrubs and trees for browse, fuel-wood and timber in the Tamaulipan thornscrub, northeastern Mexico. *Forest Ecology and Management* **36**:61–79.

Ruelle J., Yoshida M., Clair B., Thibaut B. **2007**. Peculiar tension wood structure in *Laetia procera* (Poepp.) (Flacourtiaceae). *Trees* **21**:345–355.

Stamm A.J. **1972**. Maximum effective vessel diameters of harwoods. *Wood Science and Technology* **6**(4):263–271.

Stienen H., Smits M.P., Reid N., Landa J., Boerboom J.H.A. **1989**. Ecophysiology of 8 woody multipurpose species from semiarid northeastern Mexico. *Annales des Sciences Foestières* **46**:454–458.

Tsuchiya R., Furukawa I. **2009**. Radial variation in the size of axial elements in relation in *Quercus serrata*. *IAWA Journal* **30**(1):15–26.

van Leeuwen M., Hilker T., Coops N.C., Frazer G.W., Wulder M.A., Newnham G.J., Culvenor D.S. **2011**. Assessment of standing wood and fiber quality using ground and airborne laser scanning: A review. *Forest Ecology and Management* **261**(9):1467–1478.

Velásquez J., Jiménez B., Monagas P., Terzo F.M. Toro M.E., Ruiz Y. **2014**. Aspectos morfológicos en las fibras de la madera de *Erisma uncinatum* Warm. *Interciencia* **39**(5):344–349.

Wimmer R. **2002**. Wood anatomical features in tree-rings as indicators of environmental change. *Dendrochronologia* **20**(1/2):21–36.

Zwieniecki M.A., Melcher P.J., Holbrook N.M. **2001**. Hydrogel control of xylem hydraulic resistance in plants. *Science* **291**:1059–1062.

CHAPTER 8

Phenology

8.1 Introduction

The forest contains a great diversity of species which vary in their time of floral initiation, flowering, fruiting and seed dispersal. Different species flower at different times of the year, depending on its photoperiod requirements, and add to a beautiful landscape, a galaxy with flowers of different colours at different stages of blooming. A study on the duration of the different stages in a plant, such as germination, leaf emergence, the duration of flowering, fruiting, fruit maturity and seed dispersal, is what we call phenology. In a forest one can notice flowers of different sizes and colours, very attractive to the pollinators visiting flowers and sucking nectar and at the same time helping in pollination. I have noticed that the mangrove species produce big white, hanging flowers which bloom at night to attract the insects for pollination. In the afternoon, many flower petals drop down to the ground, indicating quick pollination and completion of the fertilisation process. We can observe massive viviparous germination which falls to the ground and produces a large number of seedlings. Mangroves are so efficient in the fertilisation and propagation process.

Phenology determines the different phases of the life cycle of a plant/tree. It can be classified as:

- *Germination and seedling:* In this phase, the seeds of a plant germinate in the soil, emerge and grow as seedlings. The seeds of a woody plant which fall on the soil possess dormancy. Exposure to the edaphic and atmospheric conditions breaks the dormancy and germination occurs with the advent of rain showers. The seeds/fruits of a few species are eaten by birds, passed through their digestive system and liberated through excreta, thereby breaking dormancy.

- *Vegetative:* The seedlings after germination pass through different phases and finally reach the vegetative stage. At this phase, new leaves are derived from the meristematic activity of the lateral meristem, finally emerge as leaves, expand and finally reach a mature stage. Then, the mature leaves undergo the phase of senescence. This sequence of phases is called leaf phenology. It consists of the time taken from initiation to final expansion of the leaves. The apical meristem of the shoot at the same time leads to elongation of the stem. With the growth of the plant, the intercalary meristem (cambium) produces the secondary

Autoecology and Ecophysiology of Woody Shrubs and Trees: Concepts and Applications, First Edition.
Edited by Ratikanta Maiti, Humbero Gonzalez Rodriguez and Natalya Sergeevna Ivanova.
© 2016 John Wiley & Sons, Ltd. Published 2016 by John Wiley & Sons, Ltd.

xylem and wood of a timber plant. Before that phase, the plant reached its reproductive phase. Leaf phenology involves the time for the emergence of leaves, their expansion and sencescence.

- *Reproductive phase:* This phase determines the different stages, such as flowering, fruiting and seed dispersal. Apart from the other phases mentioned, the reproductive phase is a very important phase related to the productivity of the plant. The photoperiod and temperature determine the initiation of flower buds. Annual plants initiate flowering every year, while perennial plants take one or more years to initiate flowering. A knowledge of flowering and seed dispersal is important for efficient forest management by forest managers.

Various studies have been undertaken on the phenology of trees.

Sarquís et al. (2009) studied the physiology of photosynthesis in chimalacate (*Viguiera dentata*) in Zapotititlan of Las Salinas valley of the Tehuacán reserve biosphere in Puebla, Mexico. Growing interest has been generated in understanding ecological deterioration phenomena; and the natural restoration of native landscapes led us to initiate a research effort on a wild species which may be instrumental in the recovery of severely disturbed ecosystems. Chimalacate plants grew almost twice as tall, developed twice as many tillers, showed 54% greater dry weight, 40% greater leaf expansion rate and over five times as many flowers when exposed to full sun as compared to plants growing in the shade. However, shaded plants developed 48% more leaf area per plant mainly due to more leaves per plant. Optimum leaf

temperature for photosynthesis at high light intensity and ambient CO_2 concentration was 34 °C. Light and CO_2 compensation points under a controlled environment ranged between 23 and 48 μmol m^{-2} s^{-1} and between 22 and 32 μl l^{-1} CO_2, respectively. The photosynthesis data presented indicates chimalacate is a C3 plant well adapted for growth in hot dry conditions.

We undertook two studies on the phenology of different woody species, described in the following sections.

8.2   Reproductive phenology (flowering and fruiting) of ten woody plants, north-eastern Mexico

8.2.1   Study area

The study was undertaken during the period March–December 2013, with 12 species of the Tamaulipan thorn scrub, in an undisturbed area of Nuevo Leon, Mexico. The study site was located at the Experimental Field of the Forest Science School, Universidad Autonoma de Nuevo León (24°47′N, 99°32′E, 350 m asl), 8 km south of Linares county. The climate is subtropical and semiarid, with a warm summer. Monthly mean air temperature ranges from 14.7 °C in January to 22.3 °C in August, although daily high temperatures of 45 °C are common during summer. Average total annual precipitation ranges from 600 to 805 mm with a bimodal distribution. The peak rainfall months are May, June and September (González Rodríguez et al., 2004). The dominant soils are deep, dark-grey, lime-clay vertisols, with montmorillonite, which shrink and swell noticeably in response to changes in soil moisture content.

8.2.2   Vegetation of the study area

The main type of vegetation at the study site is characterised by shrubs and semi-trees, with heights between 4 and 6 m. The perennials mainly have spines; the deciduous mainly have small leaves. The most representative species are: *Prosopis laevigata* (Humb. and Bonpl. ex Willd.). M.C. Johnst., *Ebenopsisebano* (Berland.) Barneby and J.W. Grimes, *Acacia amentacea* D.C., *Castela erecta* Turpin subsp. *texana* (Torr. and A. Gray) Cronquist, *Celtis pallida* Torr., *Parkinsonia texana* var. *macra* (I.M. Johnst.) Isely, *Forestiera angustifolia* Torr., *Cordia boissieri* A. D.C., *Leucophyllum frutescens* (Berland.) I.M. Johnst., *Guaiacum angustifolium* Engelm., *Cylindropuntia leptocaulis* (D.C.) F.M. Knuth, *Opuntia spp*, *Zanthoxylum fagara* Sarg., *Sideroxylon celastrinum* (Kunth) T.D. Penn, *Helietta parvifolia* (A. Gray ex Hemsl.) Benth., among others. The deciduous species lose their leaves in autumn or in the beginning of winter, while the leaf fall of the perennials occurs in a constant manner throughout the year, although the greater leaf abscission occurs between summer and autumn (Moro, 1992). The floristic structure composition at research site has been previously documented (Domínguez Gómez et al., 2013), as shown in Table 8.1.

8.2.3   Selection of specimens

The species were selected on the basis of their ecological and nutritional value for livestock and wild ruminant animals (González-Rodríguez et al., 2010; Domínguez Gómez et al., 2012) and the

Table 8.1 General characteristics of plant species identified at research site, north-eastern Mexico. A, D, F and IV refer to abundance, dominance, frequency and importance value, respectively.

Species	Phenology	No. of Individuals	Height (m)	Crown cover (m²)	A (%)	D (%)	F (%)	IV (%)
Acacia amentacea	Deciduous	73	2.5	142.7	6.1	6.0	5.6	17.8
A. farnesiana	Deciduous	4	6.3	86.1	0.3	3.6	1.6	5.7
A. schaffneri	Deciduous	11	2.5	61.8	0.9	2.6	2.8	6.3
Castela erecta	Deciduous	23	1.5	16.6	1.9	0.7	4.5	7.1
Celtis pallida	Deciduous	28	3.4	149.4	2.3	6.3	5.0	13.8
Condalia hookeri	Perennial	36	2.7	131.0	3.0	5.5	2.2	10.8
Cordia boissieri	Deciduous	19	3.1	127.5	1.5	5.4	3.3	10.4
Diospyros texana	Perennial	34	2.7	123.3	2.8	5.2	0.5	8.6
Eysenhardtia texana	Deciduous	33	3.5	183.2	2.7	7.8	5.6	16.2
Forestiera angustifolia	Deciduous	19	1.8	37.4	1.5	1.5	4.5	7.7
Havardia pallens	Deciduous	40	3.1	124.6	3.3	5.3	1.1	9.8
Lantana macropoda	Deciduous	230	0.9	94.3	19.2	4.0	4.5	27.8
Leucophyllum frutescens	Perennial	32	1.5	41.4	2.6	1.7	3.9	8.4
Zanthoxylum fagara	Perennial	38	2.2	67.0	3.1	2.8	5.0	11.1

Table 8.2 Native species selected for the monitoring of phenological phases.

Scientific name	Family	Type	Growth habit
Bumelia celastrina Kunth.	Sapotaceae	Tree	Perennial
Zanthoxylum fagara Sarg.	Rutaceae	Shrub	Perennial
Diospyros texana Sheele	Ebenaceae	Tree	Deciduous
Forestiera angustifolia Torr.	Oleaceae	Shrub	Deciduous
Prosopis laevigata M.C. Johnst.	Fabaceae	Tree	Deciduous
Parkinsonia texana Torr.	Fabaceae	Tree	Deciduous
Acacia berlandieri Benth.	Fabaceae	Shrub	Deciduous
Acacia rigidula Benth.	Fabaceae	Tree	Deciduous
Lantana macropoda Torr.	Verbenaceae	Tree	Perennial
Leucophyllum frutescens I.M. Johnst.	Scrophulariaceae	Shrub	Deciduous
Acacia farnesiana Willd.	Fabaceae	Tree	Deciduous
Dichondra argentea Willd.	Convolvulaceae	Tree	Perennial

multiple uses of shrubs in north-eastern Mexico (Reid et al., 1990). In the study area, a total of 12 species, including trees and shrubs, in a previously defined, undisturbed plot of 100×100 m representative of the native vegetation. Five plants of each species were selected at random within the plot, being tagged to record the species and replication. The scientific name and family of each species is shown in Table 8.2. Data were taking on a weekly visit basis and were monitored at the study area during 10 months (March to December, 2013).

8.2.4 Study on phenology

The phenology studied were the flowering and fruiting phases. The observations were made at weekly intervals. The flowering included flowers and buds, while the fruiting phenology is considered from the appearance of young fruit (Morellato and Leitão-Filho, 1992). Each phenological phase was studied in terms of percentages, by comparing the branches with the crown of each individual selected. This method

was suggested by Fournier (1974) and has one scale of semiquantitavie intervals of four different categories with intervals of 25% within each, which allows one to estimate the intensity of each phenological phase of each individual. The percentage of species in each phenological phase was calculated in order to evaluate the interspecific synchrony and identify the proportion of species within each phenological phase determined in each phase. This variable is utilised mostly in phenological phases, which allows comparison with other studies (Bencke y Morellato, 2002). On the other hand, the percentage of individuals of each species in each phenological phase allows the evaluation of interspecific synchrony, as suggested by Bencke and Morellato (2002). The percentage of Fournier is a variable for which the phenological peaks may be refined with great precision and better represents the phenologic response of each species. In this way, our study on the vegetative and reproductive phenology is shown in terms of species percentage, followed with respect to individuals and, finally, Fournier's percentage is provided.

8.2.5   Environmental variables

Air temperature (°C) and relative humidity (%) were registered on a daily basis using a HOBO Pro Data Logger (HOBO Pro Temp/RH Series; Forestry Suppliers, Inc., Jackson, MS, USA). Daily precipitation (mm) was obtained from a Tipping Bucket Rain Gauge (Forestry Suppliers, Inc.). Figure 8.1 illustrates the monthly mean air temperature, monthly mean relative humidity and cumulative monthly rainfall at the research site.

8.2.6   Results and discussion

The species included in this study were selected on the basis of their major economic importance. They include shrubs and trees belonging to nine families: Sapotaceae, Rutaceace, Ebenaceae, Oleaceae, Fabaceae, Verbenaceae, Scrophulariace and Convolvulaceae.

The phenological stages (flowering and fruiting) are graphically represented in Figure 8.2.

The duration of the flowering and fruiting phases varied widely between the species. We describe here the phenological phases of each species.

- *Bumelia celastrina*: Showed variation in percentage of flowering in different months, such as March (<50%), but during the period from April to August did not produce flowers. The plant again started flowering from September to December (but less than 50%). On the other hand, similarly it showed variation in fruiting percentage, for example less than 50% fruiting occurred during the period from March to June. In the months of October and December less than 50% fruiting occurred, but in November it produced more than 50% fruiting.
- *Acacia berlandieri*: Produced more than 50% flowering in March, but it produced less than 50% from April to June and from November to December.

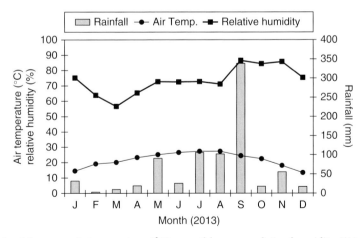

Figure 8.1 Monthly mean air temperature (°C), monthly mean relative humidity (%) and monthly rainfall (mm) during 2013 at the research site.

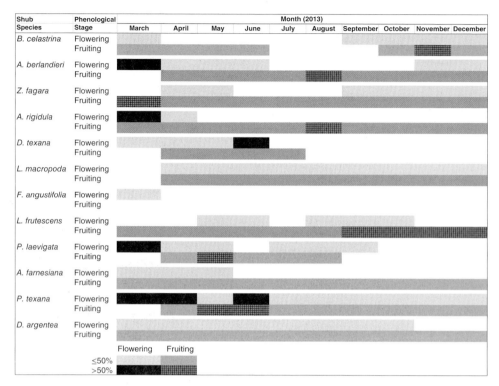

Figure 8.2 Flowering and fruiting phases in shrubs and tree species between March and December 2013, north-eastern Mexico.

From July to October it did not produce flowers. With respect to fruit phenology, it produced less than 50% fruiting from April to July, in August it produced more than 50%, then produced less than 50% again from September to December.

- *Zanthoxylum fagara*: Produced less than 50% flowering from April to May, then again from September to December less than 50%. On the other hand, fruiting phenology attained more than 50% in March, subsequently from April to December it produced less than 50% fruiting.
- *Acacia rigidula*: Varied in flowering percentage in different months; in March more than 50% while in April less than

50%. The rest of the year it did not produce flowers. The plant produced less than 50% fruiting from March to December with an exception of more than 50% in August.
- *Diospyros texana*: Kept the flowering phase with less than 50% from March to May, then more than 50% in June. With respect to the fruiting phase, from April to July it remaineds less than 50% in the fruiting phase.
- *Lantana macropoda*: Kept flowering and fruiting from April continuously up to December, with less than 50%.
- *Forestiera angustifolia*: Produced less than 50% flowering in March, but during the rest of the year it did not

produce flowers. Data on fruiting was not available.

- *Leucophyllum frutescens*: Produced less than 50% flowering from May to June, again from August to October it produced less than 50%. Produced more than 50% fruiting from September to December.
- *Prosopis laevigata*: Produced more than 50% flowering during in March, but attained less than 50% from April to May and the same percentage from July to September. The plant produced less than 50% fruiting during April to August and more than 50% only in May.
- *Acacia farnesiana*: Attained more than 50% flowering from March to May. It produced less than 50% fruit from March to December.
- *Parkinsonia texana*: Produced more than 50% flowering from March to April, followed by differing percentages (in May <50%, in June >50%, from July to December <50%). It produced less than 50% fruiting in April, 50% May to June (varying to more than 50%), then from July to December less than 50%.
- *Dichondra argentea*: Flowering less than 50% from March to October. Similarly, it produced less than 50% fruits from March to December.

An analysis of the flowering sequence revealed that different species had different times of flowering and fruiting. Almost all the species started flowering in the month of March. With respect to fruiting almost all the species were in a fruiting stage from March to December with few exceptions. In this study, it was observed that there were two phases of flowering: the first phase occurred during March to July and then in the second half of the dry season. Few species showed more than one flowering event during the study period. There was overlapping between flowering and fruiting in a few species.

Several studies on the flower phenology of trees (Morellato and Leitão-Filho, 1992; Talora and Morellato, 2000; Spina et al., 2001) reported the role of biotic factors, such as pollination and dispersal mode. Unfortunately we did not take account of this aspect. In the present study, it seems that environmental conditions did not influence flowering time, which coincides with the findings of De Medeiros et al. (2007) in the coastal vegetation of Brazil. In the present study, during the hottest months, a few species did not flower. This may be due to low pollen viability and low stigma receptibility in the prevailing high temperature during this period. On the other hand, some researchers reported in different ecosystems that the flowering time is influenced by several factors such as climatic variation (Debussche et al., 2004), soil moisture (Struck, 1994) and the availability of pollinators (Waser, 1979; Johnson and Bond, 1994; Johnson, 1992; Dreyer and Makgakga, 2003). On the other hand, excellent studies were undertaken on the phenology of different woody species in different countries. In this respect, Krishnan (2004) in a study on the reproductive phenology of endemic species in a wet forest of the Western Ghats, south India, reported that the peak flowering was observed during the dry and post-monsoon seasons for the endemic species, while the non-endemic species flowered during the dry season. Zhang et al. (2014) reported in China that low temperature reduced the activity of pollinators in figs owing to which very few figs flower in winter seasons. Similarly, Salinas-Peba and Parra-Tabla (2007) working on the phenology and pollination of *Manilkara zapota* in forest in Yucatan, Mexico, reported significant differences between environments both in the temporal distribution of flowers and mature fruit production, as well as in the proportion of mature fruits. In the present study, fruiting showed a more or less similar pattern, although no synchrony was found between the flowering and fruiting of a few species, showing a lack of relation with rainfall. Besides, at the study site, from March to August a drought prevailed, with insignificant rains. High rainfall was received only in September, followed by a drought period up to December. The presence of drought did not show any significant effect on phenology, although variations did show between species. We could not study in detail the fruit phenology of the woody species with special reference to seed dispersal syndrome, but several excellent studies have been undertaken on the fruit phenology of a few tree species in different countries.

Chmielewski et al., (2004) reported the impact of climate changes on the phenology of fruit trees and field crops in Germany. The late spring and summer phases also reacted to increased temperatures, but they usually showed lower trends. An interesting study was undertaken by Selwyn and Parthasarathy (2007) on fruiting phenology in a tropical dry evergreen forest on the Coromandel coast of India and its relation to plant life-forms, physiognomic groups, dispersal modes and climatic constraints. Although no significant differences were observed in the frequency of species at three fruiting stages

across the life-form categories and many species of upper and lower canopy trees, it was reported that the fruit maturation period was much longer for the wet season fruiting brevi-deciduous species than for the evergreen and deciduous species that fruited during the dry season. Fruiting in anemochorous species peaked during the driest months and dryness favoured the dissemination of seeds.

Borges et al. (2009) studied the phenology, pollination and breeding system of the threatened tree *Caesalpinia echinata* Lam. (Fabaceae). *Caesalpinia echinata* (brazilwood) is a threatened tree endemic to the Brazilian Atlantic forest. Flowering occurred mainly in the dry season and the peak of seed dispersal started at the beginning of the wet season. The effective pollinators were mainly bees of the genera *Centris* and *Xylocopa*. Results under the controlled pollinations and analysis of pollen tube growth revealed the predominance of bee pollination and SI system, with the occurrence of late-acting, self-incompatibility mechanisms in some species. We could not look into this aspect. Genini et al. (2009) studied the fruiting phenology of palms and trees in an Atlantic rain forest land-bridge island, stating its important role in the maintenance of frugivore populations in isolated, disturbed environments with a high density of vertebrate frugivores and a low diversity of fruiting species and fruit biomass such as those found on Anchieta Island. Hart et al. (2013), studying the fruiting phenology of the fragmented Ngele Forest Complex, KwaZulu-Natal, assessed that fruiting in three of the forest fragments did not show seasonal fruiting trends and had increased fruiting in late summer and autumn months. Fruiting varied significantly between months for all fragments and, where annual variation was observed, trends were insignificant.

Cortés-Flores et al. (2013), working on the fruiting phenology of 133 plant species in a Mexican sub-humid temperate forest located in the tropical zone, recorded monthly the presence/absence of ripe fruits for each plant species. They classified each species according to its main seed dispersal syndrome (anemochorous, autochorous, zoochorous) and its growth form (tree, shrub, herb). Zoochory was most prominent among the tree species. Shrubs showed uniform distribution of the syndromes. The three dispersal syndromes showed a negative correlation between precipitation and the number of species fruiting. The autochorous and zoochorous species also showed inverse relationships with temperature. The results suggested a complex relationship between fruiting phenology and dispersal syndromes with abiotic (precipitation, temperature) and biotic factors (growth form, seed dispersal) in the forest studied. The phenological stages of 12 species during March–October 2014 are shown in the charts in Figure 8.3.

It was observed that most of the species produced flower buds at two flushes, although varying in period (months), for example *A. rigidula*, *Z. fagara* and *D. texana* (first flush from March to April; second flush from mid-July to October). *L. macropoda* was in flower bud condition starting from the last of April to October. *P. laevigata*, *A. farnesiana* and *P. texana* had the first flush from March to May, second flush from July to September. Similarly most of the species showed two periods varying between species in the duration

of the period (months), as can be noted in Figure 8.3. In this regard, there was a large diversity in the periods of flowering. Some species flowered during the summer months, but only a few from July to October. With respect to the fruit initiation stage, the species showed variability in the period and duration of the fruit initiation stage. A few species were in a continuous fruiting stage, starting from March to October (e.g. *F. angustifolia*, *B. myricaefolia*). A few species were in the fruit initiation stage from May to September. The species showed a large diversity in the period of mature fruit.

8.2.7 Conclusions

The results of the two studies are: (a) the detailed flowering and fruiting phenology of 12 woody species from March to December, 2013, and (b) the large variability in the timings of flowering, fruiting phenology and different phenological stages of woody species shown by the study on flower bud initiation, flowering, fruit initiation, mature fruiting and seed dispersion phase of 12 species during March to October, 2014.

An analysis of the flowering sequence revealed that different species have different times of flowering. Some species possess two peaks of flowering, while few others continue flowering from April to December. A few species only flower during March. Almost all the species start flowering in the month of March. In the case of fruiting almost all the species were in fruiting stage from March to December, with few exceptions. There was overlapping between flowering and fruiting in a few species; and great variations occurred

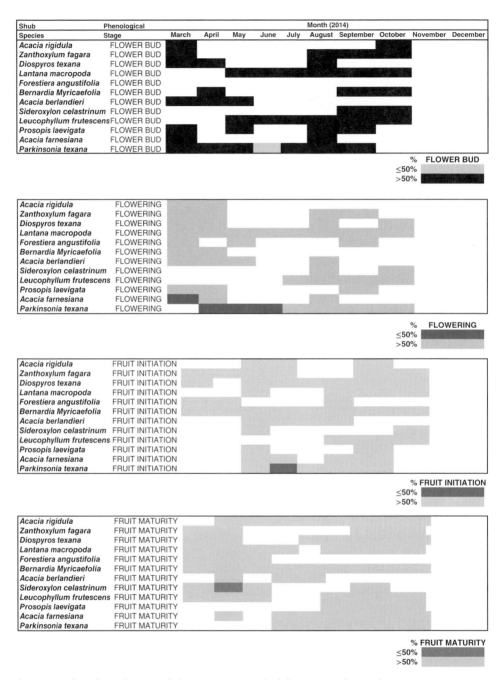

Figure 8.3 Phenological stages of the 12 species studied during March–October 2014.

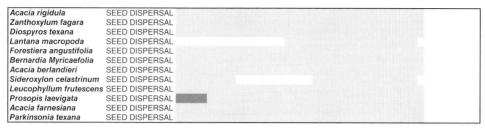

Acacia rigidula	SEED DISPERSAL
Zanthoxylum fagara	SEED DISPERSAL
Diospyros texana	SEED DISPERSAL
Lantana macropoda	SEED DISPERSAL
Forestiera angustifolia	SEED DISPERSAL
Bernardia Myricaefolia	SEED DISPERSAL
Acacia berlandieri	SEED DISPERSAL
Sideroxylon celastrinum	SEED DISPERSAL
Leucophyllum frutescens	SEED DISPERSAL
Prosopis laevigata	SEED DISPERSAL
Acacia farnesiana	SEED DISPERSAL
Parkinsonia texana	SEED DISPERSAL

% SEED DISPERSAL
≤50%
>50%

Figure 8.3 *(continued)*

in the timing of phenophases between the species,

The second study showed a remarkable variation in the initiation of flower buds, flowering stage, fruit initiation, fruit maturity and seed dispersal. The results showed that some species had bimodal peaks of flowering but a few flowered from April to December. A few flowered in the summer season. The periods of fruiting varied between the species. Although a few species did not flower in the heat of summer, climate did not seem to have an effect on phenological events, but the rates may differ and need to be studied in future. Concerted future studies need to be directed to fruit maturity, seed dispersal syndromes and seed dormancy.

On the basis of the results of these two studies it may be concluded that there exists a large variability between species in their phenological stages. The first study was concentrated on the basis of the perecentage of different phenological stages. The second study was mainly concentrated on the time of appearance of each phenological stage in individual species. This gives clear cut information about the phenological time schedule of each species. Very little attempt has been undertaken in this aspect in the literature. Therefore, concerted research input needs to be directed to evaluate the phenology of each individual species. This knowledge will help forest managers for efficient management of the forest.

Bibliography

Bencke C.S.C., Morellato L.P.C. **2002**. Estudo comparativo da fenologia de nove espécies arbóreas em três tipos de Floresta Atlântica no Sudeste do Brasil. *Revista Brasileira de Botânica* **2**:237–248.

Borges L.A., Sobrinho M.S., Lopes A.V. **2009**. Phenology, pollination, and breeding system of the threatened tree *Caesalpinia echinata* Lam. (Fabaceae), and a review of studies on the reproductive biology in the genus. *Flora – Morphology, Distribution, Functional Ecology of Plants* **204**(2):111–130.

Chmielewski F.M., Müller A., Bruns E. **2004**. Climate changes and trends in phenology of fruit trees and field crops in Germany, 1961–2000. *Agricultural and Forest Meteorology* **121**(1–2):69–78.

Cortés-Flores J., Andresen E., Cornejo-Tenorio G., Ibarra-Manríquez G. **2013**. Fruiting phenology of seed dispersal syndromes in a Mexican Neotropical temperate forest. *Forest Ecology and Management* **289**: 445–454.

De Medeiros D.P.W., Lopes A.V., Zickel C.S. 2007. Phenology of woody species in tropical coastal vegetation, northeastern Brazil. *Flora – Morphology, Distribution, Functional Ecology of Plants* **202**(7):513–520.

Debussche M., Garnier E., Thompson J.D. 2004. Exploring the causes of variation in phenology and morphology in Mediterranean geophytes: a genus-wide study of *Cyclamen*. *Botanical Journal of the Linnean Society* **145**(4):469–484.

Domínguez Gómez T.G., González Rodríguez H., Ramírez Lozano R.G., Estrada Castillón A.E., Cantú Silva I., Gómez Meza M.V., Villarreal Quintanilla J.A., Alvarado, M.S., Alanís Flores G. 2013. Diversidad estructural del matorral espinoso tamaulipeco durante las épocas seca y húmeda. *Revista Mexicana de Ciencias Forestales* **4**(17):106–123.

Domínguez Gómez T.G., Ramírez Lozano R.G., Estrada Castillón A.E., Scott Morales L.M., González Rodríguez H., Alvarado M. del S. 2012. Importancia nutrimental en plantas forrajeras del matorral espinoso tamaulipeco. *Ciencia UANL* **15**(59):77–93.

Dreyer L.L., Makgakga M.C. 2003. Oxalidaceae. In: Germishuizen, G., Meyer, N.L. (eds.), *Plants of Southern Africa: An Annotated Checklist, Strelitzia, Vol. 14.* National Botanical Institute, Pretoria, pp. 174–194.

Fournier L.A. 1974. Un método cuantitativo para la medición de características fenológicas en árboles. *Turrialba* **24**:422–423.

Genini J., Galetti M., Morellato L.P.C. 2009. Fruiting phenology of palms and trees in an Atlantic rainforest land-bridge island. *Flora – Morphology, Distribution, Functional Ecology of Plants* **204**(2):131–145.

González Rodríguez H., Cantú Silva I., Gómez Meza M.V., Ramírez Lozano R.G. 2004. Plant water relations of thornscrub shrub species, northeastern Mexico. *Journal of Arid Environments* **58**(4):483–503.

González-Rodríguez H., Cantú-Silva I., Ramírez-Lozano R.G., Gómez-Meza M.V., Uvalle-Sauceda J.I., Maiti R.K. 2010. Characterization of xylem water potential in ten native plants of north-eastern Mexico. *International Journal of Bio-resource and Stress Management* **1**(3):219–224.

Hart L.A., Grieve G.R.H., Downs C.T. 2013. Fruiting phenology and implications of fruit availability in the fragmented Ngele Forest Complex, KwaZulu-Natal, South Africa. *South African Journal of Botany* **88**: 296–305.

Johnson S.D. 1992. Plant–animal relationships. In: Cowling, R. (ed.), *The Ecology of Fynbos. Nutrients, Fire and Diversity.* Oxford University Press, Cape Town, pp. 175–205.

Johnson S.D., Bond W.J. 1994. Red flowers and butterfly pollination in the fynbos of South Africa. In: Arianoutsou, M., Groves, R.M. (eds.), *Plant Animal Interactions in Mediterranean-Type Ecosystems.* Kluwer, Dordrecht, pp. 137–148.

Krishnan R.M. 2004. Reproductive phenology of endemic understorey assemblage in a wet forest of the Western Ghats, south India. *Flora – Morphology, Distribution, Functional Ecology of Plants.* **199**(4): 351–359.

Morellato L.P.C., Leitão-Filho H.F. 1992. Padrões de frutificaço e disperso na Serra do Japi. In: Morellato, L.P.C. (ed.), *História natural da Serra do Japi: ecologia e preservação de uma área florestal no sudeste do Brasil.* Unicamp, Campinas. Pp. 112–140.

Moro M.J. 1992. Desfronde, descomposición y fijación de nitrógeno en una microcuenca con repoblación joven de coníferas y matorral de *Cistus laurifolius* y *Adenocarpus decorticans* en la Sierra de los Filabres (Almería). Doctoral thesis. Facultad de Ciencias. Universidad de Alicante. 463 pp.

Rathcke B., Lacey E.P. 1985. Phenological patterns of terrestrial plants. *Annual Review of Ecology and Systematics* **16**:179–214.

Reïd N., Marroquín J., Beyer-Münzel P. 1990. Utilization of shrubs and trees for browse, fuelwood and timber in the Tamaulipan thornscrub, northeastern Mexico. *Forest Ecology and Management* **36**:61–79.

Salinas-Peba L., Parra-Tabla V. 2007. Phenology and pollination of *Manilkara zapota* in forest and homegardens. *Forest Ecology and Management* **248**(3):136–142.

Sarquís J.I., Coria N., González-Rodríguez H. **2010**. Physiology of photosynthesis in chimalacate (*Viguiera dentata*) in the Zapotitlan de Las Salinas Valley of the Tehuacan biosphere reserve in Puebla, Mexico. *Tropical and Subtropical Agroecosystems* **12**(2):361–371.

Selwyn M.A., Parthasarathy N. **2007**. Fruiting phenology in a tropical dry evergreen forest on the Coromandel coast of India in relation to plant life-forms, physiognomic groups, dispersal modes, and climatic constraints. *Flora – Morphology, Distribution, Functional Ecology of Plants* **202**(5):371–382.

Spina A.P., Ferreira W.M., Leitão-Filho H.F. **2001**. Floração e síndromes de dispersão de uma comunidade de floresta de brejo na região de Campinas (SP). *Acta Botanica Brasilica* **15**:349–368.

Struck M. **1994**. Flowering phenology in the arid winter rainfall region of southern Africa. *Bothalia* **24**(1):77–90.

Talora D.C., Morellato P.C. **2000**. Fenologia de espécies arbóreas em floresta de planície litorânea do sudeste do Brasil. *Revista Brasileira de Botânica* **23**:13–26.

Waser N.M. **1979**. Pollinator availability as a determinant of flowering time in ocotillo (*Fouquieria splendens*). *Oecologia* **39**:107–121.

Zhang L.-S., Compton S.G., Xiao H., Lu Q., Chen Y. **2014**. Living on the edge: Fig tree phenology at the northern range limit of monoecious *Ficus* in China. *Acta Oecologica* **57**:135–141.

CHAPTER 9

Phenology, morphology and variability in pollen viability of four woody species (*Cordia boissieri, Parkinsonia texana, P. aculeata* and *Leucophyllum frutescens*) exposed to environmental temperature, north-eastern Mexico

9.1 Background

Phenology determines the different phases of the plant life cycle and plant productivity. Pollen plays an important role in the productivity of a plant. A plant requires sufficient viable pollen for pollination. Meiosis during microsporogenesis in the anther leads to the formation of pollen containing male gametophyte. The pollen grains after pollination on the stigma produce pollen tubes through pollen apertures and these pollen tubes grow down through the style tissue to fertilise the egg and form an embryo. The morphology of pollen grains, particularly the openings permitting pollen tube germination (apertures), is crucial for determining the outcome of pollen competition in flowering plants. Numerous studies have been undertaken on pollen morphology in various trees in relation to taxonomy (Chalak and Legave, 1997; Till-Bottraud

et al., 2012; Neuffer and Paetsch, 2013). The large variations observed in pollen morphology with respect to size, shape, surface structure and germ spores, are of great importance in taxonomic delimitation of species and their evolutionary sequence and its use in applied fields such as criminology and geology. In this respect, studies have been undertaken on various species, for example *Acalypha* (Sagun et al., 2006). Similar studies have been undertaken on the subtribe Nepetinae (Moon et al., 2007), on vegetation communities in arid regions of China (Luo et al., 2009) and on the morphology and ultrastructure of species of Magnoliaceae depicting an early trend of specialisation; and they supports the view that Magnoliaceae are not one of the earliest lines in the phylogeny of flowering plants (Xu and Kirchoff, 2008). Kovacik et al. (2009) studied the three-dimensional structure of channels and bacula cavities in the wall of

---

Autoecology and Ecophysiology of Woody Shrubs and Trees: Concepts and Applications, First Edition.
Edited by Ratikanta Maiti, Humbero Gonzalez Rodriguez and Natalya Sergeevna Ivanova.
© 2016 John Wiley & Sons, Ltd. Published 2016 by John Wiley & Sons, Ltd.

hazel pollen grains. Punekar and Kumaran (2010) studied pollen morphology and pollination biology of *Amorphophallus* species from north-western Ghats and Konkan of India. Lu et al. (2009) studied the pollen morphology of 86 samples from 84 species of *Gaultheria* and the closely related genera *Chamaedaphne*, *Craibiodendron*, *Diplycosia*, *Eubotrys*, *Gaylussacia*, *Leucothoe*, *Lyonia*, *Oxydendrum*, *Pieris*, *Satyria*, and *Vaccinium* (subfamily Vaccinioideae). Using pollen morphological characteristics, new non-molecular evidence was put forward to support the phylogenetic and taxonomic position of pollen grains for 20 populations of 16 species of Chinese *Curcuma* L. and *Boesenbergia* Kuntz (Zingiberaceae; Chen and Xia, 2011). Similar studies were undertaken by various authors: Khansari et al. (2012) studied the pollen grains of 47 taxa of Campanulaceae; Beretta et al. (2014) studied European bladderworts: *Utricularia australis*, *U. bremii*, *U. intermedia*, *U. minor*, *U. ochroleuca*, *U. stygia* and *U. vulgaris*; and Ćalić et al. (2013) studied the pollen morphology of *Prunus domestica* cv. Požegača.

9.2 Pollen viability

With respect to the technique, aceto-carmine and fluorescein diacetate was used to assess the pollen viability of four Požegača plums, which ranged from 67 to 99%. The pollen nucleus was binucleate. The highest pollen germination (96%) and a tube length (822 μm) of genotype PdP4 was observed in media with 20% PEG (Ćalić et al., 2013). Mazzeo et al. (2014) used three different techniques for assessing pollen grain viability: acetic carmine, fluorescein diacetate and germination. Pollen viability and pollen tube growth have profound effects on fruit and seed production. Knowledge of pollen biology is therefore necessary to increase the productivity of a tree. In this context, a concise review of literature is reported hereunder. A study was undertaken on meiosis in mandarin, examining the meiotic behavior and pollen viability in an open-pollinated population of the "Lee" mandarin [*Citrus clementina* × (*C. paradise* × *C. tangerina*)] in Southern Brazil. In most plants, the microsporogenesis was regular, with meiotic indexes over 90% and pollen viability over 83%. Several meiotic abnormalities were observed. It was concluded that most plants have the ability to produce viable gametes and could be used as pollen donors in programmed crosses. The mono-embryonic unreduced-gamete producer plants are potential progenitors in crosses aimed at producing triploid seedless fruits in mandarin (Cavalcante et al. 2000). The environment has effects on the reproductive development and pollen biology of a plant. Sukhvibul et al. (1999) reported the effects of temperature on inflorescence and floral development in four mango (*Mangifera indica* L.) cultivars. Day/night temperatures of 15/5°C severely inhibited the emergence and elongation of inflorescences of all cultivars, with inflorescence development only occurring on trees that were maintained at warmer temperatures (20/10, 25/15 and 30/20°C). Generally, warmer temperatures increased the inflorescence size of all cultivars. At 20/10°C, the pollen viability of "sensation" was significantly lower than the other culti-vars, but there was no significant difference between the cultivars held at 25/15 and 30/20 °C. Huang et al. (2010) studied the effect of low temperature on the sex-ual reproduction process of "Tainong 1" mango at a diurnal maximum temperature of < 20 °C. The results indicated that low temperatures prevailing during reproduc-tive development significantly affected pistil and male gametophyte development, resulting in poor pollen viability, owing to a few meiotic chromosomal irregularities such as univalents, multivalents, laggards, bridges, micronuclei and a higher inci-dence and significantly greater proportion of nucleolus fragmentation. Pollen tube growth was delayed under low tempera-ture stress. This clearly indicated the need for taking proper care to protect fruit-ing plants from environmental hazards. Wang et al. (2010) studied the pollen flow and mating patterns of a subtropi-cal canopy tree (*Eurycorymbus cavaleriei*) in a fragmented agricultural landscape amidst China's subtropical forests. They determined pollen dispersal and mating patterns in two physically isolated stands of *E. cavaleriei* within fragmented forests using six highly polymorphic microsatel-lite loci. Although substantial amounts of pollen travelled less than 100 m, pater-nity analysis revealed that a large extent of long-distance pollination occurred in both stands. It is assumed that high genetic diversity was found in both adults and offspring, and habitat fragmenta-tion did not have a negative impact on *E. cavaleriei*.

A study was undertaken on the pollen viability of obeche trees (*Triplochiton*

scleroxylon K. Schum.) obtained from different flowers located in Ibadan (7°17′N, 3°30′E). Pollen viability varied significantly between the flowers from the three sources. Significant differences were also observed in pollen tube growth between the flowers from each of the trees (Oni, 1990).

The viability of pollen was studied in a widely distributed anemophilous species, Scots pine (*Pinus sylvestris* L.), along a south–north gradient from 60 to 69°N in Finland. The results showed marked variations in pollen viability both between and within latitudes and between different temperature treatments. Pollen germina-tion at the lower temperatures was low across all the study areas, but it was very good at the highest temperature, which was 20 °C. The variation in pollen viability between stands might cause differential success of pollen from different origins, suggesting the possibility of gene flow from south to north (Parantainen and Pulkkinen, 2002).

Acacia mearnsii (black wattle) is a com-mercially important forestry species in South Africa. In vitro agar media germi-nation tests in ACIAR and Brewbaker and Kwack media were used together with vital stain tests (Sigma® DAB peroxi-dase and p-phenylendiamine) on pollen germination and viability of *A. mearnsii* pollen. Vital stain tests had significantly ($p < 0.05$) higher pollen viability than the agar germination tests (Beck-Pay, 2012). A study was done on the pollen viability, in vitro pollen germination and in vivo pollen tube growth in the biofuel seed crop *Jatropha curcas* using 2,3,5-triphenyltetrazolium chloride. Pollen germination was significantly higher in an agar-based medium composed of sucrose, boric acid and calcium nitrate compared with the control and indole-3-acetic acid treatment. Pollen tubes from both self- and cross-pollinated flowers entered the ovary within eight hours after pollination (Abdelgadir et al., 2012). An investigation on the pollen viability of three polliniser mango cultivars, under four storage conditions (room temperature, –4, –20 and –196 °C), used three methods of pollen viability testing (in vitro germination, fluorescein diacetate and acetocarmine staining). Storage methods and interaction between storage methods and days of storage had a highly significant effect on pollen viability ($p \leq 0.0001$). Irrespective of mango genotype, cryo-stored (–196 °C) pollens showed significantly higher viability than all the other storage conditions. The pollen viability assays confirmed that in vitro germination testing was more reliable. The authors suggested storage of pollen at –20 °C for pollination among cultivars having non-synchronised flowering in a season (Dutta et al., 2013). Olive pollination is anemophilous and an adult olive tree can produce large amounts of pollen grains spread in the air during the flowering period. The viability of pollen grains can be checked by using different methods, such as cytoplasmic stains, enzymatic reactions or germination. Among the three techniques, acetic carmine gave the highest values but staining also heat-killed the pollen grains. The number of pollen grains significantly varied among the cultivars and between the two years. The variation between pollens in pollen viability was observed between two years (Mazzeo et al., 2014).

9.3 Methodology

This study was carried out at the Experimental Research Station of the Facultad de Ciencias Forestales, University Autonomous of Nuevo León (24°47′N, 99°32′W; elevation 350 m) located 8 km south of Linares County, in Nuevo Leon State, Mexico. The climate is typically subtropical and semi-arid with a warm summer. Mean monthly air temperature ranges from 14.7 °C in January to 22.3 °C in August. Average annual precipitation is about 800 mm. The main type of vegetation is known as the Tamaulipan thornscrub or subtropical thornscrub woodland (SPP–INEGI, 1986). The dominant soils are deep, dark-grey, lime-clay vertisols, with montmorillonite, which shrink and swell noticeably in response to changes in soil moisture content. We selected four native species (trees/shrubs) of high economic importance in the semi-arid regions of north-east Mexico, namely *Cordia boissieri*, *Parkinsonia texana*, *P. aculeata* and *Leucophyllum frutescens.*

The scope of the study was a brief and selective review of research advances pertaining to general pollen morphology and viability in relation to plant productivity. We studied the phenology and general morphology of the pollen of these four species. We studied general pollen morphology by squashing the anthers and staining the pollen with 1% safranin in water, with observation at 40×. The study on pollen viability lasted from 20 May up to the end of June 2014. We collected flowers from 11 a.m. to 12 a.m. This study was undertaken to determine the effect of the prevailing environmental temperatures (23, 28, 30 and 32 °C), on the percentage pollen viability of four species of woody trees, at the time of collection. We adopted two techniques: (a) staining with 1% safranin and (b) using 3% iodine in potassium iodide (KI). The latter was found to be better. The fresh anthers were separated, squashed and stained with 3% iodine in KI for 10 min and then we counted the number of pollen grains stained as viable at 40× under a light microscope, each with 10 replications. We calculated the average percentage of viable pollens.

9.4   Results and discussion

9.4.1   Phenology

All four species started flowering from March to June 2014. At the end of June some of them started to fruit, that is *Cordia boissieri*, *Parkinsonia texana*, *P. aculeata*, and *Leucophyllum frutescens*.

9.4.2   Pollen morphology

We could not study detailed pollen morphology owing to the non-availability of facilities. The general morphology of the

pollens are depicted and described below (Figures 9.1 and 9.2)

Among the four studied species, *Cordia boissieri* possessed longest (420 µm), followed by *Leucophyllum frutescens* (330 µm), *Parkinsonia aculeata* (270 µm) and *P. texana* (250 µm). In *Cordata boissieri*, the pollen is more or less ovoidal with surface regular, smooth with little undulations, two visible germ pores are seen. Both sexine and intine are thick-walled. The pollen grain of *Leucophyllum frutescens* is oval with a regular, smooth surface, sexine thick-walled and intine thin-walled, bi-partite, germpore one visible. In *P. texana* the pollen is more or less pentangular in shape, surface shows little undulations. Sexine is mediumly thick-walled, intine thin-walled. In *P. aculeata*, the pollen is more or less round with irregular surface. Sexine is composed of ovoidal cells with surface undulations and projections. Two to three germpores are visible. Kovacik et al. (2009) undertook electron tomography of structures in all hazel pollen grains

In the present study, the pollen of the four species showed large variations in size, shape, and surface ornamentations,

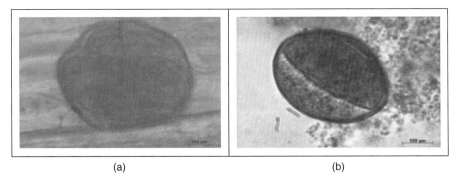

(a)   (b)

Figure 9.1 (a) Photograph of pollen grain of *Cordia boissieri* at 10×, length of pollen grain 420 µm. (b) Photograph of pollen grain of *Leucophyllum frutescens* at 10×, length of pollen grain 330 µm.

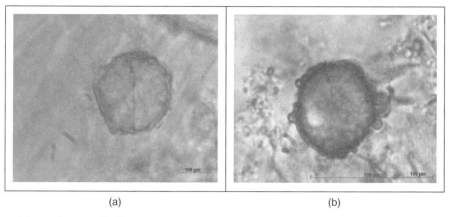

(a)   (b)

Figure 9.2 (a) Photograph of pollen grain of *Parkinsonia texana* at 10×, length of pollen grain 250 μm. (b) Photograph of pollen grain of *P. aculeata* at 10×, length of pollen grain 270 μm.

which coincides with the findings of other authors on various plant species (Chalak and Legave, 1997; Till-Bottraud et al., 2012; Neuffer and Paetsch, 2013). Most of these authors used the morphological characteristics in taxonomic delimitations and the evolutionary relationships of the species and a few for its utility in criminology.

9.4.3   Pollen viability

Depending on the air temperature, we took data on percentage pollen viability (Table 9.1).

It was observed that, at 23 °C, the percentage pollen viability in all four species was 80%. With an increase in temperature the responses of the four species with respect to percentage pollen viability differed. In the case of *Cordia boissieri*, the pollen viability reduced drastically from 81 to 54%, thereby showing susceptibility to an increase in environmental temperature, while in the case of the other three species, *P. aculeata*, *P. texana* and *L. frutescens*, the pollen viability did not reduce drastically. From 23 to 30 °C, the percentage viability in these three species was more or less

Table 9.1 Effect of air temperature on pollen viability (%) of four shrub species, north-eastern Mexico.

	Shrub species			
Temperature (°C)	*Cordia boissieri*	*Parkinsonia aculeata*	*Parkinsonia texana*	*Leucophyllum frutescens*
23	81.22	80.18	79.49	73.72
28	75.13	73.51	73.81	75.0
30	67.27	77.83	79.33	79.34
32	54.05	56.94	66.67	64.34

stable/or fluctuating, but at 32 °C their percentage viability was reduced to some extent, except *P. texana* showed a drastic decrease in pollen viability at 32 °C, from 79 to 64%. It is expected that this variability would have a direct impact on fertility and fruitification (not attempted in this present study). Similar to the present study, the variability in pollen viability in response to temperature has been documented for various trees. The pollen viability of obeche trees (*Triplochiton scleroxylon* K. Schum) obtained from different flowers located in Ibadan (7°17′N, 3°30′E) varied significantly between the flowers from the three sources and significant differences were also observed in pollen tube growth between the flowers from each of the trees (Oni, 1990).

Similarly, in a study on the pollen viability of the anemophilous species, Scots pine (*Pinus sylvestris* L.), along a south–north gradient from 60 to 69°N in Finland, marked variations in pollen viability were noted, both between and within latitudes and also between different temperature treatments. Pollen germination at the lower temperatures was low across all the study areas, but it was very good at the highest temperature, which was 20 °C. The variation in pollen viability between stands might cause a differential success of pollen of different origins, suggesting the possibility of a gene flow from south to north (Parantainen and Pulkkinen, 2002). It is reported that, at low temperatures (20/10 °C), the pollen viability in a few mango cultivars was reduced drastically, while at 25/15 °C no significant differences were observed between the mango cultivars (Sukhvibul et al., 1999). In another study, low temperature affected pistil and

male gametophyte development, resulting into poor pollen viability in mango (Huang et al., 2010). Low temperature affected pollination in mango (Dag et al., 2000), pollen germination in *Pistacia vera* (Acar and Kakani 2010) and in vitro pollen germination and pollen tube growth of Almond (*Prunus dulcis* Mill.; Sorkheh et al., 2011). Akoroda (1983) reported long-term storage of yam pollen. Alcaraz et al. (2010) studied in vitro pollen germination in avocado (*Persea americana*) and optimisation of the method and effect of temperature.

9.5 Conclusions

The present study was undertaken on the phenology, pollen morphology and the effect of environmental temperature on the pollen viability of four woody species (*Cordia boissieri, Parkinsonia texana, P. aculeata* and *Leucophyllum frutescens*) in north-eastern Mexico. These four species started flowering from March to June, 2014, and then entered the fruiting/dispersion phase. Large variations were observed on pollen morphology with respect to the form, size and surface ornamentation of pollen grains. Among the four species studied, *C. boissieri* possessed the longest (420 μm), followed by *L. frutescens* (320 μm), *P. aculeata* (270 μm) and *P. texana* (250 μm). The species varied widely in percentage pollen viability in response to the prevailing environmental temperatures. The temperature of 23 °C was found to be favourable for optimum pollen viability (80%) in all the species studied. With an increase in temperature, *C. boissieri* showed a drastic decrease in pollen viability, while the pollen viability of the others remained more or less stable, but decreased at 32 °C.

Bibliography

Abdelgadir H.A., Johnson S.D., Van Staden J. 2012. Pollen viability,pollen germination and pollen tube growth in the biofuel seed crop *Jatropha curcas* (Euphorbiaceae). *South African Journal of Botany* **79**:132–139.

Acar I., Kakani V.G 2010. The effects of temperature on *in vitro* pollen germination and pollen tube growth of *Pistacia* spp. *Scientia Horticulturae* **125**(4):569–572.

Akoroda M.O. 1983. Long-term storage of yam pollen. *Scientia Horticulturae* **20**(3):225–230.

Alcaraz M.L., Montserrat M., Hormaza J.M. 2010. *In vitro* pollen germination in avocado (*Persea Americana* Mill.): Optimization of the method and effect of temperature. *Scientia Horticulturae* **130**(1):152–156.

Beck-Pay S.L. 2012. The effect of temperature and relative humidity on *Acacia mearnsii* polyad viability and pollen tube development. *South African Journal of Botany* **83**:165–171.

Beretta M., Rodondi G., Adamec L., Andreis C. 2014. Pollen morphology of European bladderworts (*Utricularia* L., Lentibulariaceae). *Review of Palaeobotany and Palynology* **205**:22–30.

Cavalcante H.C., Schifino-Wittmann M.T., Dornelles A.L.C. 2000. Meiotic behaviour and pollen fertility in an open-pollinated population of 'Lee' mandarin [*Citrus clementina*×(*C. paradisi*×*C. tangerina*)]. *Scientia Horticulturae* **86**(2):103–114.

Chalak L., Legave J.M. 1997. Effects of pollination by irradiated pollen in Hayward kiwifruit and spontaneous doubling of induced parthenogenetictrihaploids. *Scientia Horticulturae* **68**(1/4):83–93.

Xu F.-X., Kirchoff B.K. 2008. Pollen morphology and ultrastructure of selected species of Magnoliaceae. *Review of Palaeobotany and Palynology* **150**(1/4):140–153.

Chen J., Xia N.H. 2011. Pollen morphology of Chinese *Curcuma* L. and *Boesenbergia* Kuntz (Zingiberaceae): Taxonomic implications. *Flora – Morphology, Distribution, Functional Ecology of Plants* **206**(5):458–467.

Dag A., Eisenstein D., Gazit S. 2000. Effect of temperature regime on pollen and the effective pollination of 'Kent' mango in Israel. *Scientia Horticulturae* **86**(1):1–11.

Dutta S.K., Srivastav M., Chaudhary R., Lal K., Patil P., Singh S.K., Singh A.K. 2013. Low temperature storage of mango (*Mangifera indica* L.) pollen. *Scientia Horticulturae* **161**:193–197.

Huang J.-H. Ma W.-H., Liang G.-L., Zhang L.-Y., Wang W.-X., Cai Z.-J., Wen S.-X. 2010. Effects of low temperatures on sexual reproduction of 'Tainong 1' mango (*Mangifera indica*). *Scientia Horticulturae* **126**(2):109–119.

Khansari E., Zarre S., Alizadeh K., Attar F., Aghabeigi F. 2012. Pollen morphology of *Campanula* (Campanulaceae) and allied genera in Iran with special focus on its systematic implication. *Flora – Morphology, Distribution, Functional Ecology of Plants* **207**(3):203–211.

Kovacik L., Plitzko J.M., Grote M., Reichelt R. 2009. Electron tomography of structures in the all of hazel pollen grains. *Journal of Structural Biology* **166**(3):263–271.

Lu L., Fritsch P.W., Wang H., Li H.-T., Li D.-Z., Chen J.Q. 2009. Pollen morphology of *Gaultheria* L. and related genera of subfamily Vaccinioideae: Taxonomic and evolutionary significance. *Review of Palaeobotany and Palynology* **154**(1/4):106–123.

Luo C., Zheng Z., Tarasov P., Pan A., Huang K., Beaudouin C., An F. 2009. Characteristics of the modern pollen distribution and their relationship to vegetation in the Xinjiang region, northwestern China. *Review of Palaeobotany and Palynology* **153**(3/4):282–295.

Ćalić D., Devrnja N., Kostić I., Kostić M. 2013. Pollen morphology, viability, and germination of *Prunus domestica* cv. Požegača. *Scientia Horticulturae* **155**:118–122.

Mazzeo A., Palasciano M., Gallotta A., Camposeo S., Pacifico A., Ferrara G. 2014.

Amount and quality of pollen grains in four olive (*Olea europaea* L.) cultivars as affected by 'on' and 'off' years. *Scientia Horticulturae* **170**:89–93.

Moon H.-K., Vinckier S., Smets E., Huysmans S. **2007**. Comparative pollen morphology and ultrastructure of Mentheae subtribe Nepetinae (Lamiaceae). *Review of Palaeobotany and Palynology* **149**(3/4):174–186.

Neuffer B., Paetsch M. **2013**. Flower morphology and pollen germination in the genus *Capsella* (Brassicaceae). *Flora – Morphology, Distribution, Functional Ecology of Plants* **208**(10/12):626–640.

Oni O. **1990**. Between-tree and floral variations in pollen viability and pollen tube growth in obeche (*Triplochiton scleroxylon*). *Forest Ecology and Management* **37**(4):259–265.

Parantainen A., Pulkkinen P. **2002**. Pollen viability of Scots pine (*Pinus sylvestris*) in different temperature conditions: high levels of variation among and within latitudes. *Forest Ecology and Management* **167**(1/3):149–160.

Sorkheh K., Shiran B., Rouhi V., Khodambashi M., Wolukau J.N., Ercisli S. **2011**. Response of *in vitro* pollen germination and pollen tube growth of almond (*Prunus dulcis* Mill.) to temperature, polyamines and polyamine synthesis inhibitor. *Biochemical Systematics and Ecology* **39**(4/6):749–757.

Sukhvibul N., Whiley A.W., Smith M.K., Hetherington S.E., Vithanage V. **1999**. Effect of temperature on inflorescence and floral development in four mango (*Mangifera indica* L.) cultivars. *Scientia Horticulturae* **82**(1/2):67–84.

Punekar S.A., Kumaran K.P.N. **2010**. Pollen morphology and pollination ecology of *Amorphophallus* species from North Western Ghats and Konkan region of India. *Flora – Morphology, Distribution, Functional Ecology of Plants* **205**(5):326–336.

Sagun V.G., Levin G.A., van der Ham R.W.J.M. **2006**. Pollen morphology and ultrastructure of *Acalypha* (Euphorbiaceae). *Review of Palaeobotany and Palynology* **140**(1/2):123–143.

SPP-INEGI. **1986**. *Síntesis Geografía del Estado del Nuevo León*. Instituto Nacional de Geografía Estadística e Información, México, D.F.

Till-Bottraud I., Gouyon P.H., Ressayre A., Godelle B. **2012**. Gametophytic vs. sporophytic control of pollen aperture number: a generational conflict. *Theoretical Population Biology* **82**(3):147–157.

Wang J., Kang M., Gao P., Huang H. **2010**. Contemporary pollen flow and mating patterns of a subtropical canopy tree *Eurycorymbus cavaleriei* in a fragmented agricultural landscape. *Forest Ecology and Management* **260**(12):2180–2188.

CHAPTER 10

Pollen biology and plant productivity: A review

Ratikanta Maiti[1], Humberto Gonzalez-Rodriguez[1], and Ek Raj Ojha[2]

[1] Facultad de Ciencias Forestales, Universidad Autonoma de Nuevo Leon, Linares, Mexico
[2] Climate Change and Sustainable Development Program, Tribhuvan University, Kathmandu, Nepal

10.1 Introduction

Pollen biology plays an important role in the productivity of a plant and thus crops. A plant should produce sufficient viable pollens to pollinate stigma and fertilise ovules to form embryos after fertilisation and finally seeds inside its fruits. In flowering plants, the haploid phase is produced through meiosis the from diploid phase, leading to the formation of haploid pollen grains and haploid eggs in the embryo sac. These reproductive tissues (male and female gametophytes) are actually distinct from the genome of the plant (sporophyte) and are more or less independent. The pollen grains after pollination on the stigma produce pollen tubes through pollen apertures and these pollen tubes grow down through the style tissue to fertilise the egg and form an embryo. Many species of flowering plants simultaneously produce pollen grains with different aperture numbers in a single individual through heteromorphism.

Meiosis during microsporogenesis in the anther leads to the formation of pollens. A study has been done on the meiotic behaviour and pollen fertility in an open-pollinated population of the "Lee" mandarin [*Citrus clementina* × (*C. paradise* × *C. tangerina*)] in Southern Brazil. In most plants, microsporogenesis was regular, with meiotic indexes over 90% and pollen viability over 83%. Several meiotic abnormalities were observed, including univalents and stickiness in metaphase I, laggards and bridges in anaphase and telophase I and II and microcytes at the tetrad stage. That led to the conclusion that most plants have the ability to produce viable gametes and could be used as pollen donors in programmed crosses. The monoembryonic unreduced-gamete producer plants are potential progenitors in crosses aimed at producing triploid seedless fruits (Cavalcante et al., 2000).

The environment can have effects on the reproductive development and pollen biology of a plant. For instance, the prevailing

Autoecology and Ecophysiology of Woody Shrubs and Trees: Concepts and Applications, First Edition.
Edited by Ratikanta Maiti, Humbero Gonzalez Rodriguez and Natalya Sergeevna Ivanova.
© 2016 John Wiley & Sons, Ltd. Published 2016 by John Wiley & Sons, Ltd.

ambient temperature during the reproductive phase is one of the important factors affecting the setting of seeds and fruits in different plant species. Sukhvibul et al. (1999) studied the effects of temperature on inflorescence and floral development in four mango (*Mangifera indica* L.) cultivars. Day/night temperatures of 15/5 °C severely inhibited the emergence and elongation of inflorescences of all cultivars, with inflorescence development only occurring on trees that were maintained at warmer temperatures (20/10, 25/15 and 30/20 °C). Generally, warmer temperatures increased the inflorescence size of all cultivars. At 20/10 °C, the pollen viability of "sensation" was significantly lower than the other cultivars, but there was no significant difference between the cultivars held at 25/15 and 30/20 °C. Low temperatures caused morphological changes in the styles, stigmas, ovaries and anther size in all cultivars.

Huang et al. (2010) studied the effect of low temperature on the sexual reproduction process of "Tainong 1" mango at a diurnal maximum temperature of < 20 °C. The results indicated that low temperatures prevailing during reproductive development significantly affected pistil and male gametophyte development, resulting in pollen grains with low viability. Low temperature caused such meiotic chromosomal irregularities as univalents, multivalents, laggards, bridges and micronuclei, with higher incidences and significantly greater proportions of nucleolus fragmentation and dissolution. Pollen tube growth was delayed under low temperature stress. This clearly indicated the need for taking proper case to protect fruiting plants from environmental hazards.

Wang et al. (2010) carried out a study on the pollen flow and mating patterns of a subtropical canopy tree (*Eurycorymbus cavaleriei*) in a fragmented agricultural landscape amidst China's subtropical forests. The genetic effect of habitat fragmentation depends largely on the level of gene flow within and between population fragments. Contemporary pollen dispersal and mating patterns were estimated in two physically isolated stands of *E. cavaleriei* within fragmented forests using six highly polymorphic microsatellite loci. High genetic diversity (HE = 0.670–0.754) was observed in both adults and offspring in the fragmented agricultural landscape, suggesting that habitat fragmentation did not necessarily erode the genetic diversity of *E. cavaleriei*. Although a substantial amount of pollens travelled less than 100 m, paternity analysis revealed that a large extent of long-distance pollination occurred, with the average pollen dispersal distance of 1107 m and 325 m for the two stands, respectively. It may be concluded from these findings that high genetic diversity was found in both adults and offspring. Paternity analysis reflected long-distance pollen dispersal. Extensive pollen immigration revealed effective genetic connectivity between the stands. Thus, habitat fragmentation did not have a negative impact on *E. cavaleriei*.

The morphology of pollen grains, particularly the openings permitting pollen tube germination (apertures), is crucial for determining the outcome of pollen competition in flowering plants. Till-Bottraud et al. (2012) found that the heteromorphic pollen aperture pattern depends on the genetic control of pollen morphogenesis. This raises a conflict of interest between genes expressed in the sporophyte and in the gametophyte. For pollen aperture, heteromorphism has been observed in about 40% of angiosperm species, suggesting that conflicting situations are the rule. Therefore, the sporogametophytic conflict could be one of the factors that led to the reduction of the haploid phase in plants.

Neuffer and Paetsch (2013) studied flower morphology and pollen germination in the genus *Capsella* (Brassicaceae). Within that genus, which is closely related to the molecular model species pair *Arabidopsis lyrata* (SI)/*A. thaliana* (SC), they studied the self-incompatible and self-compatible species. SC species *C. rubella* and *C. bursa-pastoris* produced in comparison with the SI species *C. grandiflora*: (a) smaller petals as a result of decreased cell division and only less of decreasing cell volume, (b) less production of pollen in one flower, (c) a lesser incision between the two valves of the fruits, in combination with a shorter style, and (d) a much quicker fertilisation of SC pollen after pollination. Crossing success was proven between the diploid species, between different provenances of the tetraploid *C. bursa-pastoris*, and between the two diploid species and particular individuals of the self-incompatible in *Magnolia grandiflora*.

Parthenogenesis induced by irradiated pollen was investigated for the main cultivar Hayward of *Actinidia deliciosa* as a possible means of genetic improvement. Pollinations were carried out for two years involving two different sources of pollen which were previously irradiated with gamma rays in the dose range of 200–1500 Gy. The best efficiency was obtained with the male genotype M2 at dosages of 500–1500 Gy. Hexaploid seedlings were also found but mainly at the lowest dosage of 200 Gy (Chalak and Legave, 1997).

10.2 Materials, methods and scope of the study

This study was entirely based on a review of literature relating to the subject under study. The authors' own intuitions by virtue of their academic and professional orientations in related subjects have naturally been put to use in searching, gleaning, organising, analysing, and arranging materials in accomplishing the study. Websites constituted the chief sources of literature reviewed for and used in carrying out and completing the study. The scope of the study is limited within a brief and selective review of research advances pertaining to pollen morphology and pollen biology in relation to plant productively.

10.3 Elaboration of the review

The review was subdivided into various topical sections, as presented below. Specific sections focused on specific aspects of the subject under study are expected to provide useful insights into the subject under the review and could have an impetus for further and more important works and findings on this specialised subject.

10.4 Pollen morphology

There exist large variations in pollen morphology with respect to size, shape, surface structure and germ spores, which are of great importance in the taxonomic delimitation of species and their evolutionary sequence and its use in applied fields such as criminology and geology. A study was done on the pollen morphology of 73 species of *Acalypha*, representing most of the taxonomic diversity in the genus, using light and scanning electron microscopy, and 14 species with transmission electron microscopy. *Acalypha* pollen is small (9–22 μm), sub-isopolar to isopolar or apolar, and oblate-spheroidal to sub-oblate in meridional outline. The aperture system is two- to eight-colporate, with the ectocolpi short (1–5 μm) and almost equal in length to the endopore diameter. The exine ornamentation is rugulate to microrugulate or areolate with distinct scabrae localised on the margins or scattered over the pollen surface. It was concluded that *Acalypha* pollens show large variation in size, aperture number, aperture distribution and exine ornamentation, and are not homogeneous. This coincides with previous reports. Though pollen morphology is inconsistent with a phylogeny based on ITS and *ndh*F sequence data, it does provide support for the subgenera *Linostachys* and *Androcephala* and several small clades in the molecular phylogeny (Sagun et al., 2006).

Moon et al. (2007) studied the pollen morphology and ultrastructure of 52 representative species belonging to all 12 genera in the current classification of the subtribe Nepetinae, using light, scanning electron and transmission electron microscopy. According to these, Nepetinae pollen is small to large (P = 16–65 μm, E = 17–53 μm), oblate to prolate (P/E = 0.7–1.6) in shape, mostly hexacolpate and sometimes octocolpate.

The exine stratification in all taxa is similar and characterised by unbranched columellae and a continuous, granular endexine. Sexine ornamentation in the Nepetinae is bireticulate, microreticulate or perforated. In perforate and microreticulate pattern a tendency towards a bireticulum could be observed due to a trace of secondary tectal connections. The bireticulate pattern is most common with variations of primary muri and secondary reticulum. In *Hymenocrater* and *Schizonepeta* the variation in sexine ornamentation is valuable at the generic level. Pollen morphology supports that *Lophanthus* and *Nepeta* are very closely allied and *Lallemantia* is quite distinct from *Dracocephalum*. The formerly suggested infrageneric relationships within *Dracocephalum* and *Nepeta* are only partly corroborated by palynological characters. Orbicules are absent in the Nepetinae.

Similarly, Luo et al. (2009) studied modern pollen distribution and its relationship to vegetation communities in the Xinjiang region, an arid area of China. Pollen concentrations did not appear to be linearly related to the vegetation cover. The results of cluster analysis and principal-components analysis showed vegetation groups that reflected the relationships among the pollen taxa, and this provided a basis for subdividing xeromorphic pollen assemblages in arid areas.

Xu and Kirchoff (2008) examined the pollen morphology and ultrastructure of 20 species, representing eight genera of the Magnoliaceae, based on observations with light, scanning and transmission electron microscopy. Their findings indicated that the family represents a homogeneous group on the basis of pollen morphological characteristics. The pollen grains are boat-shaped with a single elongate aperture on the distal face. The tectum is usually microperforate, rarely slightly or coarsely rugulose. Columellae are in general irregular, but well-developed columellae are present in some taxa. The endexine is distinct in 14 species, but difficult to discern in the genera *Parakmeria*, *Kmeria* and *Tsoongiodendron*. Within the aperture zone the exine elements are reduced to a thin foot layer. The intine possesses three layers with many vesicular-fibrillar components and tubular extensions in intine 1. The symmetry of the pollen grains and the shape and type of aperture and ultrastructure of the intine show a remarkable uniformity in the family. Nevertheless, they vary widely in pollen size, ornamentation and the ultrastructure of the exine. The pollens of Magnoliaceae are an example of an early trend of specialisation, and this supports the view that Magnoliaceae are not one of the earliest lines in the phylogeny of flowering plants.

Kovacik et al. (2009) studied the three-dimensional (3-D) structure of channels and bacula cavities in the wall of hazel pollen grains using automated electron tomography in order to explore their role in the release of allergen proteins from the pollen grains. The 3-D reconstructions of 100–150 nm thick resin-embedded sections, stabilised by thin platinum–carbon coating, revealed that the channels aimed directly towards the surface of the grain and that the bacula cavities were randomly sized and merged into larger ensembles. The analysis of positions where the hard sphere fits into the resolved channels and bacula cavity structures revealed that unbound allergens could freely traverse through the channels and that the bacula cavities support the path of the allergens towards the surface of the grain.

A study undertaken by Punekar and Kumaran (2010) on a pollen morphology and pollination biology of *Amorphophallus* species from North Western Ghats and Konkan of India revealed that pollen exine ornamentations of the genus are represented – psilate, striate, fossulate, verrucate, scabrate and the new type "pseudofossulate" of *Amorphophallus commutatus* var. *wayanadensis*. These diverse pollen exine ornamentations can be utilised as an important taxonomic tool to distinguish taxa and to resolve taxonomic problems. The pollen of all eight taxa retained their sculptured surface (exine ornamenatation type) after acetolysis. Observation of pollination ecology of the studied *Amorphophallus* taxa revealed that all are mostly visited by beetles (Bostrichidae, Cetoniidae, Hybosoridae, Lyctidae, Nitidulidae, Rutelinae, Scarabaeidae, Staphylinidae), followed by flies (Drosophilidae, Muscidae), bees (Apidae, Trigone), ants (Formicidae, Dolichoderinae) and cockroaches (Blaberidae/Panesthiinae).

Lu et al. (2009) investigated the pollen morphology of 86 samples from 84 species of *Gaultheria* and the closely related genera *Chamaedaphne*, *Craibiodendron*, *Diplycosia*, *Eubotrys*, *Gaylussacia*, *Leucothoe*, *Lyonia*, *Oxydendrum*, *Pieris*, *Satyria*, and *Vaccinium* (subfamily Vaccinioideae), using light and scanning electron microscopy (SEM). The exine ornamentation (apocolpia and mesocolpia) of pollen grains showed great variations. Six exine types of apocolpia and five exine types of mesocolpia were observed. Colpus length was correlated with both pollen size and colpus maximum width. The pollen data supported sister relationships between *Satyria* and *Vaccinium* and between *Chamaedaphne* and *Eubotrys* of a previously reported molecular phylogenetic analysis. An evolutionary trend of mesocolpium ornamentation within Vaccinioideae from granulate through granulate–rugulate to rugulate was apparent.

To establish new non-molecular evidence to support the phylogenetic and taxonomic position, pollen grains from 20 populations of 16 species of Chinese *Curcuma* L. and *Boesenbergia* Kuntz (Zingiberaceae) was studied under SEM and TEM. As indicated, the pollen grains were spherical and ovoid, non-aperturate. The pollen wall constituted a very thin exine and a thick intine. The exine was psilate or echinate. The intine had two layers, that is a thick, channelled layer (exintine) and an inner homogenous layer (endintine). A morphological similarity was found between the pollen grains of species of *Curcuma*, which according to DNA sequence data appears to be a polyphyletic genus. However, the uniform pollen morphology in *Curcuma* did not show any evidence to divide this genus into separate taxonomic entities. The results on pollen morphology do not provide any additional evidence to either unite or segregate *Boesenbergia albomaculata* and *Curcumorpha longiflora* in the same genus, suggesting that more taxonomic data on the genus *Boesenbergia* and its relatives are needed before a final conclusion can be drawn in this regard (Chen and Xia, 2011).

Khansari et al. (2012) studied the pollen grains of 47 taxa of Campanulaceae, including 35 taxa of *Campanula* that represent its five subgenera and nine sections. In addition, five species and three subspecies representing three sections in *Asyneuma* and one species of each genera *Legousia*, *Michauxia*, *Zeugandra* and *Theodorovia* were also studied, using light and scanning electron microscopy. The ornamentation pattern of exine was rugulate-echinate, rugulate-microechinate or in a few species rugulate-microreticulate and microechinate. The most valuable characters for sub-generic classification were the length and density of echini. The length of echini showed significantly long (> 2 μm) in *C. sclerotricha*, *Legousia falcate* and *Michauxia laevigata*. However, the pollen grains showed low variation in different species of subgen. *Rapunculus*, although there was variation among different species of some groups such as sect. *Rupestres*, probably revealing their non-monophyly despite homogeneity with respect to other morphological characters. The pollen morphology did not support recognition of *Asyneuma*, *Legousia*, *Michauxia*, *Symphyandra*, *Theodorovia*, and *Zeugandra* as separated from *Campanula* as none of them exhibited any unique characteristics.

A study was done on the pollen morphology and exine ornamentation of a plum cultivar (*Prunus domestica cv. Požegača* L.), using both light and scanning electron microscope. Acetocarmine and fluorescein diacetate were used to determine the pollen viability of four Požegača plum genotypes. The variation in pollen vaiability was observed among cultivars. The pollen nucleus morphology showed binucleate mature pollens. The effect of polyethylene glycol (PEG; 10, 15 and 20%, w/v) on pollen germination and tube growth was evaluated. Overall, the inclusion of PEG in the medium improved both pollen germination and tube growth. The genotype had a remarkable effect on germination and length of pollen tubes, regardless of the PEG concentration used (Ćalić et al., 2013).

Beretta et al. (2014) conducted a study on the pollen morphology of the seven known European bladderworts: *Utricularia australis, U. bremii, U. intermedia, U. minor, U. ochroleuca, U. stygia* and *U. vulgaris,* using both light microscopy and scanning electron microscopy to determine the size, shape (P/E ratio), number of colpori and exine ornamentation as important diagnostic characteristics of *Utricularia* pollens. In general, the pollen grains were usually medium-sized (~30 μm), sub-isopolar, radially symmetric and zonocolporate. For the non-fruiting species *U. bremii, U. stygia* and *U. ochroleuca*, the grains were often malformed, asymmetric or in the form of gigapollen or micropollen. A remarakable variation in number of gigapollen grains were found in *U. stygia* while micropollen was observed in *U. ochroleuca*. The shape of the normal grains was variable from suboblate to prolate spheroidal and they were (10)-11-18-(19)-zonocolporate. The prevalent ornamentations were psilate (on mesocolpi) and fossulate (on apocolpium) except for *U. bremii*, which had a somewhat perforate ornamentation.

Olive pollination is anemophilous and an adult olive tree can produce large amounts of pollen grains spread in the air during the flowering period. The viability of pollen grains can be checked by using different methods, such as cytoplasmic stains, enzymatic reactions or germination. In a study by Mazzeo et al. (2014), the pollen grain viability was estimated by using three different techniques: acetic carmine, fluorescein diacetate and germination. The three techniques came up with statistically different data: (a) acetic carmine always showed the highest values but staining also heat-killed pollen grains; (b) fluorescein diacetate and germination were significantly correlated with a high R^2 (0.862). The study also indicated that the viability of the pollen grains (with all techniques) was significantly higher in the 'off' year compared to the 'on' one, with important consequences for the fertilisation process.

10.5   Pollen dispersal

Sufficient pollen load and pollen dispersion contribute greatly in fruit production. Pacini et al. (2004) studied the cytological, physiological, chemical and ecological characteristics of pollen and nectar offered by male and female flowers of the dioecious plant, *Laurus nobilis*. The various phases of floral phenology and the insect pollinators were observed. Cytological methods were applied to determine anther, pollen and nectary structures. The nectar sugar composition was evaluated by HPLC. The pollen viability in time was compared with cytoplasmic and intine water content. The pollen dispersion was found to be reversible by opening and closing the anther valves, as determined by hydration of the mechanical layer of the anther. The pollen, covered by pollenkitt, was ready for dispersal for three consecutive days and during this time the intine and cytoplasm lost water and the pollen viability diminished. At germination, exine ruptured together with the outermost layer of the intine. The nectaries of male flowers were observed on the anther filament and on the staminodes of female flowers. The nectar consisted almost entirely of sucrose and was more concentrated in male flowers. Secreted through stomata, nectar was presented in a thin layer. In the study area, the main pollinators (about half the total number of all visits) were hymenopterans. Insects were attracted by both male and female flowers. Similarly, males offered nectar and pollen, whilst females only nectar.

A study was conducted on pollen biology and stigma receptivity in the long blooming species, *Parietaria judaica* L. (Urticaceae family). It was found that the pollen was dispersed by rapid uncurling of the filament and anther opening. The filament and anther lacking lignified wall thickenings were usually responsible for anther opening and ballistic pollen dispersal, whereas dispersal was the result of a sudden movement of the filament. The pollen was of the partially hydrated type and readily lost water and hence was viable at low relative humidity. Pollen carbohydrate reserves differed with season. Starchless grains germinated quickly and were less subject to water loss. Flowers were protogynous, pollen release occurred only after complete cessation of the female phase within an inflorescence. Dispersal and pollination were adapted to pollen features (Franchi et al., 2007).

10.6   Pollen germination

In a favourable environment, pollens after pollination start germinating and go down through the stylar tissue and finally fertilise the egg, leading to the formation of zygote and embryo. However, unfavourable conditions such as heat stress and low temperature affect pollen germination. Several studies have been undertaken on the pollen germination of plants, and some of them are reviewed hereunder.

A few techniques have been developed for the study of pollen germination. Luza and Polito (1985) developed a method for in vitro germination and tube growth of English walnut (*Juglans regia* L.) pollens. The medium contained 20% sucrose, 1.0 mM $CaCl_2$ and 0.16 mM boric acid solidified with 0.65% agar, and was suitable for freshly collected pollen from each of the 21 clones tested. Pollen collection and storage conditions were evaluated using pollen germination on this medium as an indicator. There was considerable variation in results among the clones examined. When the relative humidity of the −20 °C storage environment was maintained near 30%, the germinability was retained, although at reduced levels, for up to one year. Cryopreservation using liquid nitrogen (−196 °C) was satisfactory for the pollen of two cultivars which retained the ability to germinate and grow in vitro, but not for two others which did not.

According to Okusak and Hiratsuka (2009), the pollen of Japanese pear (*Pyrus pyrifolia* Nakai) germinated well on agar medium containing 10% sucrose or glucose, but not on agar containing fructose. Sucrose enhanced pollen tube growth much more effectively than glucose. The addition of 5% fructose to 5 or 10% sucrose or glucose media suppressed germination completely. When pollen was transferred onto fructose medium after culturing it on glucose or sucrose medium for 1–2 h, germination was completely inhibited. On the other hand, pollen transferred to sucrose or glucose medium from the fructose medium germinated at almost the same ratio as pollen on sucrose or glucose medium without transfer. Therefore, pollen inhibition by fructose is reversible. Germinated pollen on sucrose and glucose media contained sucrose and glucose, but ungerminated pollen on fructose medium contained only trace levels of these sugars, suggesting that pollen on fructose medium predominantly used sucrose and glucose as respiration substrates and could not maintain a constant level of these sugars.

Jiang et al. (2009) studied the germination and growth of sponge gourd (*Luffa cylindrical* L.) pollen tubes, with Fourier transform infrared (FTIR) analysis of the pollen tube wall. The emergence of multiple pollen tubes from single pollen grains occurred both in vitro and in vivo in sponge gourd (*L. cylindrical* L.). The frequency with which pollen grains produced multiple pollen tubes in vivo (7.2%) was lower than that under in vitro conditions (14.9%). In addition, the growth of the single pollen tubes continued for a longer period in vitro than that of the multiple tubes. The results of FTIR microspectroscopy revealed that abnormal cell wall components (peaks at 800–1000 cm^{-1}) were more frequent in multiple pollen tubes lacking nuclei, and the pectin content (1733 cm^{-1}) in multiple pollen tubes was much lower than that in single pollen tubes. These findings indicated that there were significant differences in pollen tube growth and wall composition between single and multiple pollen tubes, and that multiple pollen tubes had much less opportunity than single pollen tubes to reach the embryo sac and achieve double fertilisation.

A modified bilinear model best described the response to temperature of pollen germination and pollen tube length. The pollen germination of the genotypes ranged from 83 to 97% and pollen tube length ranged from 697 to 1270 µm. The genotype variation was found for cardinal temperatures (T_{min}, T_{opt} and T_{max}) of pollen germination percentage and pollen tube growth (Acar and Kakani, 2010).

Sabrine et al. (2010) studied the effects of cadmium and copper on pollen germination and fruit set in pea (*Pisum sativum* L). While cadmium concentrations did not affect in vivo pollen germination, only higher copper concentrations rendered a significant reduction. This is in contrast with the clear negative effect on pollen germination in vitro and might be explained by the different dynamic and bioavailability of both metals. A clear effect of Cd and Cu was however observed on two important yield components, that is fruit weight and seed set.

Alcaraz et al. (2010) developed an improved in vitro pollen germination method for avocado (*Persea americana* Mill.) using different concentrations of sucrose, PEG, Mg and Ca on pollen germination. Once the germination medium was optimised, the effect of temperature on in vitro pollen germination and tube growth in different cultivars from the three botanical varieties of avocado was studied, which differed in their adaptation to environmental conditions. Significant differences in percentage of pollen germination and in pollen tube growth were observed among the cultivars. These results could have implications not only for optimising pollen management in avocado but also for selecting the best pollinisers for a particular cultivar.

An improved in vitro pollen germination assay was developed to assess the viability of stored *Hedychium* pollen. The effect of PEG (10, 15, and 20%, w/v) on pollen germination and tube growth was evaluated for *H. longicornutum* and two commercial *Hedychium* cultivars, "Orange Brush" and "Filigree". Overall, the inclusion of PEG 4000 in the medium improved both pollen germination and tube growth for the three different genotypes tested and the results varied depending on genotype. In vitro germination was used to assess the viability of *Hedychium* pollen stored up to two months. Large variation in pollen nucleus status was observed among the four *Hedychium* cultivars (Sakhanokho and Rajasekaran, 2010).

The in vitro germinability of pollens collected from young mature cultivars of walnut (*Juglans regia*), pistachio (*Pistacia vera*) and kiwifruit (*Actinidia deliciosa*) was evaluated at sub-optimal, optimal and supra-optimal temperatures. Throughout the temperature range that permitted pollen germination, pollens from young walnut and kiwifruit trees/vines gave germination percentages consistently lower than those from the older trees/vines growing in the same location. At 20–26 °C, germination of pistachio pollen collected from young trees was lower than that collected from older trees, but no differences were evident at higher temperatures. The physiological basis for the influence of tree age on intraclonal variation in pollen germinability awaits further experimentation (Polito and Weinbaum, 1988).

Rosell et al. (1999) studied the in vivo characterisation and optimisation of in vitro germination of cherimoya (*Annona cherimola* Mill.). Cherimoya is a subtropical fruit-tree highly dependent on pollination. In most parts of the world hand pollination is a common farm practice. In this work, the pollen germination was first characterised in vivo and the temperature effect on pollen germination evaluated. Subsequently, a medium for pollen germination was optimised and the conditions for in vitro pollen germination determined. Mature pollens were trinucleated and were shed from the anther in tetrads. Upon pollen hydration, either on the stigma or in the germination medium, the pollens were liberated from the tetrad. While the kinetics of the process was quicker in vitro, both processes occurred rapidly and were completed within 2 h in vitro and within 5 h in vivo. Pollen requires prehydration prior to in vitro germination and it only germinates when the stamens are also added to the medium, which requires the provision of calcium and boron as well as sucrose at 5–10%.

Cherimoya, a subtropical fruit tree, is cultivated in a good range of subtropical regions. Its often erratic fruit set could be related to problems in pollen handling. Very little however is known of the time the pollen remains viable and the best stage to collect the anthers or pollen from the flower. The aim was to evaluate: (a) pollen germinability prior and after anther dehiscence and (b) how the age of pollen affects pollen vigour, understood as speed of germination. Pollen samples at different times following anther dehiscence were germinated in vitro and in vivo. Pollen taken up to 90 min following dehiscence performed as well as freshly dehisced pollen. However, pollen taken 120 min following dehiscence showed a clear reduction in vigour and germinated much slower in vivo. To overcome this short pollen germinability, the pollen was taken from anthers 30.5 h prior to natural anther dehiscence and compared with pollen taken at anther dehiscence and 20 h later. A reduction in germination rate was obtained in pollen taken prior to anther dehiscence. The narrow stage at which pollen could be collected, together with its ephemeral germinability explained the erratic results obtained following hand-pollination in this crop; and these results provided clues for adequate pollen handling (Rosell et al., 2006).

A study was undertaken on the influence of different pollen sources on the nut and kernel characteristics of hazelnut. Pollination is very important to obtain economical yield in hazelnut. The selection of a polliniser with a suitable quality and quantity of pollen is an essential practice in hazelnut orchards. The result showed that the percentage of pollen germination was more than 60% for all cultivars at the time of application, indicating that all studied cultivars have a high viable pollen ratio and pollen germination capacities. The dichogamy type in the selected cultivars was all the same, and protandry was common. Cross-pollination with the pollen of different pollinisers had a significant effect on increasing the values of final fruit setting, nut and kernel weight and kernel percentage, with decreasing blank nuts in all pollinated cultivars. While self-pollination reduced the values of nut and kernel traits, with an increased percentage of blank nuts, it was concluded that the pollen sources had xenia and metaxenia effects in hazelnut (Fattahi et al., 2014).

10.7 Pollen load, pollination and seed production

Pollination is a very important determinant of fruit set in a tree and fruit crop for economical yields. The selection of a polliniser with a suitable quality and quantity of pollen is therefore an essential practice to adopt.

In many Douglas-fir (*Pseudotsuga menziesii* Mirb.) seed orchards, supplemental mass pollination is practised to improve yields and genetic quality. Douglas-fir pollen can be successfully stored at low pollen moisture content and low storage temperature. A reciprocal time of pollination experiment suggested that pollen must be applied early to ensure successful pollination and results were explained using pollination mechanisms (Webber, 1987).

Quamar and Bera (2003) studied pollen production and depositional behaviour of teak (*Tectona grandis* Linn. F.) and sal (*Shorea robusta* Gaertn. F.) in tropical deciduous forests of Madhya Pradesh, India. Teak and sal were found to be high pollen producers, but very difficult to retrieve from samples of sediment profiles/cores, which could be attributed to poor pollen preservation in the sediments as well as low dispersal efficiency. Species in association with variations in environmental factors could govern the pollen production of a particular taxon/variety.

Habitat fragmentation has great effects on plants with highly specialised pollination systems. The sub-canopy tree *Oxyanthus pyriformis* sub-sp. *pyriformis* has deep tubular flowers (˜90 mm in length) that are specialised for pollination by long-tongued hawk-moths. Fruit set in natural condition was very low (0.04–0.07%), from an examination of more than 150 000 flowers, but increased to more than 70% of flowers that were hand-pollinated with pollens from another tree. This > 1000-fold increase in the likelihood of fruit production following pollen-supplementation indicates that trees experience severe pollen-limitation of fruit set. Natural fruit set is so low that only a few dozen fruits were recorded in each population during two years of monitoring. Regular hand-pollination and, in some instances, further planting of saplings, may be the only way to safeguard these populations (Johnson et al., 2004).

Silva et al. (2008) studied genetic effects of selective logging and pollen gene flow in a low-density population of the dioecious tropical tree *Bagassa guianensis* in the Brazilian Amazon. According to them, forest logging reduces population density and increases the distance between co-specifics and so can cause the loss of alleles and affect the genetic diversity, spatial genetic structure (SGS), mating system and pollen flow of the population. Non-significant differences were observed between these samples. The effective number of pollen donors was estimated as five to seven trees. The distance of pollen gene dispersal was estimated as 308–961 m, depending on the dispersal model used (normal and exponential) and the assumed population density. The estimated neighbourhood pollination area (A_{ep}) ranged from 81 to 812 ha, depending on the assumed population density.

Effects of suspension media used for spray pollination on pollen grain viability were investigated in Japanese pear [*Pyrus pyrifolia* (Burm. f.) Nakai "Kosui"]. The suspension media tested in this study consisted of pectin methylesterase (PME) and polygalacturonase (PG) combined with either 0.1% agar or 0.04% xanthangum (XG). The media containing XG combined with either PME or PG seemed to show better results for pollen grain viability and fruit set, although the results were variable from year to year. With regard to fruit size, shape and other parameters for fruit quality, spray pollination and hand pollination gave comparable results, irrespective of the medium composition (Sakamoto et al., 2009).

Magnolia denudate is an excellent ornamental and ecologically important tree that exhibits low fecundity because of seed abortion. Cross-pollination, geitonogamy and self-pollination were undertaken on flowers of sample trees. It was observed that flowers produced a large number of pollen grains available for effective pollination. Pollen viability varied at different stages, but was highest during stage III, in which the stamens detach from the axis, the anthers dehisce and the gynoecium stigma starts to fade. The duration of pistil receptivity was approximately 5 h. The results revealed that poor fertilisation might occur if the optimum pollination period was missed or the stigma received poor-quality pollen grains from stages other than stage III. The anatomical analysis of ovule and seed development further revealed that fertilisation occurred in samples with geitonogamy, but that the embryo degenerated in the torpedo stage, demonstrating that aborted seeds were produced by fertilisation, rather than by pseudogamy. The results provided new insights into the mechanism of reduced seed set under natural conditions (Wang et al., 2010).

Fragmentation of tropical forest trees is a great danger to some tree populations, by reducing local population size and gene flow from other populations. Both processes can decrease out-crossing rates and genetic variation in remnant stands. A study was undertaken on both pollen flow and diversity of pollen sources in continuous forest and isolated stands of *Swietenia humilis*, a tropical tree species pollinated by small insects. Using seven nuclear microsatellite markers, the study revealed that allelic richness of seeds is lower in isolated populations (6.1 vs 8.3 alleles per locus), even though adult populations do not show this difference. Pollen pool structure is greater in isolated patches ($\Phi_{Iso} = 0.26$) than in continuous forest ($\Phi_{For} = 0.14$), which yields estimates of the average effective number of pollen donors (N_{ep}) of 1.9 and 3.6, respectively. In addition, estimates of number of sires per mother indicate that isolated trees have half the number of pollen sources (4.98) of trees in the forest (9.8). It is thus concluded that forest fragmentation has negative impacts on the genetic connectivity of tropical trees (Rosas et al., 2011).

Gallotta et al. (2014) examined flower anomalies and pollen production in several apricot cultivars. Abnormally developed pistils often occur in different apricot cultivars, as a consequence of physiological-biochemical factors and climatic conditions. Apricot is genetically prone to the formation of a high percentage of flowers morphologically hermaphroditic but physiologically unisexual due to pistil hypotrophy, atrophy and/or necrosis. Higher anomalies were found in flower buds of the basal portions of the long twigs. Anomalies frequently occurred at Baggiolini stages C and D (BBCH 55 and 57). The number of anthers per flower ranged from 23 to 33. The number of pollen grains per flower was in the range 35 000–76 000. The results revealed that cultivars both producing the greatest amount of stainable and compatible pollen grains per flower and bearing mainly on spurs or short twigs (lower flower anomalies) should be preferred either when planting new apricot orchards or for breeding programs.

Selak et al. (2014) studied pollen tube performance in an assessment of compatibility in olive (*Olea europaea* L.) cultivars. Most olive cultivars are partially self-incompatible. The flowering periods of the studied cultivars overlapped in a high degree in both years. In "Lastovka", cross-pollination was more efficient than self-pollination with regard to pollen germination, tube growth and fertilisation. It was assessed that cross-pollination resulted with higher pollen tube growth rates than self-pollination. An earlier and greater amount of fertilisation was found for cross-pollination than for self-pollination. Pollen-pistil interaction is an early and reliable mode to estimate compatibility in olive. Further, most cross-pollination treatments provided higher fertilisation levels than self-pollination, proving that cross-pollination was better. Inter-incompatibility reactions between cultivars were not observed.

Citrullus lanatus is cultivated in Côte d'Ivoire for its edible kernels, which are used as a soup thickener. Koffi et al.

(2013) studied the effect of pollen load, source and mixture on the reproduction success of four cultivars of *Citrullus lanatus* (Thunb.) Matsumara and Nakai (Cucurbitaceae). Various pollination treatments were applied during four growing seasons to assess pollen load, pollen source and diversity (cross-pollination with single source, cross-pollination with multiple sources, self-pollination and natural pollination) and pollen mixture. All cultivars were self-compatible and higher level of inbreeding depression was observed. Cross-pollinated plants gave a higher yield than natural pollination and self-pollinated plants. With manual pollination, cross-pollination with multiple sources produced more fruits and seeds, resulting in an increased yield. Pollen diversity also improved the production of fruits and seeds. Pollen mixture would increase yields when the proportion of cross pollen in the mixture was higher than the self-pollen proportion. In conclusion, fruits and seeds production of those local self-compatible cultivars were influenced by the amount of pollen load, pollen source and diversity and also by the proportion of each pollen type in the pollen mixture.

10.8   Pollen tube growth

Tangmitcharoen and Owens (1997) studied pollen viability and pollen-tube growth following controlled pollination and their relation to low fruit production in teak (*Tectona grandis* Linn. f.). Pollen released at 1100 h had the highest viability (92.2%) but was no longer viable three days (84 h) after anthesis. In vitro pollen-tube growth was fast (140 μm h^{-1}) and increased

significantly within the first 8 h. In vivo pollen tubes also grew quickly, reaching the base of the style within 2 h after pollination and entering the micropyle 8 h after pollination. Teak has late-acting gametophytic self-incompatibility. The majority of pollen tubes grow through the style but some do not continue to grow from the style towards the embryo sacs. Pollen-tube abnormalities include swollen, reversed, forked and tapered tips and irregular or spiralling tubes. These are most prevalent in self-pollination (20.4%). Drastic fruit abortion occurred within the first week following controlled pollination. Within 14 days, fruit size and fruit set from cross-pollination was generally much greater than from self-pollination.

Self-incompatibility and synchronous protandrous dichogamy have previously been reported in *Ziziphus* species. Asatryan and Tel-Zur (2013) did a comparative analysis of fluorescence microscopy observations of pollen tube growth following controlled cross-pollinations of emasculated flowers and self-pollinations of non-emasculated flowers in three *Ziziphus* species, *Z. jujuba*, *Z. mauritiana* and *Z. spina-christi*, to determine the type of the self-incompatibility system of each species. The presence of binucleate pollen grains and the cessation of pollen tube growth in the style suggested that the self-incompatibility system operating in the studied *Ziziphus* species was gametophytically controlled. Controlled self-pollination in *Z. mauritiana* produced fruits that dropped off before maturation, whereas in *Z. spina-christi* the flowers dropped off one or two days after pollination. In the cultivar Tamar of *Z. jujuba*, the relatively high percentage of seedless fruits obtained in emasculated bagged flowers without hand pollination suggested that this cultivar could set seedless fruits without any pollination stimulus.

10.9 Pollen viability

Pollen viability and pollen tube growth have profound effects on fruit and seed production. Knowledge of pollen biology is therefore necessary to increase the productivity of a tree. In this context, a concise review of literature is made hereunder.

Pollens of obeche trees (*Triplochiton sclerxylon* K. Schum.) obtained from different flowers located in Ibadan (7° 17′N, 3° 30′E) were studied. Pollen viability varied significantly between the flowers from three sources. Significant differences were observed in pollen tube growth between flowers from each of the trees. Besides the effects of pests and diseases, variation in pollen viability between trees and from flower to flower may also contribute to the low seed-set in this species (Oni, 1990).

The viability of pollen was studied in a widely distributed anemophilous species, Scots pine (*Pinus sylvestris* L.), along a south–north gradient from 60 to 69°N in Finland. The pollen was collected directly from the male strobili and germinated in vitro at different temperatures. The results showed marked variations in pollen viability both between and within latitudes and also between different temperature treatments. Pollen germination at the lower temperatures was low across all the study areas, but very good at the highest temperature, which was 20 °C. The variation in pollen viability between stands might lead to differential success of pollen of different

origin. The variation between years was also high. Thus, different pollens might be favoured in different years. Finally, the possibility of gene flow from south to north was assessed (Parantainen and Pulkkinen, 2002).

Acacia mearnsii (black wattle) is a commercially important forestry species in South Africa, grown for its timber and bark. Previous research on crosses between diploid and tetraploid parent plants to produce triploid progeny resulted in poor seed set. In vitro agar media germination tests in ACIAR and Brewbaker and Kwack media were used together with vital stain tests (Sigma® DAB peroxidase and p-phenylendiamine) to test the pollen germination and viability of *A. mearnsii* pollen and showed that the vital stain tests had significantly ($p < 0.05$) higher pollen viability than the agar germination tests (Beck-Pay, 2012).

A study was done on the pollen viability, in vitro pollen germination and in vivo pollen tube growth in the biofuel seed crop *Jatropha curcas* using 2,3,5-triphenyltetrazolium chloride. Pollen germination was significantly higher in an agar-based medium composed of sucrose, boric acid and calcium nitrate compared with the control and indole-3-acetic acid treatment. Pollen from hermaphrodite flowers showed lower viability, lower germination rates and shorter pollen tubes, with abnormal shapes, compared to the pollen from male flowers. Pollen tubes from both self- and cross-pollinated flowers entered the ovary within eight hours after pollination (Abdelgadir et al., 2012).

In an investigation on pollen viability of three polleniser mango cultivars, "Sensation", "Tommy Atkins" and "Janardan

Pasand" up to 24 weeks under four storage conditions (room temperature, −4, −20 and −196 °C) and using three methods of pollen viability testing: in vitro germination, fluorescein diacetate (FDA) and acetocarmine staining. Storage methods and interaction between storage methods and days of storage had a highly significant effect on pollen viability ($p \leq 0.0001$). Room temperature storage of pollen in the three mango cultivars showed very low pollen viability after four weeks of storage, after which the pollen was not viable. Irrespective of mango genotypes, cryo-stored (−196 °C) pollen showed significantly higher viability as compared to all the other storage conditions. It was assessed that the pollen viability of mango cultivars up to 24 weeks under four storage conditions were comparable. Pollen viability tests, that is in vitro germination, FDA and acetocarmine were compared. Pollen viability assays confirmed that the in vitro germination test was most reliable. In FDA and acetocarmine tests, pollen viability was often overestimated. Cryo-stored (−196 °C) pollens showed significantly higher viability in mango cultivars. The authors suggested the storage of pollen at −20 °C for pollination among cultivars having non-synchronised flowering in a season (Dutta et al., 2013).

Olive pollination is anemophilous and an adult olive tree can produce large amounts of pollen grains spread in the air during the flowering period. The viability of pollen grains can be checked by using different methods such as cytoplasmic stains, enzymatic reactions or germination. A study was undertaken to verify if the quality and number of pollen grains vary between years. The grain viability was estimated by using three different techniques: acetic carmine, fluorescein diacetate and germination. Among the three techniques, acetic carmine gave the highest values but staining also heat-killed pollen grains. The number of pollen grains significantly varied among the cultivars and between the two years. The variation in pollen in pollen viability was observed in between two years (Mazzeo et al., 2014).

10.10   Effects of nitrogen dioxide

Using a randomised sampling design, the pollen viability of Austrian pine (*Pinus nigra* Arnold) exposed to NO_2 pollution was measured in northern Italy by means of the tetrazolium test. It was found that the ambient levels of nitrogen dioxide (NO_2) might reduce the pollen viability of *P. nigra*. Analysis of variance in a generalised linear model showed that NO_2 was a significant factor ($P = 0.0425$) affecting pollen viability. The pollen viability was significantly related to pollen germination ($P < 0.01$) and tube length ($P < 0.01$). This suggested a possible impact of NO_2 on the regeneration of Austrian pine in polluted environments (Gottardini, 2008).

In an experiment, the pollens of *Betula pendula*, *Ostrya carpinifolia* and *Carpinus betulus* were exposed in vitro to two levels of NO_2 (about 0.034 and 0.067 ppm), both below the current atmospheric hour-limit value acceptable for human health protection in Europe (0.11 ppm for NO_2). The experiment was done under artificial solar light with temperature and relative humidity continuously monitored. The viability, germination and total soluble proteins of all the pollen samples exposed to NO_2 decreased significantly when compared with the non-exposed. The polypeptide profiles of all the pollen samples showed bands between 15 and 70 kDa and the exposure to NO_2 did not produce any detectable changes in these profiles. The common reactive bands to the three pollen samples corresponded to 58 and 17 kDa proteins (Cuinica et al., 2014).

10.11   Effects of temperature

In mango (*Mangifera indica* L.), an effective pollination rate was determined at two stages of the fertilisation process: (a) pollen germination on the stigma; (b) pollen-tube penetration into the ovule. Pollination rates were negligible during the first part of the flowering season (31 March to 18 April), reaching a high value only at the end of the season (21 May). Ovule penetration was hampered in orchard pistils, even when phytotron pollen was used for pollination. Chilling injury appeared to be responsible for the damage to the reproductive organs of the orchard flowers. The negligible rate of effective pollination found in mango orchards in Israel during a significant part of the flowering season therefore appeared to be due to the detrimental effect of cold weather on the pollination and fertilisation processes as well as on the functional viability of the male and female reproductive organs (Dag et al., 2000).

Acar and Kakani (2010) studied the effects of temperature on in vitro pollen germination and pollen tube growth of *Pistacia* spp. to identify differences in the tolerance of pollen to temperature variations. The effect of temperature on in vitro pollen germination and pollen tube growth was investigated in *Pistacia vera* (Uygur, Atli, Kaska, Sengel, Kavak), *P. atlantica, P. khinjuk, P. terebinthus* and *P. Palaestina.* Pollen was collected at anther dehiscence and was exposed to temperatures from 5 to 40 °C at 5 °C intervals. Some differences were observed between the percentage of pollen germination and pollen tube growth at different temperatures. *Pistacia* species and cultivars were found to range from most tolerant to most susceptible, depending on pollen characters.

Prunus dulcis L. "Mamaei" is grown widely in Southwest Iran. It blooms in early spring when temperatures are still low. Sorkheh et al. (2011) observed the response of in vitro pollen germination and pollen tube growth of almond (*Prunus dulcis* Mill.) to temperature, polyamines and polyamine synthesis inhibitor. The effect of three different temperatures (10, 25, 35 °C), polyamines (putrescine, spermidine, spermine) and polyamine synthesis inhibitor, methylglyoxals-bis (guanyl-hydrazone), MGBG,on in vitro pollen germination and pollen tube growth were investigated in *P. dulcis* L. "Mamaei". All temperatures and chemicals significantly affected both pollen germination percentage and pollen tube growth. In general, different polyamines stimulated the pollen germination percentage compared to the control at all temperatures, but increasing the temperature, particularly to 35 °C, demonstrated inhibitory effects on pollen germination. At a concentration of 0.05 mM putrescine and spermidine and at 0.005 and 0.025 mM spermine revealed a longer pollen tube growth than that of the control at 10 °C, while higher concentrations tended to inhibit pollen tube growth. At 25 °C, most of the treatments had an inhibitory effect on pollen tube growth except for 0.25 mM putrescine and 0.005 mM spermine, which slightly stimulated pollen tube growth. Pollen germination and pollen tube growth were inhibited by MGBG at all temperatures and all concentrations.

Temperature and pollen genotype affected pollen germination, pollen tube growth and fertilisation percentage in olive. Increased temperature reduced pollen tube growth and fertilisation percentage. Cross-pollination led to higher pollen tube growth rate and fertilisation percentage than self-pollination (Selak et al., 2013).

Nava et al. (2009) studied the effects of high temperatures in pre-blooming and blooming periods on the growth of sexual gametes and yield of "Granada" peach at Rio Grande do Sul State, Brazil. Two treatments were tested: (a) trees in the greenhouse with partial ventilation and (b) trees in the orchard. They studied the phenology, morphologic constitutions of the pollen grains, evaluating ovule growth, yield and germination percentage of pollen grains and fruit-set. High temperatures in the pre-blooming and blooming periods induced a break of dormancy and blooming. These conditions also delayed the female gametophytes (embryo sac) and promoted anomalies in the formation of male gametophytes. Those factors promoted low pollen viability and a lack of

synchrony in fertilisation, thereby leading to low fruit set percentages and yield.

10.12   Other factors

Franchi et al. (2007) studied the ecophysiological effects of different relative humidity levels, osmotic relationships and cytological methods (stigmatic receptivity, pollen viability, histology and histochemistry) to determine pollen and pollination features during the long blooming period of the species *Parietaria judaica* L. (Urticaceae). The results reveal that pollen is dispersed by rapid uncurling of the filament and anther opening. The filament and anther lack cells with lignified wall thickenings are usually responsible for anther opening and ballistic pollen dispersal. Dispersal is the result of a sudden movement by the filament. Pollen is of the partially hydrated type. Starchless grains germinate rapidly and are less subject to water loss. Flowers are protogynous, pollen release occurring only after complete cessation of the female phase within an inflorescence.

10.13   Storage of pollen

Akoroda (1983) reported on the long-term storage of yam pollen. Yam pollen stored at 0% relative humidity and −5 °C temperature remained highly viable for over one year (from one flowering season to the next). Fluctuations in storage conditions accelerated loss of pollen viability. Pollen germination in vitro was not significantly correlated with the degree of fruit-set, and pollen samples with low percentage germination did give satisfactory fruit-set.

When the papaya pollen viability, tube growth and storage were studied, pollen germination on a drop of modified Brewbaker medium at 22–26 °C closely reflected in vivo germination. The pollen could be stored successfully for six months when kept at −18 °C in a deep-freeze. Storage of pollen for the winter months, during which no or very little viable pollen is formed in the coastal area of Israel, seemed a practical means to ensure adequate fruit-set (Cohen et al., 2011).

Due to protogynous dichogamy of cherimoya and to absence of proper pollinating vectors, hand-pollination with fresh pollen is a common practice for cherimoya commercial production. To optimise the process of hand-pollination, the conservation of cherimoya pollen at −20, −80 and −196 °C for up to three months was studied. In vitro pollen germination of fresh pollen was 57.1% and progressively reduced with conservation time at the three temperatures studied, reaching a minimum after three months of storage at −20, −80 and −196 °C of 10.4, 14.2 and 13.6%, respectively. The results indicated that pollens collected and stored at sub-zero temperatures at the beginning of the cherimoya blooming season can be used along the whole blooming season, avoiding the need for collecting fresh pollen daily (Lora et al., 2006).

Masum Akond et al. (2012) studied the longevity of crapemyrtle pollen stored at different temperatures. Temperatures for the storage of crapemyrtle (*Lagerstroemia* spp.) pollen over time were studied using clones of two interspecific hybrids (*L. cheyenne* and *L.wichita*) and five species [*L.*

indica Catawba, *L. subcostata* (NA 40181), *L. limii* (SHL2004-1), *L. speciosa* (MIA 36606) and *L. fauriei* Kiowa]. Pollen samples were stored at room temperature (23 ± 1), 4, −20 and −80 °C. Overall, this study revealed that *Lagerstroemia* pollen is best maintained over time when pollen is stored under refrigeration, with a storage temperature of −20 °C often being preferable to 4 °C. Lowering the storage temperature to −80 °C is generally unnecessary, but not deleterious to pollen germination. Storage of viable crapemyrtle pollen for no longer than 75–105 days is adequate for allowing breeders to carry out efficiently the hybridisation of germplasm flowering at different times or in widely separated locations. It is assessed that pollens of all clones lost viability within seven days stored at the room temperature. *Lagerstroemia* pollen is best maintained over time when pollen is stored under refrigeration. A storage temperature of −20 °C often improved pollen longevity as compared to 4 °C. Lowering the storage temperature to −80 °C is normally unnecessary, but not deleterious. Crapemyrtle pollen remained viable for 75–105 days when refrigerated.

10.14   Seed production

Several possible factors affecting seed production were investigated by Rasmussen and Kollmann (2004) who then reported poor sexual reproduction on the distribution limit of the rare tree *Sorbus torminalis*. This study about the effects of seven pollination treatments on fruit production and on the timing of fruit abortion was done in south-east Denmark's two relatively small and isolated populations. There was evidence of a lack of pollination and spontaneous self-pollination causing particularly high fruit abortion, which indicated that apomixis was unlikely and spontaneous self-pollination not efficient. Fruit abortion was delayed after hand-pollination, which suggested limitation by pollen quantity. Self-pollination caused earlier abortion than experimental cross-pollination within or between populations, indicating inbreeding depression. There was no evidence for out-breeding depression as measured by fruit abortion. The conclusion was that generative reproduction of *S. torminalis* was reduced on its northern distribution limit and that it might be negatively affected by pollen limitation and inbreeding effects, which were not compensated for by increased self-compatibility or apomixes.

Hirayama et al. (2007) studied the genetic factors and pollen shortage affecting seed production in a small population of *Magnolia stellata*, a threatened insect-pollinated tree, using microsatellite and manual pollination techniques. Compared to the large population, the small population showed low allelic variation and had an F_{IS} significantly greater than 0 in adults, which revealed that genetic deterioration, including genetic drift and inbreeding, might have occurred in adults. Manual self-pollination led to lowered seed production compared to manual cross-pollination to a different extent between populations. However, under natural pollination, the estimated embryo mortality rates after ovules self-fertilised were similar between the two populations because the primary selfing rate was higher in the small population. The

ovule mortality rate due to pollen shortage and that due to factors independent of pollen shortage and selfing were both approximately 10% higher in the small population, suggesting that pollen transfer might have decreased and genetic deterioration effects might have increased in that population. The results suggested that seed production in the small population of *M. stellate* was highly limited by elevated pollen shortage, selfing and genetic deterioration in adults, which accelerated the risk of extinction.

Caesalpinia echinata (brazil-wood) is a threatened tree endemic to the Brazilian Atlantic forest. It bears a global importance as it is used worldwide to manufacture the most desirable bows for violins, violas and cellos. Borges et al. (2009) investigated the phenology, pollination and breeding system of *C. echinata*. The genus has a complex taxonomic history with recent attempts to reassign its limits awaiting a final resolution of the placing of some species (including *C. echinata*). Flowering occurs mainly in the dry season and the peak of seed dispersal is at the beginning of the wet season. Anthesis is diurnal, lasting one day. The flowers are zygomorphic, yellow, sweet-scented, and the effective pollinators are mainly medium-sized to large bees of the genera *Centris* and *Xylocopa*, together with the introduced *Apis mellifera*. Nectar volume and sugar concentration average at $2.9 \pm 1.0\,\mu l$ and $29.5 \pm 9.4\%$, respectively. The results of controlled pollinations and analysis of pollen tube growth revealed *C. echinata* presents late-acting self-incompatibility.

da Silva Carneiro et al. (2009) used six highly polymorphic microsatellite loci and a categorical paternity analysis approach to investigate the contemporary pollen gene flow in the neotropical tree species *Symphonia globulifera*. High levels of pollen immigration from outside of the study plot were found in both sampled seed-years ($\geq 49\%$) suggesting long distance pollen gene flow. Low levels of self-fertilisation were also detected ($\leq 2\%$). The analysis showed long-distance pollen dispersal occurring within the study area in both 2002 ($\delta = 907 \pm 652$ m SD) and 2003 ($\delta = 963 \pm 542$ m SD). Patterns of pollen dispersal distance within the plot were also found to be shorter than the distances between potential male parents and seed-trees. This result revealed that the distance between trees did not explain the identified pollen dispersal pattern. The results supported the hypothesis that animal-pollinated species occurring in low-density populations could disperse pollens over long distances, despite the very dense nature of the forest.

Protogynae and poor pollen quality are factors causing poor seed production. In this context, Li et al. (2008) stated that protogynous pollen limitation and low seed production were causal factors for the dieback of *Spartina anglica* in coastal China. The flower was protogynous. Owing to regular pollen grains, the pollen viability in two sites was 28.2 ± 1.9 and $20.8 \pm 0.9\%$, respectively. In the field, pollen grains reached neighbours 3 m away at the most. Because of an irregular pollen tube, only about one or two pollen tubes could penetrate the micropyle in a cross-pollinated flower and self-pollination hardly occurred. The proportion of seed production per inflorescence was as low as 5.9% in a natural coastal area in China. Thus, protogynous, poor pollen quality, abnormal pollen grains and pollen tube were the main causes of low seed production.

Jatropha curcas Lisa, a species from the euphorbiaceae family, native to Mexico and Central America, has been the focus of recent research because its seeds can be used as a source for biodiesel production. A study was done to evaluate the morphological characteristics of development from ovary to fruit, and from ovule to seed, as well as early stages of embryo development. The results showed that pollen grains had hexagonal and polygonal patterns. At 15 days post-anthesis (DPA), the early globular and cotyledonal embryo was observed. Similarly, the caruncle around the micropyle exhibited an increase of body oils in the first 15 DPA. Ripening of fruits started at 40 DPA and the testa of the seeds became hard at 35 DPA. The research findings generated new insights into the development of *J. curcas* fruits and seeds, and this could be useful for genetic improvement of the species to produce more and better oil for biodiesel production (Catzín-Yupit et al., 2014).

10.15 Conclusions

Various techniques have been developed for testing pollen viability, pollen germination and pollen tube growth. A substantially high level of research advances has been achieved in this direction. Pollen viability, pollen germination, pollen tube growth and fertilisation contribute to the productivity of crops. For the effective productivity of a plant, the anthers should produce a sufficient pollen load with a higher percentage of viable pollens, higher pollen germination and growth of pollen tubes leading to the fertilisation of an egg, embryo development and finally seed production. Adverse environmental conditions such as air pollution, high, low temperature and heat stress greatly affect all these parameters, leading to the poor productivity of seeds. Causal factors for the poor productivity of some fruits have been discussed and precautions have been suggested to protect crops from these menaces. A critical synthesis of the factors affecting plant productivity has also been discussed. A good knowledge on various aspects of pollen biology and seed production could help in the efficient management and adoption of measures to protect plants from adverse environmental conditions.

Bibliography

Abdelgadir H.A., Johnson S.D., Van Staden J. 2012. Pollen viability,pollen germination and pollen tube growth in the biofuel seed crop *Jatropha curcas* (Euphorbiaceae). *South African Journal of Botany* 79:132–139.

Acar I., Kakani V.G. 2010. The effects of temperature on *in vitro* pollen germination and pollen tube growth of *Pistacia* spp. *Scientia Horticulturae* 125(4):569–572.

Akoroda M.O. 1983. Long-term storage of yam pollen. *Scientia Horticulturae* 20(3):225–230.

Alcaraz M.L., Montserrat M., Hormaza J.M. 2010. *In vitro* pollen germination in avocado (*Persea Americana* Mill.): Optimization of the method and effect of temperature. *Scientia Horticulturae* 130(1):152–156.

Asatryan A., Tel-Zur N. 2013. Pollen tube growth and self-incompatibility in three *Ziziphus* species (Rhamnaceae). *Flora – Morphology, Distribution, Functional Ecology of Plants* 208(5/6):390–399.

Beck-Pay S.L. 2012. The effect of temperature and relative humidity on *Acacia mearnsii* polyad viability and pollen tube development. *South African Journal of Botany* 83:165–171.

Beretta M., Rodondi G., Adamec L., Andreis C. 2014. Pollen morphology of European bladderworts (*Utricularia* L., Lentibulariaceae). *Review of Palaeobotany and Palynology* 205:22–30.

Borges L.A., Sobrinho M.S., Lopes A.V. 2009. Phenology, pollination, and breeding system of the threatened tree *Caesalpinia echinata* Lam. (Fabaceae), and a review of studies on the reproductive biology in the genus. *Flora – Morphology, Distribution, Functional Ecology of Plants* 204(2):111–130.

Ćalić D., Devrnja N., Kostić I., Kostić M. 2013. Pollen morphology, viability, and germination of *Prunus domestica* cv. Požegača. *Scientia Horticulturae* 155:118–122.

Catzín-Yupit C.N., Ramírez-Morillo I.M., Barredo Pool F.A., Loyola-Vargas V.M. 2014. Ontogenic development and structure of the embryo, seed, and fruit of *Jatropha curcas* L. *South African Journal of Botany* 93:1–8.

Cavalcante H.C., Schifino-Wittmann M.T., Dornelles A.L.C. 2000. Meiotic behaviour and pollen fertility in an open-pollinated population of 'Lee' mandarin [*Citrus clementina*×(*C. paradisi*×*C. tangerina*)]. *Scientia Horticulturae* 86(2):103–114.

Chalak L., Legave J.M. 1997. Effects of pollination by irradiated pollen in Hayward kiwifruit and spontaneous doubling of induced parthenogenetictrihaploids. *Scientia Horticulturae* 68(1/4):83–93.

Chen J., Xia N.H. 2011. Pollen morphology of Chinese *Curcuma* L. and *Boesenbergia* Kuntz (Zingiberaceae): Taxonomic implications. *Flora – Morphology, Distribution, Functional Ecology of Plants* 206(5):458–467.

Cohen E., Lavi U., Spiegel-Roy P. 2011. Papaya pollen viability and storage. *Scientia Horticulturae* 40(4):317–324.

Cuinica L.G., Abreu I., da Silva J.E. 2014. Effect of air pollutant NO_2 on *Betula pendula*, *Ostrya carpinifolia* and *Carpinus betulus* pollen fertility and human allergenicity. *Environmental Pollution* 186:50–55.

da Silva Carneiro F., Degen B., Kanashiro M., Biscaia de Lacerda A.E., Sebbenn A.M. **2009**. High levels of pollen dispersal detected through paternity analysis from a continuous *Symphonia globulifera* population in the Brazilian Amazon. *Forest Ecology and Management* **258**(7):260–1266.

Dag A., Eisenstein D., Gazit S. **2000**. Effect of temperature regime on pollen and the effective pollination of 'Kent' mango in Israel. *Scientia Horticulturae* **86**(1):1–11.

Dutta S.K., Srivastav M., Chaudhary R., Lal K., Patil P., Singh S.K., Singh A.K. **2013**. Low temperature storage of mango (*Mangifera indica* L.) pollen. *Scientia Horticulturae* **161**:193–197.

Fattahi R., Mohammadzedeh M., Khadivi-Khub A. **2014**. Influence of different pollen sources on nut and kernel characteristics of hazelnut. *Scientia Horticulturae* **173**:15–19.

Franchi G.G., Nepi M., Matthews M.L., Pacini E. **2007**. Anther opening, pollen biology and stigma receptivity in the long blooming species, *Parietaria judaica* L. (Urticaceae). *Flora – Morphology, Distribution, Functional Ecology of Plants* **202**(2):118–127.

Gallotta A., Palasciano M., Mazzeo A., Ferrara G. **2014**. Pollen production and flower anomalies in apricot (*Prunus armeniaca* L.) cultivas. *Scientia Horticulturae* **172**:199–205.

Gottardini E., Cristofori A., Cristofolini F., Maccherini S., Ferretti M. **2008**. Ambient levels of nitrogen dioxide (NO_2) may reduce pollen viability in Austrian pine (*Pinus nigra* Arnold) trees correlative evidence from a field study. *Science of the Total Environment* **402**(2/3):299–305.

Hirayama K., Ishida K., Setsuko S., Tomaru N. **2007**. Reduced seed production, inbreeding, and pollen shortage in a small population of a threatened tree, *Magnolia stellata*. *Biological Conservation* **136**(2):315–323.

Huang J.-H. Ma W.-H., Liang G.-L., Zhang L.-Y., Wang W.-X., Cai Z.-J., Wen S.-X. **2010**. Effects of low temperatures on sexual reproduction of 'Tainong 1' mango (*Mangifera indica*). *Scientia Horticulturae* **126**(2):109–119.

Jiang B., Shen Z., Shen J., Yu D., Sheng X., Lu H. **2009**. Germination and growth of sponge gourd (*Luffa cylindrica*) pollen tubes and FTIR analysis of the pollen tube wall. *Scientia Horticulturae* **122**(4):638–644.

Johnson S.D., Neal P.R., Peter C.I., Edwards T.J. **2004**. Fruiting failure and limited recruitment in remnant populations of the hawkmoth-pollinated tree *Oxyanthus pyriformis* subsp. *pyriformis* (Rubiaceae). *Biological Conservation* **120**(1):31–39.

Khansari E., Zarre S., Alizadeh K., Attar F., Aghabeigi F. **2012**. Pollen morphology of *Campanula* (Campanulaceae) and allied genera in Iran with special focus on its systematic implication. *Flora – Morphology, Distribution, Functional Ecology of Plants* **207**(3):203–211.

Koffi A., Séverin B.B., Guillaume K.K., Bertin Y.K., Clémence K.L., Sylvère Y.K., Yao D. **2013**. Effect of pollen load, source and mixture on reproduction success of four cultivars of *Citrullus lanatus* (Thunb.) Matsumara and Nakai (Cucurbitaceae). *Scientia Horticulturae* **164**:521–531.

Kovacik L., Plitzko J.M., Grote M., Reichelt R. **2009**. Electron tomography of structures in the all of hazel pollen grains. *Journal of Structural Biology* **166**(3):263–271.

Li H., An S., Zhi Y., Yan C., Zhao L., Zhou C., Deng Z., Su W., Liu Y. **2008**. Protogynous, pollen limitation and low seed production reasoned for the dieback of *Spartina anglica* in coastal China. *Plant Science* **174**(3):299–309.

Lora J., Pérez de Oteyza M.A., Fuentetaja P., Hormaza J.I. **2006**. Low temperature storage and in vitro germination of cherimoya (*Annona cherimola* Mill.) pollen. *Scientia Horticulturae* **108**(1):91–94.

Lu L., Fritsch P.W., Wang H., Li H.-T., Li D.-Z., Chen J.Q. **2009**. Pollen morphology of *Gaultheria* L. and related genera of subfamily Vaccinioideae: Taxonomic and evolutionary significance. *Review of Palaeobotany and Palynology* **154**(1/4):106–123.

Luo C., Zheng Z., Tarasov P., Pan A., Huang K., Beaudouin C., An F. **2009**. Characteristics of the modern pollen distribution and their relationship to vegetation in the Xinjiang region, northwestern China. *Review of Palaeobotany and Palynology* **153**(3/4):282–295.

Luza J.G., Polito V.S. **1985**. *In vitro* germination and storage of English walnut pollen. *Scientia Horticulturae* **27**(3/4):303–316.

Masum Akond A.S.M.G., Pounders C.T., Blythe E.K., Wang X. **2012**. Longevity of crapemyrtle pollen stored at different temperature. *Scientia Horticulturae* **139**:53–57.

Mazzeo A., Palasciano M., Gallotta A., Camposeo S., Pacifico A., Ferrara G. **2014**. Amount and quality of pollen grains in four olive (*Olea europaea* L.) cultivars as affected by 'on' and 'off' years. *Scientia Horticulturae* **170**:89–93.

Moon H.-K., Vinckier S., Smets E., Huysmans S. **2007**. Comparative pollen morphology and ultrastructure of Mentheae subtribe Nepetinae (Lamiaceae). *Review of Palaeobotany and Palynology* **149**(3/4):174–186.

Nava G.A., Dalmago G.A., Bergamaschi H., Paniz R., dos Santos R.P., Marodin G.A.B. **2009**. Effect of high temperatures in the pre-blooming and blooming periods on ovule formation, pollen grains and yield of 'Granada' peach. *Scientia Horticulturae* **122**(1):37–44.

Neuffer B., Paetsch M. **2013**. Flower morphology and pollen germination in the genus *Capsella* (Brassicaceae). *Flora – Morphology, Distribution, Functional Ecology of Plants* **208**(10/12):626–640.

Okusak K., Hiratsuka S. **2009**. Fructose inhibits pear pollen germination on agar medium without loss of viability. *Scientia Horticulturae* **122**(1):51–55.

Oni O. **1990**. Between-tree and floral variations in pollen viability and pollen tube growth in obeche (*Triplochiton scleroxylon*). *Forest Ecology and Management* **37**(4):259–265.

Pacini E., Sciannandrone N., Nepi M. **2004**. Floral biology of the dioecious species *Laurus nobilis* L. (Lauraceae). *Flora – Morphology, Distribution, Functional Ecology of Plants* **209**(3/4): 153–163.

Parantainen A., Pulkkinen P. **2002**. Pollen viability of Scots pine (*Pinus sylvestris*) in different temperature conditions: high levels of variation among and within latitudes. *Forest Ecology and Management* **167**(1/3):149–160.

Polito V.S., Weinbaum S.A. **1988**. Intraclonal variation in pollen germinability in kiwifruit pistachio and walnut as influenced by tree age. *Scientia Horticulturae* **36**(1/2):97-102.

Punekar S.A., Kumaran K.P.N. **2010**. Pollen morphology and pollination ecology of *Amorphophallus* species from North Western Ghats and Konkan region of India. *Flora – Morphology, Distribution, Functional Ecology of Plants* **205**(5):326–336.

Quamar M.F., Bera S.K. **2003**. Pollen production and depositional behaviour of teak (*Tectona grandis* Linn. F.) and sal (*Shorea robusta* Gaertn. F.) in tropical deciduous forests of Madhya Pradesh, India: An overview. *Quaternary International* **325**:111–115.

Rasmussen K.K., Kollmann J. **2004**. Poor sexual reproduction on the distribution limit of the rare tree *Sorbus torminalis*. *Acta Oecologica* **25**(3):211–218.

Rosas F., Quesada M., Lobo J.A., Sork V.L. **2011**. Effects of habitat fragmentation on pollen flow and genetic diversity of the endangered tropical tree *Swietenia humilis* (Meliaceae). *Biological Conservation* **144**(12):3082–3088.

Rosell P., Herrero M., Saúco V.G. **1999**. Pollen germination of cherimoya (*Annona cherimola* Mill.): In vivo characterization and optimization of in vitro germination. *Scientia Horticulturae* **81**(3):251–265.

Rosell P., Saúco V.G., Herrero M. **2006**. Pollen germination as affected by pollen age in cherimoya. *Scientia Horticulturae* **109**(1):97–100.

Sabrine H., Afif H., Mohamed B., Hamadi B., Maria H. **2010**. Effects of cadmium and copper on pollen germination and fruit set in pea (*Pisum sativum* L.). *Scientia Horticulturae* **125**(4):551–555.

Sagun V.G., Levin G.A., van der Ham R.W.J.M. **2006**. Pollen morphology and ultrastructure of *Acalypha* (Euphorbiaceae). *Review of Palaeobotany and Palynology* **140**(1/2):123–143.

Sakamoto D., Hayama H., Ito A., Kashimura Y., Moriguchi T., Nakamura Y. **2009**. Spray pollination as a labor-saving pollination system in Japanese pear (*Pyrus pyrifolia* (Burm.f.) Nakai): Development of the suspension medium. *Scientia Horticulturae* **119**(3):280–285.

Sakhanokho H.F., Rajasekaran K. **2010**. Pollen biology of ornamental ginger (*Hedychium* spp. J. Koenig). *Scientia Horticulturae* **125**(2):129–135.

Selak G.V., Cuevas J., Ban S.G., Perica S. **2014**. Pollen tube performance in assessment of compatibility in olive (*Olea europaea* L.) cultivars. *Scientia Horticulturae* **165**:36–43.

Selak G.V., Perica S., Ban S.G., Poljak M. **2013**. The effect of temperature and genotype on pollen performance in olive (*Olea europaea* L.). *Scientia Horticulturae* **156**:38–46.

Silva M.B., Kanashiro M., Ciampi A.Y., Thompson I., Sebbenn A.M. **2008**. Genetic effects of selective logging and pollen gene flow in a low-density population of the dioecious tropical tree *Bagassa guianensis* in the Brazilian Amazon. *Forest Ecology and Management* **255**(5/6):1548–1558.

Sorkheh K., Shiran B., Rouhi V., Khodambashi M., Wolukau J.N., Ercisli S. **2011**. Response of *in vitro* pollen germination and pollen tube growth of almond (*Prunus dulcis* Mill.) to temperature, polyamines and polyamine synthesis inhibitor. *Biochemical Systematics and Ecology* **39**(4/6):749–757.

Sukhvibul N., Whiley A.W., Smith M.K., Hetherington S.E., Vithanage V. **1999**. Effect of temperature on inflorescence and floral development in four mango (*Mangifera indica* L.) cultivars. *Scientia Horticulturae* **82**(1/2):67–84.

Tangmitcharoen S., Owens J.N. **1997**. Pollen viability and pollen-tube growth following controlled pollination and their relation to low fruit production in teak (*Tectona grandis* Linn. f.). *Annals of Botany* **80**(4):401–410.

Till-Bottraud I., Gouyon P.H., Ressayre A., Godelle B. **2012**. Gametophytic vs. sporophytic control of pollen aperture number: a generational conflict. *Theoretical Population Biology* **82**(3):147–157.

Wang J., Kang M., Gao P., Huang H. **2010**. Contemporary pollen flow and mating patterns of a subtropical canopy tree *Eurycorymbus cavaleriei* in a fragmented agricultural landscape. *Forest Ecology and Management* **260**(12):2180–2188.

Wang R., Jia H., Wang J., Zhang Z. **2010**. Flowering and pollination patterns of *Magnolia denudate* with emphasis on anatomical changes in ovule and seed development. *Flora – Morphology, Distribution, Functional Ecology of Plants* **205**(4):259–265.

Webber J.E. **1987**. Increasing seed yield and genetic efficiency in Douglas-fir seed orchards through pollen management. *Forest Ecology and Management* **19**(1/4):209–218.

Xu F.-X., Kirchoff B.K. **2008**. Pollen morphology and ultrastructure of selected species of Magnoliaceae. *Review of Palaeobotany and Palynology* **150**(1/4):140–153.

CHAPTER 11

Seed characteristics

11.1 Introduction

Seeds play a primordial role in the regeneration of plants and maintaining the growth cycle of trees. There exists large variability in the morphology, size, shape, weight and surface stratification. Unlike that of crop seeds, very little attention is directed to tree seed morphology and structure. Concerted research input needs to be directed to various aspects of the seed morphology, seed dormancy and seed biology of tree species. In the following we depict variability in the seed morphology of different shrub species of the Tamaulipan thorn scrub vegetation, north-eastern Mexico. There is a necessity to study the anatomical structure of seeds, such as the size of macro- and microsclereids and its relation to the dormancy and germinability of seeds of each species.

11.2 Seed productivity in trees and shrubs

The seed production capacity of tree species is an important trait for the regeneration of the species and for maintaining the life cycle. Very little attention has been directed to this aspect, owing to the non-availabilty of efficient technology for assessing the seed productivity of trees. Few studies are available on this aspect.

Greene and Johnson (1994) estimated the mean annual seed production of trees. Plant seed production is a function of both seed size (mean mass per seed) and plant size. They examined the interspecific relationship between the size of seeds and plants and the mean long-term annual seed production per tree. For canopy trees, they showed that seed production is highly (inversely) correlated with seed mass. Tree size (basal area or leaf mass) is directly proportional to seed production over a limited range. Analysis of seed production for herbaceous plants reveals a relationship similar to that for trees. With respect to seed size, the relation to seed production is >1.0. It shows that large-seeded plants produce more total annual crop mass than do small-seeded species. But this is balanced by a greater investment in ancillary reproductive tissue by small-seeded species. The results have both theoretical and applied value in, for example, stand-level simulations of population dynamics or for planning the optimal size and shape of clear-cuts intended to be regenerated naturally.

Koenig and Knops (2000) tested whether annual seed production (masting

Autoecology and Ecophysiology of Woody Shrubs and Trees: Concepts and Applications, First Edition.
Edited by Ratikanta Maiti, Humbero Gonzalez Rodriguez and Natalya Sergeevna Ivanova.
© 2016 John Wiley & Sons, Ltd. Published 2016 by John Wiley & Sons, Ltd.

or mast fruiting) in Northern Hemisphere trees is an evolved strategy or is a consequence of resource tracking by comparing their masting patterns with annual rainfall and mean summer temperatures.

The book *Applied Forest Tree Improvement* (Zobel and Talbert, 1984) gives a summary of the concepts necessary for useful and efficient operational tree improvement programmes. It discusses the biological and practical aspects but also discusses some basic statistical concepts. Much of the information comes from large co-operative tree improvement programmes carried out in south-eastern USA and many examples are about pines, although broadleaves are included. Experience from other countries (South and Central America, Canada, Australia, New Zealand and Europe) is also included, as is information on the use of exotics in tropical and subtropical regions and temperate zones. There are 16 chapters: general concepts of tree improvement; variation and its use; provenance, seed source and exotics; quantitative aspects of forest tree improvement; selection in natural stands and unimproved plantations; seed production and seed orchards; use of tree improvement in natural forests and in stand improvement; genetic testing programs; selection and breeding for resistance to diseases, insects and adverse environments; vegetative propagation; hybrids in tree improvement; wood and tree improvement; advanced generations and continued improvement; gain and economics of tree improvement; the genetic base and gene conservation; and developing tree improvement programmes.

Downs and McQuilkin (1944) studied the seed production of Southern Appalachian oaks. Management of oak stands requires an understanding of the seeding habits of individual species, the productiveness of trees of different sizes and the importance of biotic factors in reducing the number of acorns available for germination. This paper put forward quantitative evidence on these points.

A book on eucalypt domestication and breeding (Davidson and Harwood, 1994) discussed aspects of variation, selection and reproduction that are unique to eucalypts, in the context of initial generations in the domestication of "wild" eucalypts for wood production. Of the 23 chapters, 11 cover general aspects, including eucalypts natural and planted, genetic resources, matching species and provenances to site, testing species and provenances, breeding strategies and breeding plans, reproductive biology, selection and breeding, seed production, mass vegetative propagation and future prospects. The other 12 chapters describe provenance variation and base populations for selection in 13 species (*Eucalyptus camaldulensis, E. deglupta, E. delegatensis, E. fastigata, E. globulus, E. grandis, E. saligna, E. nitens, E. obliqua, E. regnans, E. tereticornis, E. urophylla, E. viminalis*). Species and subject indexes are provided.

Bawa and Webb (1984) studied flower, fruit and seed abortion in tropical forest trees and its implications for the evolution of paternal and maternal reproductive patterns. The species showed large variability in fruit and seed set and the rate at which flowers and fruits were aborted. The amount of flower and fruit abortion also varied over time within species. Small samples of open-pollinated flowers in three species showed adequate amounts of pollen on the stigma, but it could not be determined whether the pollen was compatible or incompatible.

LaDeau and Clark (2001) studied the rising of CO_2 levels and the fecundity of forest trees. They determined the reproductive response of 19 year old loblolly pine (*Pinus taeda*) to four years of carbon dioxide (CO_2) enrichment (ambient concentration plus 200 ml/l) in an intact forest. After three years of CO_2 fumigation, trees were twice as likely to be reproductively mature and produced three times as many cones and seeds as trees at ambient CO_2 concentration. A disproportionate carbon allocation to reproduction under CO_2 enrichment results in trees reaching maturity sooner and at a smaller size. This reproductive response to future increases in atmospheric CO_2 concentration is expected to change dispersal and recruitment patterns.

Silvertown (1980) reported the evolutionary ecology of mast seedlings in trees. The hypothesis that masting by trees is a defensive strategy which satiates seed predators in mast years and starves them in the intervening periods is tested in 59 sets of data on the seed production and pre-dispersal seed predation of 25 tree species. An analysis of the data sets supported the hypothesis and showed a statistically significant positive relationship between the proportion of seeds surviving the pre-dispersal stage and the $\log_{10}$ of the crop size for the same year. Evidence that pre-dispersal seed survival increases with the length of the mast interval is poor. A positive relationship between the strength of the masting habit and the maximum observed pre-dispersal seed mortality in a sample of 15 tree species recommends that the masting habit is best developed in predator-prone species. A survey of seed

crop frequencies in the woody plant flora of North America showed masting species to be under-represented amongst shrubs and trees which disperse their seeds in fleshy dispersal units. The selection pressures and evolutionary constraints which operate on the evolution of masting plants and their seed predators are emphasised.

Seed stands or seed production areas are formed to produce seed of the best provenances of forest trees (Matthews, 1964). Seed orchards are planted to produce seed of new improved cultivars. It is observed that forest trees flower and fruit most regularly and profusely when given favourable climatic conditions, sufficient growing space and adequate nutrition. Periodicity in flowering and fruiting can be further reduced by adequate protection from animals, insects and fungi which damage or destroy flowers, fruits and seeds. Seed stands are formed by selecting vigorous, healthy and well-formed trees as seed trees and releasing their crowns by removal of all other trees in the crop. Fertilisers are applied to increase seed production and the ground cover is carefully managed. Seed orchards may consist of clones of grafts, cuttings or layers derived from selected (plus) trees, or selected seedlings, derived from open or controlled pollination. Rapid early growth is essential for early onset of flowering. Thereafter production of well-filled viable seed is maintained by the use of fertilisers and careful treatment of the ground cover. Choice of rootstock and treatment with plant growth substances should eventually provide additional increases in flowering and seed production. In both seed stands and seed orchards the degree of genetic improvement depends on effective isolation of the seed trees from inferior sources of pollen and on the intensity of selection practiced. The object of seed certification procedures is to make available to the forester seed and plants that are true to name and satisfy certain minimum requirements of quality. Four categories of seed and plants are moving in national and international trade: unclassified, source-identified, selected and certified. The first two categories of seed and plants are being discarded as rapidly as possible and are being replaced by selected and certified seed and plants. Twelve national comprehensive certification schemes for forest seed and plants are analysed to bring out the essential features of national schemes for forest seed and plants. The present lack of quick growing-on tests and the difficulties of separating provenances and cultivars of some species make essential adequate field inspection and records. Six international bodies are active in matters affecting trading in forest tree seed. Nationally acceptable seed certification standards are being framed for forest tree seed and plants.

Bibliography

Bawa K.S., Webb C.J. **1984**. Flower, fruit and seed abortion in tropical forest trees: implications for the evolution of paternal and maternal reproductive patterns. *American Journal of Botany* **71**(5):736–751.

Davidson J., Harwood C. **1994**. *Eucalypt domestication and breeding*. OUP, Oxford. 308 pp.

Downs A.A., McQuilkin, W.E. **1944**. Seed production of southern Appalachian oaks. *Journal of Forestry* **42**(12):913–920.

Greene D.F., Johnson E.A. **1994**. Estimating the mean annual seed production of trees. *Ecology* **75**(3):642–647.

Koenig W.D., Knops J.M.H. **2000**. Patterns of annual seed production by northern hemisphere trees: a global perspective. *The American Naturalist* **155**(1):59–69.

LaDeau S.L., Clark J.S. **2001**. Rising CO_2 levels and the fecundity of forest trees. *Science* **292**(5514):95–98.

Matthews J.D. **1964**. Seed production and seed certification. *Unasylva* **18**(2/3):104–118.

Silvertown J.S. **1980**. The evolutionary ecology of mast seeding in trees. *Biological Journal of the Linnean Society* **14**(2):235–250.

Zobel B.J., Talbert J.T. **1984**. *Applied forest tree improvement*. The Blackburn Press, Wiley & Sons, Inc., New York. 524 pp.

CHAPTER 12

Tree mortality

12.1 Introduction

Various factors affect tree mortality in a forest. Some studies have been undertaken on the factors affecting tree mortality.

12.2 Studies on tree mortality

Horsley et al. (2002) studied the health of eastern North American sugar maple (*Acer saccharum*) forests and factors affecting its decline. Sugar maple is a keystone species in the forests of the north-eastern and mid-western United States and eastern Canada. Its sustained health is an important issue in both managed and unmanaged forests. While sugar maple generally is healthy throughout its range, decline disease of sugar maple has occurred sporadically during the past four decades. Soil moisture deficiency or excess, highway de-icing salts and extreme weather events, including late spring frosts, mid-winter thaw/freeze cycles, glaze damage and atmospheric deposition, are the most important abiotic agents. Defoliating insects, sugar maple borer (*Glycobius speciosus*), *Armillaria* root disease and injury from management activities represent important biotic factors. It is observed that nutrient deficiencies of magnesium, calcium and potassium, insect defoliation, drought and *Armillaria* were important predisposing, inciting and contributing factors in sugar maple declines.

Suarez et al. (2004) reported factors predisposing episodic drought-induced tree mortality in *Nothofagus*. Although climatic variability is a strong driving force for forest dynamics, drought-induced mortality has generally received much less attention than other types of disturbance. In 1998–1999 northern Patagonia was affected by one of the most severe droughts of the twentieth century, coinciding with a strong La Niña event, causing high mortality of *Nothofagus dombeyi* (coihue), the dominant tree species in Nahuel Huapi National Park. They studied the factors involved in determining this mortality of *N. dombeyi* at both patch and tree level. Radial growth characteristics of killed trees and survivors were compared by dendrochronological analyses. Relationships between growth and climate were investigated using response function analysis. At the tree individuals with variable growth were more prone to die from drought than trees with more regular growth. Juveniles whose growth patterns showed sensitivity to climate were particularly likely to die. Mean growth rate was a good predictor

Autoecology and Ecophysiology of Woody Shrubs and Trees: Concepts and Applications, First Edition.
Edited by Ratikanta Maiti, Humbero Gonzalez Rodriguez and Natalya Sergeevna Ivanova.
© 2016 John Wiley & Sons, Ltd. Published 2016 by John Wiley & Sons, Ltd.

of mortality in adult trees, revealing that trees with a slower growth rate were more susceptible to drought. Spatial patterns of extensive full and partial crown dieback, which are evident in many temperate forests worldwide, may reflect the superposition of these predisposing factors on strong/repeated interannual fluctuations of climate.

Nowak et al. (2004) studied tree mortality rates and tree population projections in Baltimore. It has been estimated that, within the boundaries of Baltimore, Maryland, trees are estimated to have an annual mortality rate of 6.6%, with an overall annual net change in the number of live trees of –4.2%. Tree mortality rates were significantly different based on tree size, condition, species and land use. *Morus alba*, *Ailanthus altissima* and trees in small diameter classes, poor condition or in transportation or commercial/industrial land uses exhibited relatively high mortality rates. Trees in medium- to low-density residential areas showed low mortality rates. The high mortality rate for *A. altissima* is an artifact of this species distribution between land use types (24% were in the transportation land use). Based on a new tree population projection model that incorporates Baltimore's existing tree population and annual mortality estimates, along with estimates of annual tree growth, Baltimore's urban forest is projected to decline in both number of trees and canopy area over the next century. Factors affecting urban tree mortality are discussed.

Ostertag et al. (2005) studied factors affecting mortality and resistance to damage following hurricanes in a rehabilitated subtropical moist forest. They examined the relative importance of tree size, species, biogeographic origin, local topography and damage from previous storms in long-term permanent plots in a rehabilitated subtropical moist forest in Puerto Rico following Hurricane George in order to better predict patterns of resistance. Severe damage led to uprooted trees, snapped stems or crowns with >50% branch loss. The hurricane-caused mortality after 21 months was 5.2%/year, more than seven times higher than background mortality levels during the non-hurricane periods. Species differed greatly in their mortality and damage patterns, but there was no relationship between damage and wood density or biogeographic origin. Size was also predictive of damage, with larger trees suffering more damage. Trees on ridges and in valleys received greater damage than trees on slopes. It is suggested that resistance of trees to hurricane damage is therefore not only correlated with individual and species characteristics but also with past disturbance history.

Lines et al. (2010) investigated the influence of forest structure, climate and species on the composition on tree mortality across the eastern United States. A few studies have quantified regional variation in tree mortality or explored whether species compositional changes or within-species variation are responsible for regional patterns, despite the fact that mortality has direct effects on the dynamics of woody biomass, species composition, stand structure, wood production and forest response to climate change. Using a Bayesian analysis of over 430 000 tree records from a large eastern United States forest database, they characterised tree mortality as a function of climate, soils, species and size (stem diameter). They found: (a) mortality is U-shaped versus stem diameter for all 21 species examined, (b) mortality is hump-shaped versus plot basal area for most species, (c) geographical variation inmortality is substantial and correlated with several environmental factors and (d) individual species vary substantially from the combined average in the nature and magnitude of their mortality responses to environmental variation. Regional variation in mortality is therefore the product of variation in species composition combined with highly varied mortality environment correlations within species. The results revealed that variation in mortality is a crucial part of variation in the forest carbon cycle, such that this variation is included in models of the global carbon cycle.

Ruiz-Benito et al (2013) studied patterns and drivers of tree mortality in Iberian forests in relation to the climatic effects modified by competition. A knowledge on tree mortality driven by simultaneous drivers is needed to evaluate the potential effects of climate change on forest composition. Using repeat-measure information from approximately 400 000 trees from the Spanish Forest Inventory, they quantified the relative importance of tree size, competition, climate and edaphic conditions on the tree mortality of 11 species; and they explored the combined effect of climate and competition. Tree mortality was affected by all of these multiple drivers, especially tree size and asymmetric competition; and strong interactions between climate and competition were found. All species showed L-shaped mortality patterns (i.e. showed decreasing mortality with tree size), but pines were more sensitive to asymmetric competition than broad-leaved species. Among climatic variables, the negative effect of temperature on tree mortality was much larger than the effect of precipitation. Moreover, the effect of climate (mean annual temperature and annual precipitation) on tree mortality was aggravated at high competition levels for all species, but especially for broad-leaved species. The significant interaction between climate and competition on tree mortality indicated that global change in Mediterranean regions, causing hotter and drier conditions and denser stands, could lead to profound effects on forest structure and composition. Therefore, to evaluate the potential effects of climatic change on tree mortality, forest structure must be considered, since two systems of similar composition but different structure could radically differ in their response to climatic conditions.

Mezei et al. (2014) studied host and site factors affecting tree mortality caused by the spruce bark beetle (*Ips typographus*) in mountainous conditions. They analysed the entire course of an outbreak from 1990 to 2000 in the Tatra Mountains (Western Carpathians, Central Europe). This time period represents the last complete bark beetle gradation in this area. They distinguished three outbreak phases: the incipient epidemic, epidemic and post-epidemic stages. The sampling unit was the forest subcompartment. They analysed a total of 315 forest subcompartments over more than 2000 ha. They investigated the influence of 11 environmental and stand variables on two processes in different phases of the outbreak: the initiation and the severity of spruce mortality. They used factor analysis, discriminant analysis, multiple linear regressions and boosted regression trees for the statistical analyses. The results revealed that the roles of host and site factors in the initiation and severity of spruce mortality caused by the spruce bark beetle differed during the outbreak according to the exploitation of available host resources. The initiation of tree mortality was primarily related to host factors and the severity of mortality was dependent on host size and insolation.

Allen et al. (2010) reported that a global overview of drought and heat-induced tree mortality revealed emerging climate change risks for forests. Greenhouse gas emissions have significantly changed the global climate and will continue to do so in the future. Increases in the frequency, duration and/or severity of drought and heat stress associated with climate change could fundamentally alter the composition, structure and biogeography of forests in many regions. They were concerned about potential increases in tree mortality associated with climate-induced physiological stress and interactions with other climate-mediated processes, such as insect outbreaks and wildfire. Apart from risk, existing tree mortality is based on models that lack functionally realistic mortality mechanisms and there has been no attempt to track observations of climate-driven tree mortality globally. Allen et al. (2010) presented the first global assessment of recent tree mortality attributed to drought and heat stress.

van Mantgem et al. (2009) reported that persistent changes in tree mortality rates can alter forest structure, composition and ecosystem services such as carbon sequestration. The analyses of longitudinal data from unmanaged old forests in the western United States showed that background (non-catastrophic) mortality rates have increased rapidly in recent decades, with doubling periods ranging from 17 to 29 years among regions. Increases were also pervasive across elevations, tree sizes, dominant genera and past fire histories. Forest density and basal area declined slightly, which suggested that increasing mortality was not caused by endogenous increases in competition. Because mortality increased in small trees, the overall increase in mortality rates cannot be attributed solely to the aging of large trees. Regional warming and consequent increases in water deficits are likely contributors to the increases in tree mortality rates.

Anderegg et al. (2013) discussed the consequences of widespread tree mortality triggered by drought and temperature stress. Forests provide innumerable ecological, societal and climatological benefits, yet they are vulnerable to drought and temperature extremes. Climate-driven forest die-off from drought and heat stress has occurred around the world, is expected to increase with climate change and probably has distinct consequences from those of other forest disturbances. They examined the consequences of drought- and climate-driven widespread forest loss on ecological communities, ecosystem functions, ecosystem services and land–climate interactions. Furthermore, they highlighted research gaps that warrant study. As the global climate continues to warm, understanding the implications of forest loss triggered by these events will be of increasing importance.

Brando et al. (2011) reported abrupt increases in Amazonian tree mortality due to drought–fire interaction. Interactions between climate and land-use change may drive widespread degradation of Amazonian forests. High-intensity fires associated with extreme weather events could accelerate this degradation by abruptly increasing tree mortality, but this process remains poorly understood. They presented the first field-based evidence of a tipping point in Amazon forests due to altered fire regimes. Based on the results of a large-scale, long-term experiment with annual and triennial burn regimes (B1*yr* and B3*yr*, respectively) in the Amazon, they found abrupt increases in fire-induced tree mortality (226 and 462%) during a severe drought event, when fuel loads and air temperatures were substantially higher and relative humidity was lower than long-term averages. This threshold mortality response had a cascading effect, causing sharp declines in canopy cover (23 and 31%) and aboveground live biomass (12 and 30%) and favouring widespread invasion by flammable grasses across the forest edge area (80 and 63%), where fires were most intense (e.g. 220 and 820 kW/m). During the droughts of 2007 and 2010, regional forest fires burned 12 and 5% of south-eastern Amazon forests, respectively, compared with <1% in non-drought years. These results showed that a few extreme drought events, coupled with forest fragmentation and anthropogenic ignition sources, are already causing widespread fire-induced tree mortality and forest degradation across south-eastern Amazon forests. Future projections of vegetation responses to climate change across drier portions of the Amazon require more than a simulation of global climate forcing alone and must also include interactions of extreme weather events, fire and land-use change.

Eamus (2013) reported that global change-type drought-induced tree mortality and vapour pressure deficit is more important than temperature per se in causing decline in tree health. Drought-induced tree mortality is occurring across all forested continents and is expected to increase worldwide during the coming century. Regional-scale forest die-off influences terrestrial albedo, carbon and water budgets and land-surface energy partitioning. Although increased temperatures during drought were expected to exacerbate tree mortality associated with "global-change-type drought", corresponding changes in vapour pressure deficit (D) have rarely been considered explicitly and have not been disaggregated from that of temperature per se. In this study, they applied a detailed mechanistic soil–plant–atmosphere model to examine the impacts of drought, increased air temperature (+2°C or +5°C) and increased vapour pressure deficit (D; +1 kPa or +2.5 kPa), singly and in combination, on net primary productivity (NPP) and transpiration and forest responses, especially soil moisture content, leaf water potential and stomatal conductance. They showed that increased D exerts a larger detrimental effect on transpiration and NPP than increased temperature alone, with or without the imposition of a three-month drought. Combined with drought, the effect of increased D on NPP was substantially larger than that of drought plus increased temperature. Thus, the number of days when NPP was zero across the two-year simulation was 13 or 14 days in the control and increased temperature scenarios, but increased to approximately 200 days when *D* was increased. Drought alone increased the number of days of zero NPP to 88, but drought plus increased temperature did not increase the number of days. In contrast, drought and increased *D* increased the number of days when NPP=0 to 235 (+1 kPa) or 304 days (+2.5 kPa). They concluded that correct identification of the causes of global change-type mortality events requires explicit consideration of the influence of *D* as well as its interaction with drought and temperature.

Bibliography

Allen C.D., Macalady A.K., Chenchouni H., Bachelet D., McDowell N., Vennetier M., Kizberger T., Rigling A., Breshears D.D., Hogg E.H., Gonzalez P., Fensham R., Zhang Z., Castro J., Demidova N., Lim J.H., Allard G., Running S.W., Semerci A., Cobb N. **2010**. A global overview of drought and heat-induced tree mortality reveals emerging climate change risks for forests. *Forest Ecology and Management* **259**(4):660–684.

Anderegg W.R.L., Kane J.M., Anderegg L.D.L. **2013**. Consequences of widespread tree mortality triggered by drought and temperature stress. *Nature Climate Change* **3**:30–36.

Brando P.M., Balch J.K., Nepstad D.C., Morton D.C., Putz F.E., Coe M.T., Silvério D., Macedo M.N., Davidson E.A., Nóbrega C.C., Alencar A., Soares-Filho B.S. **2014**. Abrupt increases in Amazonian tree mortality due to drought–fire interactions. *Proceedings of the National Academy of Sciences of the United States of America* **111**(17):6347–6352.

Eamus D., Boulain N., Cleverly J., Breshears D.D. **2013**. Global change-type drought-induced tree mortality: vapor pressure deficit is more important than temperature per se in causing decline in tree health. *Ecology and Evolution* **3**(8):2711–2729.

Horsley S.B., Long R.P., Bailey S.W., Hallett R.A., Wargo P.M. **2002**. Health of eastern north american sugar maple forests and factors affecting decline. *Northern Journal of Applied Forestry* **19**(1):34–44.

Lines E.R., Coomes D.A., Purves D.W. **2010**. Influences of forest structure, climate and species. composition on tree mortality across the Eastern US. *Plos One* **5**(10):e13212.

Mezei P., Grodzki W., Blaženec M., Škvarenina J., Brandýsová V., Jaku R. **2014**. Host and site factors affecting tree mortality caused by the spruce bark beetle (*Ips typographus*) in mountainous conditions. *Forest Ecology and Management* **331**(1):196–207.

Nowak D.J., Kuroda M., Crane D.E. **2004**. Tree mortality rates and tree population projections in Baltimore, Maryland, USA. *Urban Forestry and Urban Greening* **2**(3):139–147.

Ostertag R., Silver W.L., Lugo A.E. **2005**. Factors affecting mortality and resistance to damage following hurricanes in a rehabilitated subtropical moist forest. *Biotropica* **37**(1):16–24.

Ruiz-Benito P., Lines E.R., Gómez-Aparicio L., Zavala M.A., Coomes D.A. **2013**. Patterns and drivers of tree mortality in iberian forests: Climatic effects are modified by competition crossmark. *Plos One* **8**(2):e56843.

Suarez M.L., Chermandi L., Ktzerger T. **2004**. Factors predisposing episodic drought-induced tree mortality in *Nothofagus* – site, climatic sensitivity and growth trends. *Journal of Ecology* **92**(6):954–966.

van Mantgem P.J., Stephenson N.L., Byrne J.C., Daniels L.D., Franklin J.F., Fulé P.Z., Harmon M.E., Larson A.J., Smith J.M., Taylor A.H., Veblen T.T. **2009**. Widespread increase of tree mortality rates in the Western United States. *Science* **323**(5913):521–524.

CHAPTER 13

Plant traits related to the productivity of trees

13.1 Background

Studies have been undertaken to identify the plant traits related to the productivity of trees, among which basal diameter of the trunk at breast height, total plant height and canopy-bole cover are important.

13.2 Basal diameter

According to Wikipedia (the free encyclopedia), the diameter at breast height, or DBH, is a standard method of expressing the diameter of the trunk or bole of a standing tree. DBH is one of the most common dendrometric measurements. Electronic calipers can measure diameter at breast height (DBH) and send measured data online via a Bluetooth field computer. Tree trunks are measured at the height of an adult's breast, which is defined differently in different countries and situations.

Ek and Monserud et al. (1979) studied performance and a comparison of stand growth models based on individual trees and diameter-class growth. A distance-dependent individual tree-based growth model (FOREST) was compared with a diameter-class growth model (SHAF) for describing changes in stand density and structure. Projections of the Lake States' northern hardwood stand development were made by each model for 5–26 years over a range of stand conditions and harvest treatments. Results from numerous performance tests and comparisons of actual and predicted diameter distributions, basal areas and numbers of trees, revealed that the individual tree model was considerably more sensitive to harvest treatments and reproduction response than the diameter-class model. Conversely, the latter was much less expensive to operate. Prediction of species and individual tree growth with the individual tree model appeared to provide sensitivity nearly equal to that observed for predictions of the stand as a whole. Long-term projections (120 years) for reserve (no cut) and clear-cut stand conditions recommended the potential and limitations of the models for management analyses.

Leak and Filip (1977) reported that a 38-year period of group selection in northern hardwoods resulted in a permanent composition of one-quarter to one-third intermediate and intolerant species. The diameter distribution follows the inverse J-shaped form that shows uneven-aged stands. Uneven-aged management based

Autoecology and Ecophysiology of Woody Shrubs and Trees: Concepts and Applications, First Edition.
Edited by Ratikanta Maiti, Humbero Gonzalez Rodriguez and Natalya Sergeevna Ivanova.
© 2016 John Wiley & Sons, Ltd. Published 2016 by John Wiley & Sons, Ltd.

on a combination of group and single-tree selection appears to be silviculturally sound.

Borders and Patterson (1990) made a comparison of the Weibull diameter distribution method, a percentile-based projection method and a basal area growth projection method. The first model uses the Weibull probability density function, the second a percentile-based algorithm and the last is a modification of an individual tree distance independent basal area projection model. It was observed that, for most criteria evaluated, the model which makes use of the distance-independent basal area projection algorithm is superior.

Vanclay (1991) developed diameter increment equations for tropical rainforests. Pairwise F-tests offered an efficient approach for aggregating large numbers of species into a manageable number of groups for developing diameter increment functions. The first stage of the two-stage procedure identified the number of groups required and the species defining these groups; the second stage aggregated all the remaining species into the most appropriate group. This approach is readily automated and computationally efficient. An analysis of diameter increments of 237 species from the rainforests of north Queensland revealed 41 species groups, each with increment functions significantly different at $P < 0.01$. These provided a substantially better model than the previous model based on subjectively formed groups.

Zhao et al. (2004) developed individual-tree diameter growth and mortality models for the bottomland mixed-species hardwood stands in the Lower Mississippi Alluvial Valley (LMAV). Data were obtained from 5-year remeasurements of continuous forest inventory plots. Six species groups were formed to diameter structure, tree growth, mortality, recruitment and light demand of species. A 5-year basal area increment model and logistic mortality model was developed for species groups. Potential predictor variables at tree-level and stand-level were selected based on the available data and their biological significance to tree growth and mortality. The developed models possess desirable statistical properties and model behaviors and can be used to update short-term inventory.

Johnson and Abrams (2009) investigated basal area increment trends across age classes for two long-lived tree species in the eastern United States. Basal area increment (BAI) was used to determine forest growth and modeling studies because it provides an accurate quantification of wood production due to the ever-increasing diameter of a growing tree (Rubino and McCarthy, 2000). The BAI growth of individual trees typically follows a sigmoidal pattern: BAI increases rapidly from young to middle age, plateaus and remains level during a protracted period of middle age and then declines as trees become old age (Weiner and Thomas 2001). This BAI trend may be related to an increasing tree canopy during early age, a constant canopy volume during middle age and then a physiological decline in old trees (Spiecker et al., 1996). The sigmoidal growth model using BAI may represent an excellent means for detecting post-European settlement changes in tree growth in eastern North America. BAI is often ignored in dendrochronological studies in lieu of raw or standardised ring width indices (Briffa et al., 1998). In this study, they took a unique approach of examining BAI changes over time across all age classes from young to old trees, rather than just studying the oldest individuals of each species where a sample bias may exist (Cherubini et al., 1998; Voelker et al., 2006). Two tree species were chosen for the study: hemlock (*Tsuga canadensis*) and blackgum (*Nyssa sylvatica*). The primary objectives of the study were to examine tree growth rate (including BAI) variation from young to old age classes for contrasting species and the relationship between tree growth rate and maximum longevity for species and for individuals.

Poage and Tappeiner (2002) studied long-term patterns of diameter and basal area growth of old-growth Douglas-fir trees in western Oregon. Diameter growth and age data collected from stumps of 505 recently cut old-growth Douglas-fir (*Pseudotsuga menziesii* (Mirb.) Franco) trees at 28 sample locations in western Oregon (USA) revealed that rapid early and sustained growth of old Douglas-fir trees were extremely important in terms of attaining large diameters at ages 100–300 years. The diameters of the trees at ages 100–300 years (D100–D300) showed a highly significant positive correlation and related to their diameters and basal area growth rates at age 50 years. Average periodic basal area increments (PAI_{BA}) of all trees increased for the first 30–40 years and then plateaued, remaining relatively high and constant from age 50 to 300 years. Average PAI_{BA} of the largest trees at ages 100–300 years were significantly greater by age 20 years than were those of smaller trees at ages 100–300 years. The site factors (province,

site class, slope, aspect, elevation and establishment year) accounted for little of the variation observed in basal area growth at age 50 years and D100–D300. The mean age range for old-growth Douglas-fir at the sample locations was wide (174 years). The hypothesis that large-diameter old-growth Douglas-fir developed at low stand densities was supported by these observations.

Cao (2000) made prediction of annual diameter growth and survival for individual trees from periodic measurements. It is difficult to fit annual tree survival and diameter growth models to data that were measured, not every year, but at some interval. This study was undertaken to determine suitable methods to obtain parameter estimates for such a system from periodic measurements. Given a system consisting of a tree survival model and a tree diameter growth model, the result presented an iterative method for estimating system parameters. The method involves sequentially updating the parameters of both models within the growth period. Data from 111 plots from the Southwide Seed Source Study of loblolly pine (*Pinus taeda* L.) were then utilised to evaluate this iterative approach against the averaging method that assumes a constant tree survival probability and diameter growth rate during the remeasurement interval. Results revealed that the iterative method out-performed the averaging method in predicting future individual tree survival, diameter growth and stand basal area. The iterative method was superior because it accounted for the variable rate of diameter growth and tree survival probability as functions of ever-changing stand and tree attributes.

According to Wikipedia, the basal area is the area of a given section of land that is occupied by the cross-section of tree trunks and stems at their base. The term is used in forest management and forest ecology.

In most countries, this is usually a measurement taken at the diameter at breast height (1.3 m or 4.5 ft) of a tree above the ground and includes the complete diameter of every tree, including the bark. Measurements are usually made for a plot and this is then scaled up for 1 ha of land for comparison purposes to examine a forest's productivity and growth rate.

To estimate a tree's basal area BA, use the tree's diameter at breast height DBH in inches with the following formula:

$$BA = \frac{\pi \times (DBH/2)^2}{144}$$

(Note: The factor of 144 is there to convert from square inches to square feet.)

This formula simplifies to:

BA = $0.005454 \times DBH^2$.

The result will be in square feet.

For the DBH in cm use:

BA = $0.00007854 \times DBH^2$.

The result will be in square metres.

The basal area of a forest stand can be found by adding the basal areas (as calculated above) of all of the trees in an area and dividing by the area of land in which the trees were measured. Basal area is generally expressed as ft^2/acre or m^2/ha.

A wedge prism can be used to quickly estimate the basal area per hectare. To find basal area using this method, simply multiply the basal area factor (BAF) by the number of "in" trees in the variable radius plot. The BAF will vary based on the prism used, common BAFs include 5/8/10 and all "in" trees are those trees, when viewed through a prism from plot centre, that appear to be in-line with the standing tree on the outside of the prism.

13.2.1 Worked example

If yu carry out a survey using Angle Count Sampling (wedge prism) and it is necessary to select a basal area factor (BAF) of four and you measured the diameter breast height (DBH) of your first tree as 14 cm, then the standard way of calculating how much of 1 ha was covered by tree area (scaling up from that tree to the hectare) would be: $4/[(DBH + 0.5)^2 \times \pi/4)]$.

- BAF would be 4 – this is the BAF selected for that sampling technique.
- DBH would be 14 – this uses an assumed diameter, when actually used it is the radius perpendicular to the tangent line.
- The +0.5 allows under and over measurement to be accounted for.
- The $\pi/4$ converts the rest to the area.

Finally, to see the figure in m^2, it needs to be multiplied by 10 000. In this case, this means in every hectare there are 242 m^2 of tree area, according to this sampled tree being taken as representative of all the unmeasured trees.

13.3 Plant height

King (1981) measured tree dimensions and the rate of height growth in dense stands. To determine the effect of tree dimensions on the rate of height growth a model was constructed relating the tree weight to total height and R, the ratio of crown weight to trunk weight. The model was constructed on the the assumption that the trunk buckling safety factor is constant. If trees also maintain a constant R as they grow, then the rate of height growth is maximised by $R = 0.17$. In addition, the height growth rate increases as the buckling safety factor decreases. These predictions of optimal form for height growth are appropriate for shade-intolerant, successional species growing in dense stands. Dimensional measurements of self-thinning *Populus tremuloides* indicate near optimal dimensions for height growth. Trees ranging from seven to 19 m in height had trunks which were only 50% thicker than the minimum required to prevent them from buckling under their own weight and had a mean R of 0.13. This ratio of crown weight to trunk weight is significantly lower than the optimal value, but the predicted height growth rate for $R = 0.13$ is 99% of that predicted for $R = 0.17$.

Biging and Matthia (1992) made a comparison of distance-dependent competition measures for height and basal area of individual conifer trees The effect of competition on height and diameter^2 growth of individual conifer trees was examined. In this paper, they modelled growth of individual trees as a product of potential growth reduced by competition. They also investigated a number of competition indices that incorporated tree sizes and distances from neighbours, evaluated over varying competition zones. For these indices, the reduction in mean square error relative to the no competition index was used to judge performance. The performance of these indices varied by species and growth component (height or diameter^2 growth), but performance was generally better for tolerant white fir than for intolerant ponderosa pine. Additional competition measures that incorporated estimated crown parameters were developed and shown to improve on many of the traditional competition measures, especially for ponderosa pine and Douglas-fir. For white fir, basal area performed well, but not as well as the individual tree competition indices that incorporate estimated crown parameters. They also found that expanding the search zone did not improve our ability to estimate competitive effects. Choosing competitors with a height angle gauge was generally superior to selecting them with a DBH angle gauge. Results were summarised by species, type of competition index and competition evaluation zone.

Bradshaw and Stettler (1995) have mapped quantitative trait loci (QTLs) for commercially important traits (stem growth and form) and an adaptive trait (spring leaf flush) in a *Populus* F_2 generation derived from a cross between interspecific F_1 hybrids (*P. trichocarpa* × *P. deltoides*). Phenotypic data were collected over a tw-year period from a replicated clonal trial containing ramets of the parental, F_1 and F_2 trees. Contrary to the assumptions of simple polygenic models of quantitative trait inheritance, one to five QTLs of large effect are responsible for a large portion of the genetic variance in each of the traits measured. For example, 44.7% of the genetic variance in stem volume after two years of growth is controlled by just two QTLs. QTLs governing stem basal area were found clustered with QTLs for sylleptic branch leaf area, sharing similar chromosomal position and mode of action and suggesting a pleiotropic effect of QTLs ultimately responsible for stem diameter growth.

13.4 Bole diameter growth of trees

Bole-canopy diameter, canopy cover, is considered as a parameter related to the the productivity of trees. Brunner and Nigh (2000) studied light absorption and bole volume growth of individual Douglas-fir trees. Empirical growth and yield models for forest management were developed toward individual-tree models that are capable of simulating the growth of mixed and uneven-aged stands. Spatially explicit (i.e., distance-dependent) models usually modified the growth of trees by means of competition indices. They used tree growth data from an even-aged, unthinned, 50-year-old Douglas-fir [*Pseudotsuga menziesii* (Mirb.) Franco] stand in British Columbia to test the hypothesis that the amount of absorbed light is a good predictor of diameter at breast height, height and bole volume growth of an individual tree. They also explored the relationships between these variables. A spatially light model was developed to simulate photosynthetically active radiation absorbed by individual trees during a growth period (APAR) based on detailed canopy architecture information. For the purpose, they used a weighted leaf area (WLA) that was linearly related to APAR. Because of the integration of light absorption by a tree crown, estimates of WLA were highly correlated with leaf area for dominant trees. It was assessed that leaf area was a poor estimator of WLA. The relationship between WLA and bole volume growth was non-linear, revealing a higher light-use efficiency in suppressed trees than in dominant trees. This relationship was strong enough to be useful for growth modelling. Only height growth of suppressed trees was affected by WLA. They concluded that single-tree WLA can be used as a process-oriented competition index in growth models for forest management.

Troxel et al. (2013) studied relationships between bole and crown size for young urban trees in the north-eastern USA. Knowledge of allometric equations can enable urban forest managers to meet desired economic, social and ecological goals. This study was to address this research gap and examine interactions between age, bole size and crown dimensions of young urban trees in New Haven (CT, USA) to identify allometric relationships and generate predictive growth equations useful for the region. They studied the 10 most common species from a census of 1474 community planted trees (ages 4–16). Regressions were analysed to relate diameter at breast height (dbh), age (years since transplanting), tree height, crown diameter and crown volume. Across all 10 species each allometric relationship was statistically ($p < 0.001$) significant at a level of 0.05. Consistently, shade trees demonstrated stronger relationships than ornamental trees. Crown diameter and dbh showed the strongest fit with eight of the 10 species having an $R^2 > 0.70$. Crown volume gave a good fit for each of the shade tree species ($R^2 > 0.85$), while the coefficients of determination for the ornamentals varied ($0.38 < R^2 < 0.73$). In the model predicting height from dbh, ornamentals displayed the lowest R^2 ($0.33 < R^2 < 0.55$) while shade trees represented a much better fit ($R^2 > 0.66$). These correlations will better equip forest managers to predict the growth of urban trees, thereby improving the management and maintenance of New England's urban forests.

Foli et al. (2003) developed a model for growing space requirement for some tropical tree specied. They reported crown diameter–bole diameter relationships for five mixed tropical species by regression analysis. The regression explained 77% of the variation in crown diameter growing space is associated with crown size.

13.5   Regeneration

After attaining maturity, each species dispersed seeds on soil which are exposed to the climatic conditions (or are consumed by birds) and, with the advent of favourable conditions, the seeds start germination and seedlings emerge and establish in the forest soil. This phase is called regeneration. Studies have been undertaken on this aspect.

Guariguata and Pinard (1998) discussed our ecological knowledge of regeneration from seed in neotropical forest trees, with implications for natural forest management. They discussed the main ecological factors that influence tree recruitment in neotropical moist and wet forests within the context of timber management based on selective logging. They observed that setting aside protection areas in managed forests as a way to preserve ecological processes may not be sufficient to ensure sustainable levels of tree regeneration and that a thorough understanding and application of tree seed ecology can help to refine management prescriptions. They made a review on relevant aspects of tree reproductive biology, seed production and

dispersal, spatial and temporal constraints on seed availability, disperser behaviour and the potential consequences of hunting and forest fragmentation on tree regeneration; and they discussed their implications for biological sustainability in managed forests. Tree seed production is influenced by the selective removal of neighbours of the same species (due to insufficient pollen transfer), flowering asynchrony and attributes of the species' sexual system. The extent to which an area is supplied by seed can affect dispersal mechanism, spatio-temporal limitations to seed dispersal and tree size-dependent levels of seed production at the species level. Studies of vertebrate-disperser behaviour and tree seed deposition in logged forests are rare. Therefore further attention should be given to define our understanding of the dependency of sustained timber production on vertebrate fauna. Although much remains to be learned about tree seed ecology in neotropical logged forests, the baseline information presented in this study may offer a starting point for developing ecological criteria for seed tree retention. Furthermore, it may contribute in improving ecologically-based management prescriptions in order to enhance or at least maintain sufficient levels of natural regeneration without the need to rely on artificial regeneration.

Brokaw (1985) described gap-phase regeneration of trees for the first five or six years of regrowth in 30 treefall gaps (20–705 m^2) in tropical moist forest on Barro Colorado Island, Panama. Trees were classified as pioneers (saplings found only in gaps) or primary species (saplings found in gaps and in the understory of mature forest). In most of the gaps studied, stem densities rose rapidly after gap formation, then levelled off or declined by years 3–6. This pattern was particularly marked in some large gaps (>150 m^2), where pioneers attained high densities, then experienced heavy mortality. Stem density of primary species did not vary with gap size. In large gaps the mean rate of growth in height was greater for pioneers than for primary species, size-class distribution broadened more for pioneers than for primary species and early recruits of both regeneration types grew faster than later ones. Gap formation fosters regeneration of pioneer and primary species and, in this forest, produces patches that differ markedly in tree population dynamics, species composition and growth rate.

Hawthorne (1985) discussed eological profiles of Ghanaian forest trees. Textual summaries are presented of the ecology of most forest trees which in Ghana attain 5 cm diameter at breast height. For rarer or smaller trees, the text is reduced to little more than name and brief ecological group, but for others information is presented under various headings relating to tree distribution and regeneration ecology, taxonomy and miscellaneous other notes. Some 1260 charts are included, summarising aspects of the ecology in Ghana of 210 of the most important species. Five charts summarise abundance across Ghana's forest zone, stratified into 21 broad landscape and forest type categories. A chart summarises trends of crown exposure for each of the 210 important species, potentially of value as objective indices of the extent to which each species is a light-demander or shade-bearer and how this changes with tree size.

Anderson (1990) mentioned alternatives to deforestation: steps toward sustainable use of the Amazon rainforest. Some of the possible sustainable uses of the Amazon rainforest in Latin America are explored. This book comprises a number of essays, drawn from an international conference held in Belem, Brazil, that was attended by scientists and policy makers. The essays present innovative approaches and technologies that will permit simultaneous use and conservation of the rainforest and that will benefit the population of the Amazon rainforest as a whole, rather than just a small rural minority. The 17 articles are arranged under five headings: background, natural forest management, agroforestry, landscape recovery and implications for regional development.

Alexander et al. (1992) discussed the role of mycorrhizas in the regeneration of some Malaysian forest trees. An investigation was made into the availability of mycorrhizal inoculum and the response of tree seedlings to mycorrhizal infection in West Malaysian forests. Spores of vesicular arbuscular (VA) mycorrhizal fungi in the soil were reduced by 25% after selective logging and by 75% after heavy logging. VA infection in the roots of plants persisting on, or colonising, a heavily logged site was reduced by up to 75%. The most probable number (MPN) of VA propagules in sieved soil was up to ten times greater than spore density, but was also greatly reduced by heavy logging. This resulted in reduced infectivity of soil from the heavily logged site, as demonstrated by reduced VA infection of bioassay plants. The infectivity of soil declined following sun-drying, but sun-dried soil devoid of vegetation retained some infectivity even after 12 months storage. Overall the data suggest that root and hyphal fragments are more important than spores as inoculum in disturbed forest and that, in undisturbed forest, living roots and hyphae are likely to be important sources of infection. In a pot experiment, shoot growth of two test species, *Albizia falcataria* (L.) Becker and *Parkia speciosa* Hassk. responded more to VA mycorrhizal infection than to P fertilisation over the range 0–6 g triple superphosphate per 8 kg of soil. The response to inoculation with a cocktail of "introduced" VA fungi propagated in pot cultures was greater than the response to inoculation with "indigenous" fungi propagated in pot cultures from roots and soil collected in undisturbed forests. Another test species, *Intsia palembanica* Miq., also responded better to mycorrhizal infection than to P fertilisation and better to VA mycorrhizal infection than to ecto-mycorrhizal infection. *Intsia palembanica* seedlings growing around mature diptero-carps quickly became ectomycorrhizal, suggesting that at least some ectomyc-orrhizal fungi infect both dipterocarps and *Intsia*. *Shorea leprosula* Miq. seedlings growing naturally in the forest had ecto-mycorrhizas 20 days after germination, that is before they had true leaves; and within seven months they supported up to 11 different ectomycorrhizal fungi. However, seedlings isolated from contact with the roots of mature *Shorea* trees remained uninfected in the field for up to six months. This shows the importance of contact with living ectomycorrhizal roots for early infection of dipterocarp seedlings, a point which should be recognised in logging operations and forest regeneration programmes.

To assess how the decimation of large vertebrates by hunting alters recruitment processes in a tropical forest, Terborgh et al. (2008) compared the sapling cohorts of two structurally and compositionally similar forests in the Rio Manu floodplain in south-eastern Peru. Large vertebrates were severely depleted at one site, Boca Manu (BM), whereas the other, Cocha Cashu Biological Station (CC), supported an intact fauna. At both sites we sampled small (≥ 1 cm tall, < 1 cm dbh) and large (≥ 1 m and < 10 cm dbh) saplings in the central portion of 4-ha plots within which all trees ≥ 10 cm dbh were mapped and identified. This design ensured that all conspecific adults within at least 50 m (BM) or 55 m (CC) of any sapling would have known locations. They used the Janzen–Connell model to make five pre-dictions about the sapling cohorts at BM with respect to CC: (a) reduced over-all sapling recruitment, (b) increased recruitment of species dispersed by abiotic means, (c) altered relative abundance of species, (d) prominence of large-seeded species among those showing depressed recruitment and (e) little or no ten-dency for saplings to cluster closer to adults at BM. Their results affirmed each of these predictions. Interpreted at face value, the evidence suggested that few species were demographically stable at BM and that up to 28% were increasing and 72% decreasing. Loss of dispersal function allows species dis-persed abiotically and by small birds and mammals to substitute for those dis-persed by large birds and mammals. Their results suggested that the best and per-haps only way to prevent compositional change and probable loss of diversity in tropical tree communities is to prohibit hunting.

Bibliography

Alexander I., Ahmad N., See L.S. **1992**. The role of mycorrhizas in the regeneration of some Malaysian forest trees. *Philosophical Transactions Biological Sciences* **335**(1275):379–388.

Anderson A.B. **1990**. *Alternatives to deforestation: steps toward sustainable use of the Amazon rain forest.* Springer, Berlin, 293 pp. ISBN 0-231-06892-1.

Biging G.S., Matthia D. **1992**. A comparison of distance-dependent competition measres for height and basal area of individual conifer trees. *Forest Science* **38**(3):695–720.

Borders B.E., Patterson W.D. **1990**. Projecting stand tables: a comparison of the Weibull diameter distribution method, a percentile-based projection method and a basal area growth projection method. *Forest Science* **36**(2):413–424.

Bradshaw D. Jr, Stettler R.F. **1995**. Molecular genetics of growth and development in populus. IV. Mapping QTLs with large effects on growth, form and phenology traits in a forest tree. *Genetics* **139**(2):963–973.

Briffa K.R., Schweingruber F.H., Jones P.D., Osborn T.J., Harris I.C., Shiyatov S.G., Vaganov E.A., Grudd H., Cowie J. **1998**. Trees tell of past climates: But are they speaking less clearly today? [and Discussion]. *Philosophical Transactions: Biological Sciences* **353**:65–73.

Brokaw N.V.L. **1985**. Gap-phase regeneration in a tropical forest. *Ecology* **66**(3):682–687.

Brunner A., Nigh G. **2000**. Light absorption and bole volume growth of individual Douglas-fir trees. *Tree Physiology* **20**(5/6):323–332.

Cao V. **2000**. Prediction of annual diameter growth and survival for individual trees from periodic measurements. *Forest Science* **46**(1):127–131.

Cherubini P., Dobbertin M., Innes J.L. **1998**. Potential sampling bias in long-term forest growth trends reconstructed from tree-rings: A case study from the Italian Alps. *Forest Ecology and Management* **109**:103–118.

Ek A.R., Monserud R.A. **1979**. Performance and comparison of stand growth models based on individual tree and diameter-class growth. *Canadian Journal of Forest Research* **9**(2):231–244.

Foli E.G., Alder D., Miller H.G., Swaine M.D. **2003**. Modelling growing space requirement for some tropical tree specied. *Forest Ecology and Management* **173**(1/3):79–88.

Guariguata M.R., Pinard M.A. **1998**. Ecological knowledge of regeneration from seed in neotropical forest trees: Implications for natural forest management. *Forest Ecology and Management* **112**(1/2):87–99.

Hawthorne W.D. **1985**. Ecological profiles of Ghanaian forest trees. *Tropical Forestry* **29**, 345 pp.

Johnson S.E., Abrams M.D. **2009**. Basal area increment trends across age classes for two long-lived tree species in the eastern U.S. In: Kaczka R, Malik I, Owczarek P, Gärtner H, Helle G, Heinrich I (eds.). *TRACE – Tree Rings in Archaeology, Climatology and Ecology*, Vol. **7**. GFZ Potsdam, Scientific Technical Report STR 09/03, Potsdam, 226 pp.

King D. **1981**. Tree dimensions: maximizing the rate of height growth in dense stands. *Oecologia* **51**(3):351–356.

Leak W.B., Filip S.M. **1977**. Thirty-eight years of group selection in New England northern hardwoods. *Journal of Forestry* **75**(10):641–643.

Poage N.J., Tappeiner J.C. **2002**. Long-term patterns of diameter and basal area growth of old-growth Douglas-fir trees in western Oregon. *Canadian Journal of Forest Research* **32**(7):1232–1243.

Rubino D.L., McMarthy B.C. **2000**. Dendroclimatological analysis of white oak (*Quercus alba* L., Fagaceae) from an old-growth forest of southeastern Ohio, USA. *Journal of the Torrey Botanical Society* **127**:240–250.

Spiecker H., Mielikaeinen K., Kohl M., Skovsgaard P. (eds) **1996**. *Growth Trends in European Forests: Studies from 12 Countries.* European

Forest Institute Research Report No. 5. Springer. Berlin.

Terborgh T., Nuñez-Iturri G., Pitman N.C.A., Cornejo Valverde F.H., Alvarez P., Swamy V., Pringle E.G., Timothy Paine C.E. **2008**. Tree recruitment in an empty forest. *Ecology* **89**:1757–1768.

Troxel B., Piana M., Ashton M.S., Murphy-Dunning C. **2013**. Relationships between bole and crown size for young urban trees in the northeastern USA. *Urban Forestry and Urban Greening* **12**:144–153.

Vanclay J.K. **1991**. Aggregating tree species to develop diameter increment equations for tropical rainforests. *Forest Ecology and Management* **42**(3/4):143–168.

Voelker S.L., Muzika R., Guyette R.P., Stambaugh M.C. **2006**. Historical CO_2 growth enhancement declines with age in *Quercus* and *Pinus*. *Ecological Monographs* **76**:549–564.

Weiner J., Thomas S.C. **2001**. The nature of tree growth and the age-related decline in forest productivity. *Oikos* **94**:374–376.

Zhao D., Borders B., Wilson M. **2004**. Individual-tree diameter growth and mortality models for bottomland mixed-species hardwood stands in the lower Mississippi alluvial valley. *Forest Ecology and Management* **199**(2/3): 307–322.

PART II

CHAPTER 14

Ecophysiology

14.1 Background

Plants in the forests are our life-savers. They protect us from heavy pollution caused by emissions of CO_2 from the combustion of fossil fuels and human activities like incessant logging and expansion of agriculture. Plants absorb this toxic gas, CO_2, through stomata and use it in the process of photosynthesis by capturing solar radiation. They store the energy as potential energy in plant biomass, wood and forest products and at the same time release oxygen which we use for our respiration. The forests give us various products for human use and feed for animals. We should protect forests for the security of our lives. A vast array of ecophysiological activities occur in the forest ecosystem and adapt to the varying environmental conditions. For maintaining forest health we should understand and study the various ecophysiological processes of plant species and their responses to varying environmental and edaphic conditions necessary for their adaptation to these environments. Here, we mention research advances in these processes.

Ecophysiology dealing with environmental physiology or physiological ecology is a biological discipline that studies the adaptation of plant physiology to environmental conditions. It is closely related to comparative physiology and evolutionary physiology. The ecophysiology of tree growth can be defined in terms of the increase in size for an individual or a stand. Growth is generally expressed as a change in size per unit time and area. The growth of trees is influenced by several physiological traits and environmental conditions in the forest ecosystem.

Plant ecophysiology deals largely with two mechanisms:

The sensitivity and response of plants to environmental change require integrated responses to highly variable conditions. In many cases, animals are able to escape unfavourable and changing environmental factors, such as heat, cold, drought or floods, but plants are unable to move away and, therefore, must endure the adverse conditions or perish. Plants are therefore phenotypically plastic and have an impressive array of genes which aid in adapting to changing conditions. It is hypothesised that this large number of genes can be partly explained by the plant species' need to adapt to a wider range of conditions.

The following narrates in brief some studies on the ecophysiology of trees in different ecological regions of the world.

Autoecology and Ecophysiology of Woody Shrubs and Trees: Concepts and Applications, First Edition.
Edited by Ratikanta Maiti, Humbero Gonzalez Rodriguez and Natalya Sergeevna Ivanova.
© 2016 John Wiley & Sons, Ltd. Published 2016 by John Wiley & Sons, Ltd.

14.2 Tropical rainforest

The productivity of tropical forests is affected by several climatic and environmental factors, such as heavy rainfall, cyclonic weather, high-velocity winds, soil erosion and several edaphic factors.

A study was undertaken on the relative importance of photosynthetic traits and allocation patterns as correlates of seedling shade tolerance of 13 tropical trees. It was important to compare traits for plants at the same developmental stage in full sun, regressed against the first-year mortality rates of seedlings in shade (Kitajima, 1994).

Leaf nitrogen influences photosynthesis. In this respect, Ellsworth and Reich (1996) studied photosynthesis and leaf nitrogen in five Amazonian tree species during early secondary succession field measurements of maximum net photosynthesis (P_{max}), leaf nitrogen (N) content (leaf N per area and N %), and specific leaf area (SLA) for Amazonian tree species within and across early successional sites of known ages after abandonment from slash and burn agriculture. They investigated five species across a successional sere near San Carlos de Rio Negro, Venezuela, to test whether plasticity was associated with successional status and to determine whether changes in foliar properties during secondary succession can be attributed to shifts either in species composition, in resource availability, or in both. Average leaf N concentration was high (nearly 3%) for a pioneer species (*Cecropia ficifolia*) early in succession (1–3 years after abandonment) but was always lower for the other early and mid to late succession species, especially later in succession (1–2% at 5–10 years

after abandonment). Net photosynthetic capacity (P_{max}/area and P_{max}/mass) showed variations up to as much as sixfold, being higher in pioneer species such as *Cecropia* and *Vismia* and lower in late successional species such as *Miconia* and *Licania* on 10-year abandoned agricultural sites. Total daily light availability also varied widely (14-fold) from its peak one year after farm abandonment to low levels nine years into succession. During the first five years of secondary succession, there were significant ($P < 0.05$) differences in P_{max} and leaf N concentration among species in any given year. In most species, P_{max} values decreased with increasing time. Large differences were observed between species in photosynthetic plasticity among species: P_{max} tended to be much greater in earlier than later successional species soon after abandonment. Early successional species showed strong ($r^2 \geq 0.57$, $P < 0.0001$) mass-based photosynthesis-N relationships but weak ($r^2 = 0.40$ or lower, $P < 0.0001$) area-based relationships both across the secondary successional sere after agriculture and across sites varying in types of disturbance. Both mass- and area-based photosynthesis–N relationships were poorer or not significant ($P > 0.05$) for mid- to late successional species. The results suggest that early and late successional species may differ.

Reich et al. (1997) studied global convergence in plant functioning from tropics to tundra. Despite striking differences in climate, soils and evolutionary history among diverse biomes ranging from tropical and temperate forests to alpine tundra and desert, they observed similar interspecific relationships among leaf structure and function and plant growth in all biomes.

Their results demonstrated convergent evolution and global generality in plant functioning, in spite of the enormous diversity of plant species and biomes. For 280 plant species from two global data sets, they observed that potential carbon gain (photosynthesis) and carbon loss (respiration) increase in similar proportion with decreasing leaf life-span, increasing leaf nitrogen concentration and increasing leaf surface area to mass ratio. The productivity of individual plants and of leaves in vegetation canopies also changes in constant proportion to leaf life-span and surface area to mass ratio. These global plant functional relationships have significant implications for global-scale modelling of vegetation–atmosphere CO_2 exchange in the mode and degree of leaf-level physiological plasticity across succession.

Light intensity influences seedling growth. In this respect, Poorter (1999) compared the growth of seedlings of 15 rainforest tree species under controlled conditions, at six different light levels (3, 6, 12, 25, 50 and 100% daylight). It was observed that most plant variables showed strong ontogenetic changes; and they were highly dependent on the biomass of the plant. Growth rate was highest at intermediate light levels (25–50%), above which it declined. Most plant variables showed a curvilinear response to irradiance, with the largest changes at the lowest light levels. It was observed that plants which were fast-growing in a low-light environment were also fast-growing in a high-light environment. At low light, interspecific variation in relative growth rate was measured mainly by differences in a morphological trait, the leaf area ratio (LAR), whereas at high light it was determined mainly by differences in a physiological trait. The net assimilation rate (NAR) was a stronger determinant of growth than LAR in more than 10–15% daylight. As light availability in the forest is generally much lower than this threshold level, it demonstrates that interspecific variation in growth in a forest environment is mainly due to variations in morphology.

Light environment influences tree growth. Poorter (2001) investigated light-dependent changes in biomass allocation and their importance for the growth of rainforest tree species They determined the sapling growth of six rainforest tree species in order to know whether species respond in a similar way to a natural light gradient. Seedling height growth showed a positive relation with light environment and leaf area. A single descriptor of light environment explained sapling growth best. Direct or diffuse light could induce plant growth, depending on species. Sapling relative growth rate increased with irradiance, mainly owing to an increase in net assimilation rate. On a shoot basis, shaded plants had a smaller leaf mass fraction (LMF) and a larger specific leaf area, giving leaf area ratios (LAR) similar to those of sun plants. This is in contrast with the results of seedling studies under controlled conditions, where LMF and LAR increased with shade. Biomass partitioning to leaf growth decreased with irradiance and relative growth rate of the sapling. This leaf partitioning ratio showed better correlation with RGR than with irradiance. Species showed variation in the effect of light-dependent changes in specific leaf area (SLA) on growth. The effect of SLA was not related to the shade tolerance of the species.

It may be concluded that the varying climatic and environmental conditions, such as the light intensity prevailing in a tropical forest environment, influence the photosynthesis, growth and other physiological traits of trees showing variability among species.

14.3 Temperate forest

Tree species show variation in their adaptation to temperate environments. Low temperature and drought are limiting factors in tree growth.

Abrams (1994) observed genotypic and phenotypic variations in the adaptations of temperate tree species. Species that prevail in large geographic ranges or a variety of habitats within a limited area deal with contrasting environmental conditions through genotypic and phenotypic variations. Abrams (1994) studied these forms of ecophysiological variation in temperate tree species in eastern North America by means of a series of field and greenhouse experiments, including controlled studies with *Cercis canadensis*, *Fraxinus pennsylvanica*, *Acer rubrum* and *Quercus rubra*, in relation to drought stress. The author took measurements of gas exchange, tissue water relations and leaf morphology; and he identified genotypic variations at the biome and individual community levels. Xeric genotypes generally showed higher net photosynthesis and leaf conductance and lower osmotic and water potentials at incipient wilting during drought than mesic genotypes. Xeric genotypes also produced leaves with greater thickness, leaf mass per area and stomatal density and smaller area than the mesic genotypes,

demonstrating a general coordination between leaf morphology, gas exchange and tissue water relations. Leaf phenotypic plasticity was observed in different light environment in every species, in a wide array of ecological tolerances. In one study on the interactions of genotypes with environment, shade plants showed an osmotic adjustment during drought and shade plants had smaller reductions in photosynthesis with decreasing leaf water potential. In this study, sun (but not shade) plants showed significant genotypic differences in leaf structure but, with certain variables, phenotypic variation exceeded genotype variation. Thus, genotypic variation was not reflected in all phenotypes; and some phenotypes responded differentially to stress. Overall, these studies indicate the importance of genotypic and phenotypic variation as stress adaptations in temperate tree species among both distant and nearby sites of contrasting environmental conditions.

Theodose and Bowman (1997) investigated the long-term (five year) responses of plant absolute abundance and species diversity to N, P, and N + P fertilisation in two sedge-dominated alpine plant communities that differed in soil resource availability but not in macroclimate: a resource-poor dry meadow and a more resource-rich wet meadow. Prior to analysis, species were grouped into functional groups based on growth form, potential developmental constraints and presence or absence of mutualisms. Absolute abundance changes of functional groups were more prominent in the dry-meadow community than in the wet-meadow community. In the dry meadow, mycorrhizal forbs increased with N+P fertilisation, non-mycorrhizal forbs increased with P and N+P and N_2-fixing forbs increased with P alone. Grasses increased with N and N+P whereas sedges, the dominant functional group, were not affected by fertilisation. In the wet meadow, the dominant sedges showed abundance increases in response to N, whereas grasses increased with P and N+P. Wet-meadow forb abundance was not significantly influenced by fertilisation. Shifts in species relative abundances led to an increase in species diversity following N+P fertilisation in the dry meadow and a decrease in species diversity following N+P fertilisation in the wet meadow. This study therefore showed a direct comparison of diversity responses between communities that differed primarily in soil resource availability, substantiating the theory that plant species diversity is greatest under intermediate levels of fertility.

With predictable abiotic stress and a meta-analysis of field results in arid environments, Maestre et al. (2005) investigated the change of plant–plant interactions. They performed a quantitative meta-analysis of field and common garden studies to evaluate the effect of abiotic stress (low vs high) on the net outcome of plant–plant interactions in arid and semiarid environments. Thus, they evaluated the degree of empirical support for these models. The analyses revealed that both the selection of the estimator of plant performance and the experimental approach followed have a strong influence on both the net outcome of plant–plant interactions and the effect of abiotic stress on such an outcome. Density data demonstrated that the net effect of neighbours was positive and negative, respectively, at low and high abiotic stress levels whereas

other estimators suggested that the net effect of neighbours did not differ with stress level.

Bréda et al. (2006) made a review of ecophysiological responses, adaptation processes and long-term consequences of temperature in forest trees and stands under severe drought. The extreme drought occurring in Western Europe during 2003 emphasised the need to understand the key processes that may allow trees and stands to overcome such severe water shortages. They therefore reviewed the current knowledge available about such processes, such as the impact of drought on exchanges at soil–root and canopy–atmosphere interfaces, using water and CO_2 flux measurements. The decrease in transpiration and water uptake and in net carbon assimilation due to stomatal closure was quantified and modelled. Models were developed to compute water balance at stand level, based on the 2003 climate in nine European forest sites from the CARBOEUROPE network. Then, they reviewed the irreversible damage that could be imposed on water transfer within trees and particularly within xylem. A special emphasis was given to the inter-specific variability of these properties in a wide range of tree species. They also discussed inter-specific diversity of hydraulic and stomatal responses to soil water deficit as it might reflect a large diversity in traits potentially related to drought tolerance. Finally, tree decline and mortality due to recurrent or extreme drought events were discussed on the basis of a literature review and recent decline studies. They also discussed potential involvement of hydraulic dysfunctions or deficits in carbon storage as causes for the observed long-term (several years) decline in tree growth and development and for the onset of tree dieback.

Gebrekirstos et al. (2006) studied the adaptation of five co-occurring tree and shrub species to water stress and its implication in the restoration of degraded land. The open savannah woodlands in Ethiopia are distributed over very large areas of the Rift valley in southern and eastern parts of the country and have suffered from deforestation due to excessive tree cutting and overgrazing. This research was undertaken: (a) to assess the effect of disturbing the water status of plants and their respective sites, (b) to explain the differences between species in their survival and distribution and (c) to determine their relative suitability for the restoration of degraded lands. The plant water potentials of five common co-occurring species were estimated in the field at midday and predawn in the dry season. The water potential values of the species at sites with different biophysical settings and diurnal differences showed significant differences between the species. *Balanites aegyptiaca* and *Dichro stachyscinerea*, which showed low midday (−3.05 to −4.85 MPa), predawn (−1.98 to −3 MPa) and wide diurnal (1.16−2.25 MPa) plant water potential ranges had wider capacities to withstand water changes and, hence, can be considered as suitable candidates for reforestation in drought-prone areas. Though *Acacia senegal* and *Acacia seyal* had the highest midday (−1.55 to −2.68 MPa), predawn (−0.83 to −2.09 MPa) and narrow diurnal (0.32−1.1 MPa) plant water potential ranges and could be considered as drought avoider (sensitive) species, they

had a combination of mechanisms to avoid drought and grow in dry areas.

Mycorrhizas in roots improve plant growth. A study was made by Barea et al. (2011) on the ecological and functional roles of mycorrhizas in semiarid ecosystems of South-east Spain. Mycorrhizal symbioses in higher plants' root regions improve the resilience of plant communities against environment stresses, including nutrient deficiency, drought and soil disturbance. These investigations are reviewed here in terms of: (a) analysing the diversity of mycorrhizal fungi, (b) assessing the ecological and functional interactions among plant communities and their associated mycorrhizal fungal populations and (c) using mycorrhizal inoculation technology for the restoration of degraded semiarid areas in South-east Spain. Disturbance of the target semiarid ecosystems decreases the density and diversity of mycorrhizal populations. However, the mycorrhizal propagules do not disappear completely, suggesting a certain degree of stress adaptation; and these remaining, resilient ecotypes are being used as plant inoculants. Numerous field experiments have been carried out in revegetation projects in the semiarid Iberian South-east, using plant species from a natural succession inoculated with a community of indigenous mycorrhizal fungi. This management strategy improved both plant development and soil quality and is a successful biotechnological tool to aid the restoration of self-sustaining ecosystems.

In conclusion, several studies have been undertaken on various physiological traits, such as gas exchange, leaf conductance, leaf phenotypic plasticity, responses of species to abiotic stress under fertilisation, several other ecophysiological traits and the role of mycorrhiza in relation to adaptation to temperate environments.

14.4   Alpine forest

Low temperature limits the growth of tree species in alpine forests. Very few studies are available on this aspect.

A nitrogen reservoir contributes to plant growth. Bowman (1992) investigated the inputs and storage of nitrogen in winter snowpacks in an Alpine ecosystem. They observed that the N reservoir in snow is an important source for plant growth and may explain in part the heterogeneity of primary production in alpine ecosystems.

Environment plays an important role in the distribution of alpine species. Guisan et al. (1998) predicted the potential distribution of plant species in an alpine environment. They studied the relationships between the distribution of alpine species and selected environmental variables by using two types of generalised linear models (GLMs) in a limited study area in the Valais region (Switzerland). The empirical relationships were utilised in a predictive sense to mimic the potential abundance of alpine species over a regular grid. They presented results for the alpine sedge *Carex curvula* ssp. *curvula*. The modelling approach consisted of:

1 A binomial GLM, including only the mean annual temperature as explanatory variable, which was adjusted to species presence/absence data in the entire study area.

2 A logistic model restricted to stands occurring within the a priori defined

temperature range for the species which allowed ordinal abundance data to be adjusted.

3 The two species–response functions combined in a GIS to generate a map of the species potential abundance in the study area.

4 Model predictions filtered by the classes of the qualitative variables under which the species never occur.

The model evaluation was finally carried out with the γ-measure of association in an ordinal contingency table. It showed that abundance was satisfactorily predicted for *C. curvula*.

14.5   Conclusion

In the context of the above literature, it may be concluded that the variability in the environmental conditions, such as light intensity, temperature fluctuations, wind velocity, altitude and slope, greatly affect the ecophysiology and tree growth in tropical, temperate and alpines.

Bibliography

Abrams M.D. **1994**. Genotypic and phenotypic variation as stress adaptations in temperate tree species: a review of several case studies. *Tree Physiology* **14**(7/9):833–842.

Barea J.M., Palenzuela J., Cornejo P., Sánchez-Castro I., Navarro-Fernández C., Lopéz-García A., Estrada B., Azcón R., Ferrol N., Azcón-Aguilar C. **2011**. Ecological and functional roles of mycorrhizas in semi-arid ecosystems of southeast Spain. *Journal of Arid Environments* **75**(12):1292–1301.

Baskin C.C., Baskin J.M. **1988**. Germination ecophysiology of herbaceous plant species in a temperate region. *American Journal of Botany* **75**(2):286–305.

Bowman W.D. **1992**. Inputs and storage of nitrogen in winter snowpack in an alpine ecosystem. *Arctic and Alpine Research* **24**(3):211–215.

Bréda N., Huc R., Granier A., Dreyer E. **2006**. Temperate forest trees and stands under severe drought: a review of ecophysiological responses, adaptation processes and long-term consequences. *Annals of Forest Science* **63**(6):625–644.

Carrera A.L., Mazzarino M.J., Bertiller M.B., del Valle H.F., Carretero E.M. **2009**. Plant impacts on nitrogen and carbon cycling in the Monte Phytogeographical Province, Argentina. *Journal of Arid Environments* **73**(2):192–201.

Cunningham S., Read J. **2002**. Comparison of temperate and tropical rainforest tree species: photosynthetic responses to growth temperaturerature. *Oecologia* **133**(2):112–119.

Ellsworth D.S., Reich P.B. **1996**. Photosynthesis and leaf nitrogen in five amazonian tree species during early secondary succession. *Ecology* **77**:581–594.

Gebrekirstos A., Teketay D., Fetene M., Mitlöhner R. **2006**. Adaptation of five co-occurring tree and shrub species to water stress and its implication in restoration of degraded lands. *Forest Ecology and Management* **229**(1/3):259–267.

Guisan A., Theurillat J-P., Kienast F. **1998**. Predicting the potential distribution of plant species in an alpine environment. *Journal of Vegetation Science* **9**(1):65–74.

Huxman T.E., Snyder K.A., Tissue D., Leffler A.J., Ogle K., Pockman W.T., Sandquist D.R., Potts D.L., Schwinning S. **2004**. Precipitation pulses and carbon fluxes in semiarid and arid ecosystems. *Oecologia* **141**:254–260.

Kitajima K. **1994**. Relative importance of photosynthetic traits and allocation patterns as correlates of seedling shade tolerance of 13 tropical trees. *Oecologia* **98**:419–428.

Kitajima K. **1996**. Ecophysiology of tropical tree seedlings. In: Mulkey SS, Chazdon RL, Smith AP (eds.) *Tropical Forest Plant Ecophysiology*. Chapman and Hall, New York, pp 559–596.

Körner C. **2003**. *Alpine Plant Life: Functional Plant Ecology of High Mountain Ecosystems* (2nd Edn.). Springer, Heidelberg, 359 pp.

Maestre F.T., Valladares F., Reynolds J.F. **2005**. Is the change of plant–plant interactions with abiotic stress predictable? A meta-analysis of field results in arid environments. *Journal of Ecology* **93**(4):748–757.

Newstrom L.E., Frankie G.W., Baker H.G. **1994**. A new classification for plant phenology based on flowering patterns in lowland tropical rain forest trees at La Selva, Costa Rica. *Biotropica* **26**:141–159.

Poorter L. **1999**. Light effect on growth of seedlings of 15 rain-forest tree species. *Functional Ecology* **13**(3):396–410.

Poorter L. **2001**. Light-dependent changes in biomass allocation and their importance for growth of rain forest tree species. *Functional Ecology* **15**:113–123.

Poorter L. **2002**. Gap heterogeneity and its implications for regeneration. In: Orians GH, Deindert E (eds.) *Advanced Comparative Neotropical Ecology*. Organization of Tropical Studies **01-25**, pp. 200–214.

Reich P.B., Walters M.B., Ellsworth D.S. **1997**. From tropics to tundra: Global convergence in plant functioning. *Proceedings of the National Academy of Sciences of the United States of America* **94**(25):13730–13734.

Schlesinger W.H., Pilmanis A.M. **1998**. Plant–soil interactions in deserts. *Biogeochemistry* **42**(1/2):169–187.

Theodose T.A., Bowman W.D. **1997**. Nutrient availability, plant abundance, and species diversity in two alpine tundra communities. *Ecology* **78**(6):1861–1872.

Theurillat J.P., Guisan A. **2001**. Potential impact of climate change on vegetation in the European Alps: a review. *Climatic Change* **50**(1/2): 77–109.

Vázquez-Yanes C., Orozco-Segovia A. **1984**. Ecophysiology of seed germination in the tropical humid forests of the world. In: Medina E., Mooney, H.A., Vázquez-Yanes C. (eds.) *Physiological Ecology of Plants in the Wet*

Tropics. Dr. Junk Publishers, Dordrecht, pp. 37–50.

Vázquez-Yanes C., Orozco-Segovia A. **1993**. Patterns of seed longevity and germination in the tropical rainforest. *Annual Review of Ecology and Systematics* **24**:69–87.

Westoby M., Falster D.S., Moles A.T., Vesk P.A., Wright I.J. **2002**. Plant ecological strategies: Some leading dimensions of variation between species. *Annual Review of Ecology and Systematics* **33**:125–159.

Whitmore T.C. **1975**. *Tropical Rain Forests of the Far East*. Clarendon Press, London, 282 pp.

Yeun K., Ruseel R.M. **2002**. Carotenoid bioavailability and bioconversion. *Annual Review of Nutrition* **22**(1):483–504.

CHAPTER 15

Research advances in plant ecophysiology

In the following, we discuss various aspects of ecophysiology in semiarid, arid, tropical rainforest, temperate and alpine regions and some original research carried out in the Tamaulipan thornscrub of the semiarid regions of Mexico as practical examples, with methodology supported by extensive global literature.

15.1 Leaf pigments

We studied the variability in plant pigments (Chlorophyll a, chlorophyll b and carotenoid contents) in the leaves of ten species of trees and shrubs in Linares, Northeast of Mexico.

15.2 Background

Plants contain several pigments, such as chlorophyll a, chlorophyll b, carotenoid, anthocyanin, xanthophyll, which are related to plant photosynthesis. Chloroplasts contain chlorophyll in their thylakoids and stroma. Chlorophyll has the capacity to absorb radiant energy from the sun in a photochemical reaction as part of the process of photosynthesis, the final act of which is the storage of chemical energy as carbohydrate. Carotenoids are natural fat-soluble pigments present in plants, algae and photosynthetic bacteria, where they also play a role in photosynthesis. A deficiency of chlorophyll gives chlorotic symptoms in leaves. Various studies have been undertaken to determine the variability of leaf pigments in tree species.

Native shrubs and trees in the semi-arid region of north-east Mexico serve as important resources for range ruminants and white-tailed deer. They also provide high-quality fuel, timber for fencing and timber for construction (Reid et al., 1990; Fullbright et al., 1991). But the growth of these species is affected by climatic conditions, which probably cause differences in the production of the photosynthetic pigments.

Various leaf pigments such as chlorophyll, carotenoids, xanthophyll, flavonoids and so on can provide an insight into the physiological performance of pigment content. These can be utilised with varying leaf structural characteristics indices for the protection of leaf pigment content. Chlorophylls and carotenoids play an important role in assimilatory processes in higher plants. Chlorophyll has the capacity to absorb the radiant energy of sunlight and change it into the chemical energy

Autoecology and Ecophysiology of Woody Shrubs and Trees: Concepts and Applications, First Edition.
Edited by Ratikanta Maiti, Humbero Gonzalez Rodriguez and Natalya Sergeevna Ivanova.
© 2016 John Wiley & Sons, Ltd. Published 2016 by John Wiley & Sons, Ltd.

of organic carbon through the process of photosynthesis (Sims and Gamon, 2002).

Carotenoids are natural fat-soluble pigments found in plants, algae and photosynthetic bacteria, where they also play a role in photosynthesis. In some non-photosynthetic bacteria, they may undertake a protective function against damage by light and oxygen (Biswal, 1995; Gitelson et al., 1999). Animals appear to be unable to synthesise carotenoids and may incorporate carotenoids from their diets. In animals, carotenoids impart a bright coloration and serve as antioxidants and a source for vitamin A activity (Britton, 1995). Besides, carotenoids develop important functions in plant in plant reproduction through their role in attracting pollinators and seed dispersal (Yeun and Russell, 2002). Leaf pigments show seasonal variability in tree species. Uvalle Sauceda et al. (2008) determined seasonal trends of chlorophylls a and b and carotenoids in native trees and shrubs of north-eastern Mexico. They found seasonal trends in the foliar tissue of native trees (T) and shrubs (S) from northeastern Mexico, such as *Acacia rigidula* (S), *Bumelia celastrina* (T), *Castela texana* (S), *Celtis pallida* (S), *Croton cortesianus* (S), *Forestiera angustifolia* (S), *Karwinskia humboldtiana* (S), *Lantana macropoda* (S), *Leucophyllum frutescens* (S), *Prosopis laevigata* (T) and *Zanthoxylum fagara* (T). Pigment contents were significantly different between years and seasons and between plants within years and seasons, except for the year×plant interaction for carotenoids at China site. All plants had marginally higher chlorophyll a content at Linares (0.79 mg g^{-1} f wt) than at China (0.71) or Los Ramones (0.66). Chlorophyll b content followed a trend

similar to chlorophyll a (0.29, 0.25 and 0.23 mg g^{-1} f wt, respectively). Marginal differences in carotenoid content, in all plants, were observed among sites. The 0.20 mg carotenoids g^{-1} f wt yearly and seasonal variations in plant pigments might have been related to seasonal water deficits, excessive irradiance levels during summer and extreme low temperatures in winter that could have affected leaf development and senescence.

In a later study, Himmelsbach et al. (2010, 2011) estimated leaf pigment contents in five tree and shrub species in a Mexican pine–oak forest during periods of different water during periods of different water in order to detect the seasonal means of chlorophyll a + b content, the seasonal variations in predawn leaf water (Ψ_{wpd}) and the osmotic potentials (Ψ_{spd}) of six native species co-existing in a mixed pine–oak forest, under seasonal drought. There were large variations in the variables between seasons and species, between 0.2 and 0.3 mg g^{-1} for *J. flaccida* and *Q. canbyi*, respectively. Ψ_{wpd} varied between −0.47 (*Q. canbyi* under shade) and −4.21 MPa (*J. flaccida* under sun) and Ψ_{spd} between −1.47 (*Acacia rigidula*) and −3.79 MPa (*Q. canbyi*), both under sunny conditions. Hence, *Q. canbyi* achieved the highest leaf pigment contents at both sites. Thus, chlorophyll a + b content was significantly correlated ($p < 0.001$) with air temperature (−0.70; *J. flaccida)* and relative humidity (−0.67; *Pinus pseudostrobus)*; and carotenoid content correlated with vapour pressure deficit (−0.60; *J. flaccida*) and average soil moisture content (−0.63; *P. pseudostrobus*). Furthermore, leaf chlorophyll a + b and carotenoid content were significantly correlated ($p < 0.01$) to Ψ_{wpd}

(−0.43 *P. pseudostrobus*; −0.53 *J. flaccida*) and Ψ_{spd} (−0.53 *P. pseudostrobus*; 0.58 *J. flaccida*). *P. pseudostrobus* and *J. flaccida* were at a physiological disadvantage during periods of drought in comparison to broad-leaved species such as *A. rigidula*, *Arbutus xalapensis*, *Rhus virens* and *Q. canbyi*, as they showed lower leaf pigment contents and were sensitive to environmental variables such as evaporative demands components (air temperature, relative humidity and vapour pressure deficit) and soil moisture content.

15.3 Methodology

This study was carried out at the experimental station of Facultad de Ciencias Forestales, Universidad Autonoma de Nuevo Leon, located in the municipality of Linares (24° 47′N, 99° 32′W), at an elevation of 350 m. The climate is subtropical or semiarid with a warm summer; the monthly mean air temperature varies from 14.7 °C in January to 23 °C in August, although during summer the temperature goes up to 45 °C. The average annual precipitation is around 805 mm, with a bimodal distribution. The dominant vegetation type is the known as Tamaulipan thornscrub or subtropical thornscrub wood land SPP-INEGI (1986). The dominant soil is deep, dark gray, lime-clay, vertisol with montmorrillonite, which shrinks and swells remarkably in response to changes in moisture content. The following species were included in this study: *Leucophyllum frutescens* I.M Johnst (Scrophulariaceae, shrub), *Acacia rigidula* Benth (Mimosaceae, shrub), *Sideroxylon celastrinum* Kunth. (Sapotaceae, tree), *Acacia berlandieri* Benth

(Mimosaceae, shrub), *Cordia boissieri* DC (Boraginaceae, tree), *Celtis pallida* Torr (Ulmaceae, tree), *Ligustrium lucidum* Torr (Oleaceae, tree), *Amyris madrensis* S. Watson (Rutaceae, shrub), *Dichondra argentea* Willd (Convolvulaceae, shrub) and *Lantana macropoda* Torr (Verbenaceae, shrub).

15.3.1 Determinations of chlorophyll and carotenoids

Four samples of leaf tissue (1.0 g of fresh weight) of each plant species were used for analysis. The content of chlorophyll a + b and carotenoids were extracted in 80% (v/v) aqueous acetone and vacuum filtered through a Whatman No.1 filter paper. Pigment contents were determined spectrophotometrically using a Perkin–Elmer spectrophotometer (Model Lamda 1A). Absorbance of chlorophyll a + b and carotenoid extracts were determined at wavelengths of 669, 645 and 470 nm, respectively. Chlorrophyll a + b and carotenoid (mg/g fresh weight) were estimated according to Lichtenthaler and Wellburn (1983).

15.4 Results

It was observed that, in all the variables (i.e. chlorophyll a, chlorophyll b, carotenoids), the chlorophyll a + b and the chlorophyll to carotenoids ratio showed highly significant differences among the species studied. Table 15.1 shows a summary of the test to detect significant differences in the pigment content of these native plant species.

Figure 15.1 shows the content of chlorophyll a, chlorophyll b, chlorophyll a + b and chlorophyll ratio (a/b) in different

Table 15.1 Kruskal–Wallis test to detect significant differences in pigment content among different plant species.

Statistic	Plant pigment					
	Chlorophyll a	Chlorophyll b	Carotenoids	Chlorophyll a + b	Chlorophyll ratio (a/b)	Chlorophyll to carotenoids ratio
χ^2	29.472	18.975	32.725	25.629	29.961	19.457
P-value	0.001	0.025	0.000	0.002	0.0	0.022

plant species in north-eastern Mexico. Figure 15.2 shows the carotenoid contents.

It was observed that *L. frutescens, A. rigidula* and *B. celastrinum* contained minimum chlorophyll a (about 1 mg), while *A. berlandieri* and *L. macropoda* had maximum chlorophyll a (around 2 mg). With respect to chlorophyll b, *L. frutescens, A. rigidula* and *B. celastrinum* contained the minimum (around 0.25 mg), while *A. berlandieri* and *L. macropoda* contained the maximum (around 0.5 mg). With respect to total chlorophyll (a + b) *L. frutescens, A. rigidula* and *S. celastrinum* had minimum total chlorophyll, while *A. berlandieri* and *L. macropoda* contained the maximum (2 mg). With respect to chlorophyll a/b ratio, all the species had more or less similar amount but *A. rigidula* and *C. boissieri* had a slightly higher value. With respect to carotenoids, *L. frutescens, A. rigidula* and *S. celastrinum* had minimum carotenoids (around 0.2 mg), while *A. berlandieri* and *L. macropoda* had higher carotenoids. Others fell in the intermediate group (about 0.4 mg). With respect to chlorophyll to carotenoids all the species had more or less similar values, ranging from 5.0 to 6.0 mg.

15.5 Discussion

Chlorophyll and carotenoids are responsible for absorbing light energy and transferring it to the photosynthetic apparatus in chloroplast for the production of photosynthesis and finally biomass production in plants. Therefore, the estimation of leaf pigments content serves as a valuable tool to understand the physiological and biochemical functions of leaves (Sims and Gamon, 2002).

The results of the present study reveal that the species varied remarkably in chlorophyll a and b, total a + b and carotenoid content among the 10 species studied. These findings confirm the importance of the species in their capacity in the production of each pigment to guide the photosynthetic process in leaves and its potential values. Plant pigments may play a role in the ecosystem productivity, but it is influenced by the droughts and extreme temperatures prevailing during the winter and summer seasons (González Rodríguez et al., 2000; 2004). It is well known that the productivity of higher plants is mediated by photosynthesis in leaves and its adaptation through the leaves (Valladares et al., 2000). In the

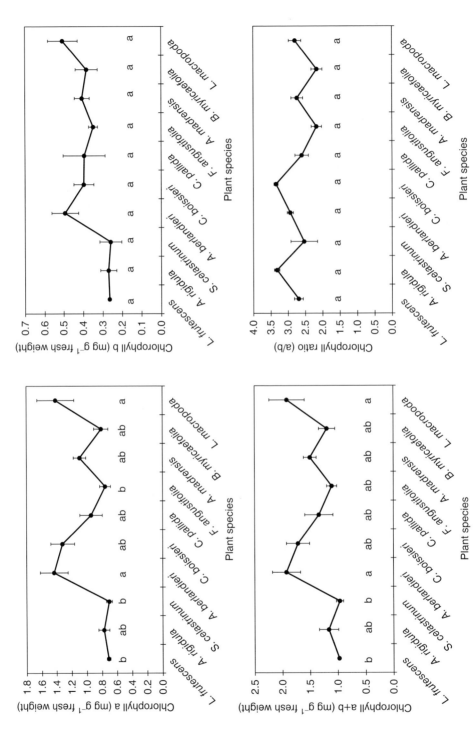

Figure 15.1 Chlorophyll a, chlorophyll b, chlorophyll a + b and chlorophyll ratio (a/b) content in different plant species, north-eastern Mexico. Means values followed by the same lowercase letter are not different at $P = 0.05$, using Tukey's HSD test.

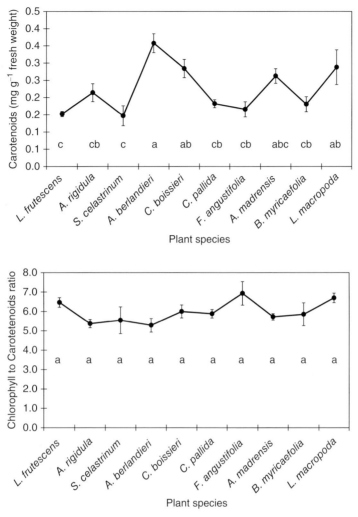

Figure 15.2 Carotenoids and chlorophyll to carotenoids ratio content in different plant species, north-eastern Mexico. Means values followed by the same lowercase letters are not different at $P = 0.05$, using Tukey's HSD test.

present study, an attempt was made to study the effects of several environmental factors in the production of plant pigments different authors have well documented the effects of these environmental factors, such as high temperature (González Rodríguez et al., 2004), low temperature during winter and high temperature in the summer thereby affecting growth of the plant species. Under such conditions, the production of photosynthates may be limited by temperature, stomatal control and light energy damage. The chlorophyll contents are affected by unfavourable temperature (Ottander et al., 1995) and also by the prevailing shade characteristics

(Castrillo et al., 2001). There exists a relationship between leaf pigment and spectral reflectance (Sims and Gamon, 2002).

The carotenoid components of the sun leaves of plants revealed that sun leaves contain a higher amount of xanthophyll components in cycles (Demmig-Adams and Adams III, 1992). The reduction of chlorophyll does not cause an adaptive response against adverse conditions in a Mediterranean summer, which may be applicable in north-eastern Mexico (Valladares et al., 2000; Kyparissis et al., 2006). This could be related to varying leaf structural characteristics indices for the protection of leaf pigment content. Sims and Gamon (2002) studied the relationship between leaf pigment content and spectral reflectance. Tattini et al. (2005) reported that the flavonoids present in the cuticle; thick cuticle and dense glandular trichomes protect *Ligustrum vulgare* against high solar radiation.

Various studies have revealed the effects of environments on leaf pigment contents. Jiang et al. (2006) reported that the chlorophyll content and photosynthesis in young leaves were much higher than in fully expanded leaves; leaf orientation, photorespiration and xanthophyll cycles protect young seedlings against high irradiation in the field. Demmig-Adams and Adams III (1992) analysed carotenoid composition in sunny and shaded plants with different forms. Giuffrida et al. (2006) studied chlorophyll and chlorophyll-derived components in pistachio kernels (*Pistacia vera* L.) and found 13 compounds. De Carvalho Gonçalves et al. (2001) studied how environmental light on a leaf affected the concentrations of photosynthetic pigments and chlorophyll fluorescence in mahogany (*Swietenia macrophylla* King) and tonka bean (*Dipteryx odorata*). Chlorophyll contents were higher in shade leaves than in sunny leaves. Kalituho et al. (2007) studied the roles of specific xanthophylls in light utilisation. Kyparissis et al. (2006) reported seasonal fluctuations in photoprotective (xanthophyll cycle) and photoselective (chlorophyll) capacity in eight Mediterranean plant species belonging to two different growth forms. Jeon et al. (2006) reported the effect of temperature on photosynthetic pigments, morphology and leaf gas exchange during ex vitro accumulation of micro-propagated CAM.

15.6 Conclusions and future research

Plant pigments play an important role in plant assimilatory systems and plant growth. A study was undertaken to determine the chlorophyll a, chlorophyll b and carotenoid contents of 10 species of trees and shrubs in Linares, north-eastern Mexico, viz. *Leucophyllum frutescens* I.M Johnst, *Acacia rigidula* Benth, *Sideroxylon celastrinum* Kunth., *Acacia berlandieri* Benth, *Cordia boissieri* DC, *Celtis pallida* Torr, *Ligustrium lucidum* Torr, *Amyris madrensis* (S. Watson), *Dichondra argentea* Willd. and *Lantana macropoda* Torr. Large variations were observed between species in their contents of chlorophyll (a, b, total a+b) and carotenoids. These variations in plant pigments could play a great role in photosynthetic activities and the productivity of different species. The contents of pigments of the studied species may vary in different regions. Therefore, there is a necessity to study these contents in different regions.

Bibliography

Biswal B. **1995**. Carotenoid catabolism during leaf senescence and its control by light. *Journal of Photochemistry and Photobiology B: Biology* **30**(1):3–13.

Britton G. **1995**. Structure and properties of carotenoids in relation to function. *FASEB J* **9**(15):1551–1558.

Castrillo M., Vizcano D., Moreno E., Lotorraca Z. **2001**. Chlorophyll content some cultivated and wild species of Lamiaceae. *Biologia Plantarum* **44**(3):423–325.

De Carvalho Gonçalves J.F., Morenco R.A., Vieira G. **2001**. Concentration of photosynthetic pigments and chlorophyll fluorescence of Mahogany and Tonka bean under two light environments. *Revista Brasileira de Fisiologia Vegetal* **13**(2):149–157.

Demmig-Adams B., Adams III W.W. **1992**. Carotenoid composition in sun and shaded plant with different forms. *Plant Cell and Environment* **15**(4):411–419.

Fulbright T.E., Reynolds J.P., Beason S.L., Demaris S. **1991**. Mineral contents of guajillo regrowth following cropping. *Journal of Range Management* **44**(5):520–522.

Gitelson A.A., Busechman C., Hartman P., Lichtenthaler K. **1999**. The chlorophyll fluorescence ratio as an accurate measure of the chlorophyll content in plants. *Remote Sensing of Environment* **69**:236–302.

Giuffrida D., Saitta M., La Torre L., Bombaci L., Dugo, G. **2006**. Chlorophyll and chlorophyll derived components in pistachio kernels (*Pistacia vera* L.) from Sicilly. *Italian Journal of Food Science* **18**(3):309–316.

González Rodríguez H., Cantú Silva I., Gómez Meza M.V., Jordan W.R. **2000**. Seasonal plant water relationships in *Acacia berlandieri*. *Arid Soil Research and Rehabilitation* **14**(4): 343–357.

González Rodríguez H., Cantú Silva I., Gómez Meza M.V., Ramírez Lozano R.G. **2004**. Plant water relations of thornscrub shrub species, north-eastern Mexico. *Journal of Arid Environments* **58**(4):483–503.

Himmelsbach W., González-Rodríguez H., Treviño-Garza E.J., Estrada Castillón A.E., Aguirre Calderón O.A., González Tagle M.A. **2010**. Leaf pigment contents in five tree and shrub species in a Mexican pine-oak forest during periods of different water availability. *International Journal of Agriculture, Environment and Biotechnology* **3**(1):5–14.

Himmelsbach W., Treviño-Garza E.J., González-Rodríguez H., González-Tagle M.A., Gómez Meza M.V., Aguirre Calderón O.A., Estrada Castillón A.E., Mitlöhner R. **2011**. Acclimation of three co-occurring tree species to water stress and their role as site indicators in mixed pine-oak forests in the Sierra Madre Oriental, Mexico. *European Journal of Forest Research* **131**(2):355–367.

Jeon M.-W., Babar Ali M., Hahn E.-J., Paek K.-Y. **2006**. Photosynthetic pigments, morphology and leaf gas exchange during *ex vitro* acclimatization of micropropagated CAM *Doritaenopsis* plantlets under relative humidity and air temperature. *Environmental and Experimental Botany* **55**(1/2):183–194.

Jiang Ch.D., Gao H.-Y., Zou Q., Jiang G.-M., Li L.-H. **2006**. Leaf orientation, photorespiration and xanthophyll cycle protect young soybean leaves against high irradiance in field. *Environmental and Experimental Botany* **55**(1/2):87–96.

Kalituho L., Rech J., Jahns P. **2007**. The roles specific xanthophylls in light utilization. *Planta* **225**(2):423–439.

Kyparissis A., Drilias P., Manetas Y. **2006**. Seasonal fluctuations in photoprotective (xanthophyll cycle) and photoselective (chlorophylls) capacity in eight Mediterranean plant species belonging to two different growth forms. *Australian Journal of Plant Physiology* **27**(3):265–272.

Lichtenthaler H.K., Wellburn A.R. **1983**. Determinations of total carotenoids and chlorophylls a and b of leaf extracts in different solvents. *Biochemical Society Transactions* **11**:591–592.

Ottander C., Campbell D., Öquist G. **1995**. Seasonal changes in photosystem II organisation and pigment composition in *Pinus sylvestris*. *Planta* **197**:176–183.

Reid N., Marroquin J., Beyer-Münzel P. **1990**. Utilization of shrubs and trees for browse, fuelwood and timber in the Tamaulipan thornscrub in northeastern Mexico. *Forest Ecology and Management* **36**(1):61–79.

Sims D.A., Gamon J.A. **2002**. Relationship between leaf pigment content and spectral reflection across a wide range of species, leaf surface structure and developmental stages. *Remote Sensing of Environment* **81**(2):337–354.

SPP-INEGI. **1986**. *Síntesis Geografía del Estado del Nuevo León, Instituto Nacional de Geografía Estadística e Información*, México, D.F.

Tattini M., Guidi L., Morassi-Bonzi L., Pinelli P., Remorini D., Degl'Innocenti E., Giordano C., Massai R., Agati G. **2005**. On the role of flavonoid in the integrated mechanisms of response of *Ligustrum vulgare* and *Phillyrea latifolia* to high solar radiation. *New Phytologist* **167**(2):457–470.

Uvalle Sauceda J.I., González Rodríguez H., Ramírez Lozano R.G., Cantú Silva I., Gómez Meza M.V. **2008**. Seasonal trends of chlorophylls *a* and *b* and carotenoids in native trees and shrubs of northeastern Mexico. *Journal of Biological Sciences* **8**(2):258–267.

Valladares F., Martinez-Ferri E., Balaguer L., Perez-Corona E., Manrique E. **2000**. Low leaf-level response to light and nutrients in mediterranean evergreen oaks: a conservative resource-use strategy? *New Phytologist* **148**(1):79–91.

Yeun K.J., Russell R.M. **2002**. Carotenoid bioavailability and bioconversion. *Annual Review of Nutrition* **22**(1):483–504.

Figure 7.1 Transverse section of *Acacia amentacea*.

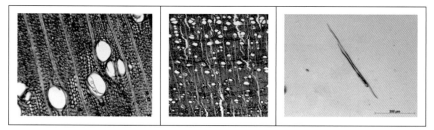

Figure 7.2 Transverse section of *Acacia berlandieri*.

Figure 7.3 Transverse section of *Acacia farnesiana*.

Autoecology and Ecophysiology of Woody Shrubs and Trees: Concepts and Applications, First Edition.
Edited by Ratikanta Maiti, Humbero Gonzalez Rodriguez and Natalya Sergeevna Ivanova.
© 2016 John Wiley & Sons, Ltd. Published 2016 by John Wiley & Sons, Ltd.

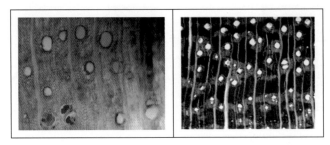

Figure 7.4 Transverse section of *Acacia shaffneri*.

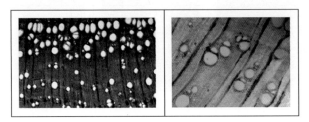

Figure 7.5 Transverse section of *Acacia wrightii*.

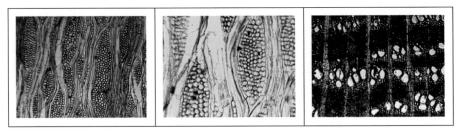

Figure 7.6 Transverse section of a stem of *Cordia boissieri* showing the organization of primary and secondary xylem.

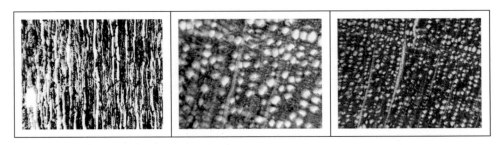

Figure 7.7 Transverse section of *Helietta parviflora*.

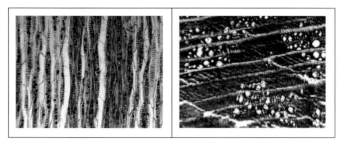

Figure 7.8 Transverse section of *Condalia hookeri*.

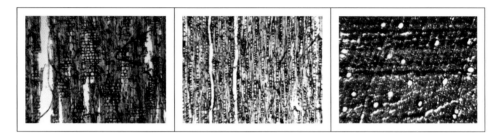

Figure 7.9 Transverse section of *Diospyros palmeri*.

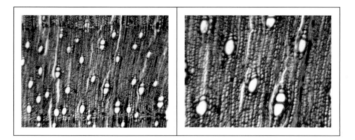

Figure 7.10 Transverse section of *Zanthoxylum fagara*.

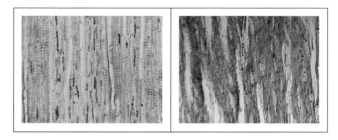

Figure 7.11 Transverse section of *Karwinskia humboldtiana*.

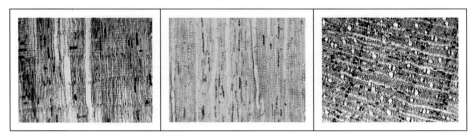

Figure 7.12 Transverse section of *Diospyros texana*.

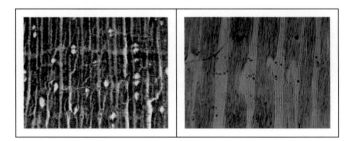

Figure 7.13 Transverse section of *Celtis pallida*.

Figure 7.14 Transverse section of *Celtis laevigata*.

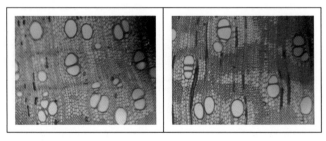

Figure 7.15 Transverse section of *Caesalpinia mexicana*.

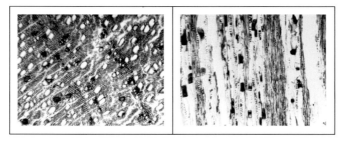

Figure 7.16 Transverse section of *Eysenthardtia polystachya*.

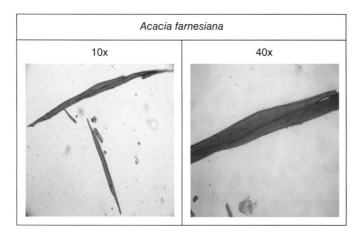

Figure 7.17 Microphotographs of wood fibre cells of *Acacia farnesiana*, seen at 10× and 40×.

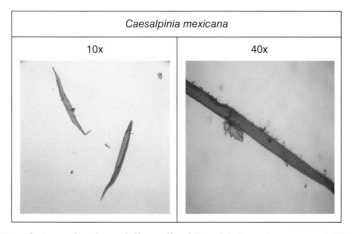

Figure 7.18 Microphotographs of wood fibre cells of *Caesalpinia mexicana*, seen at 10× and 40×.

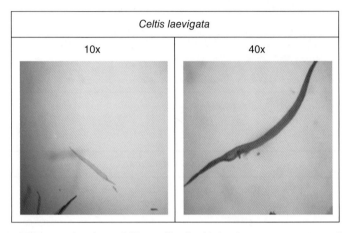

Figure 7.19 Microphotographs of wood fibre cells of *Celtis laevigata*, seen at 10× and 40×.

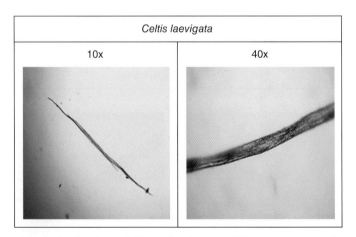

Figure 7.20 Microphotographs of wood fibre cells of *Celtis pallida*, seen at 10× and 40×.

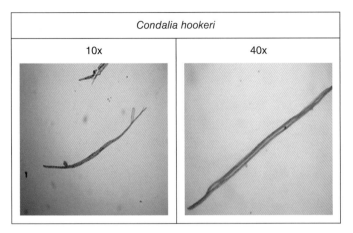

Figure 7.21 Microphotographs of wood fibre cells of *Condalia hookeri*, seen at 10× and 40×.

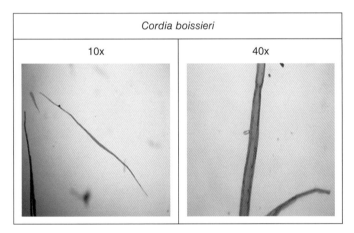

Figure 7.22 Microphotographs of wood fibre cells of *Cordia boissieri*, seen at 10× and 40×.

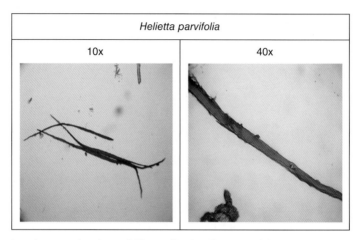

Figure 7.23 Microphotographs of wood fibre cells of *Helietta parvifolia*, seen at 10× and 40×.

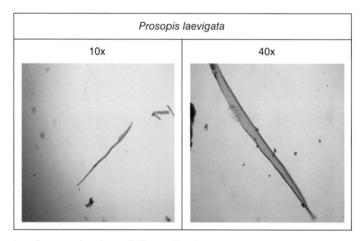

Figure 7.24 Microphotographs of wood fibre cells of *Prosopis laevigata*, seen at 10× and 40×.

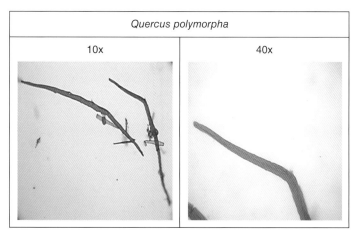

Figure 7.25 Microphotographs of wood fibre cells of *Quercus polymorpha*, seen at 10× and 40×.

Figure 25.1 Pine fir forest in the mountains of the Middle Urals (Russia).

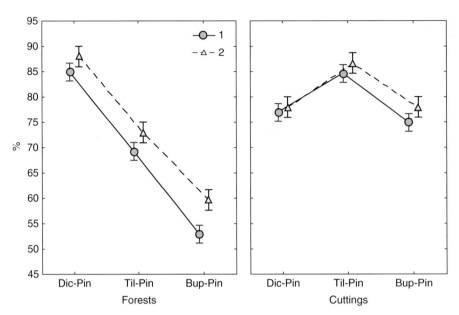

Figure 25.5 Activity rates of the seed germination of *Pinus sylvestris*. Dashed lines represent the average value, solid lines represent the 95% interval; 1 – Germination energy, 2 – seed germination; Dic-Pin – *Dicrano-Pinetum sylvestris*, Til-Pin – *Tilio-Pinetum sylvestris*, Bup-Pin – *Bupleuro longifolii-Pinetum sylvestris*.

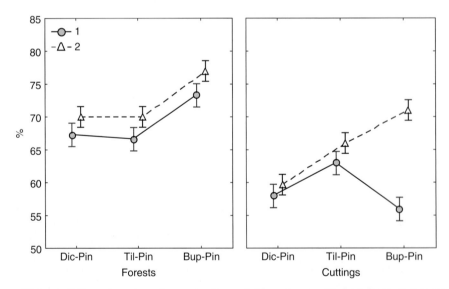

Figure 25.6 Activity rates of seed germination of *Picea obovata*. Dashed lines represent the average value, solid lines represent the 95% interval; 1 – germination energy, 2 – seed germination; Dic-Pin – *Dicrano-Pinetum sylvestris*, Til-Pin – *Tilio-Pinetum sylvestris*, Bup-Pin – *Bupleuro longifolii-Pinetum sylvestris*.

Ebenopsis ebano (Berland.) Barneby & J.W. Grimes

Cordia boissieri A. DC.

Bernardia myricifolia (Sheele). Benth. & Hook. F.

Condalia hookeri M.C. Johnst.

Eysenhardtia texana Scheele

Croton suaveolens Torr.

Karwinskia humboldtiana (Schult.) Zucc.

Leucophyllum frutenscens (Berland.) I.M. Johnst

Celtis pallida Torr.

Zanthoxylum fagara (L.) Sarg.

Caesalpinia mexicana A. Gray

Havardia pallens (Benth.) Britton & Rose

Forestiera angustifolia Torr.

Leucaena leucocephala (J. de Lamarck) H.C. de Wit

Helietta parvifolia (A. Gray) Benth.

Sideroxylon celastrinum (Kunth) T.D. Penn.

Diospyros palmeri Eastw.

Parkinsonia texana (A. Gray) S. Watson

Ehretia anacua (Terán & Berland.) I.M. John.

Guaiacum angustifolium Engelm.

Amyris texana (Buckley) P. Wilson.

Sargentia greggii S. Watson

Celtis laevigata Willd.

Berberis chococo Schlecht.

Acacia wrightii Benth.

Acacia farnesiana (L) Willd.

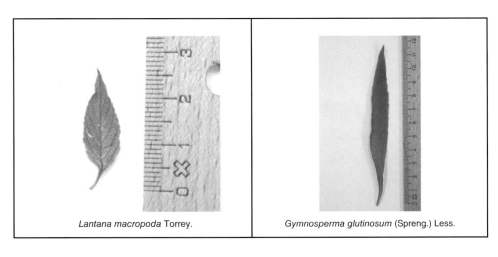

Lantana macropoda Torrey.

Gymnosperma glutinosum (Spreng.) Less.

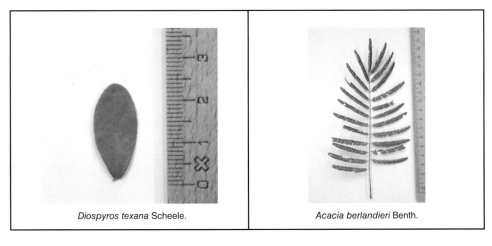

Diospyros texana Scheele.

Acacia berlandieri Benth.

Quercus virginiana Mitl.

Prosopis laevigata (H. & B.) Jonhst

Acacia rigidula (Benth.) Seigler & Ebinger.

Fraxinus greggii A. Gray

Acacia schaffneri (S. Watson) F.J. Herm.

Parkinsonia aculeata L.

CHAPTER 16

Carbon capture, carbon sequestration and carbon fixation

16.1 Introduction

Increasing global warming associated with incessant logging, illegal human activities and conversion of forest to agriculture have increased pollution and the carbon dioxide load in the atmosphere. In addition, the constant emission of CO_2 due to the combustion of fossil fuels in factories has enhanced this menace and thereby endangering our lives. This has a direct effect on climate changes, thereby reducing crop productivity and aggravating poverty. In order to reduce this CO_2 load (by carbon sequestration), concerted research activities are being directed in various countries, but with little success. Herein we narrate a few items from the literature on this aspect.

The constant emission of CO_2 by the combustion of fossil fuels from factories and burning of wood is a great menace increasing pollution in the atmosphere thereby, endangering the security of mankind and animals. To mitigate it, different technologies are adopted in different countries in relation to CO_2 capture and sequestration. In this context, plants in the forests play an important role in the reduction of carbon load by capturing CO_2 for photosynthesis and storage of carbon

in plant biomass and woods. At present, various industries in advanced countries are planting trees around their factories to reduce CO_2, in order to reduce tax through carbon credit. Otherwise, they have to pay a heavy tax to the government.

There is a great necessity to reduce the CO_2 load in the atmosphere. Plants contribute a lot to the capture of CO_2 from the atmosphere in the process of photosynthesis, synthesis of carbohydrate and liberation of oxygen, required for our respiration and other living organisms. In this gesture, various technologies have been developed and utilised in developed countries to capture carbon dioxide load from the atmosphere emitted by factories and stored underground deep in the soil, a process called carbon sequestration.

16.2 Examples of the role of lower plants in carbon fixation

According to wikipedia, cyanobacteria and several other photosynthetic plants carry out photosynthesis and absorb CO_2 thereby changing the earth's atmosphere. Carbon fixation or carbon assimilation is the conversion process of inorganic carbon (CO_2)

Autoecology and Ecophysiology of Woody Shrubs and Trees: Concepts and Applications, First Edition.
Edited by Ratikanta Maiti, Humbero Gonzalez Rodriguez and Natalya Sergeevna Ivanova.
© 2016 John Wiley & Sons, Ltd. Published 2016 by John Wiley & Sons, Ltd.

to organic compounds by living organisms. Similar to photosynthesis, chemosynthesis is another form of carbon fixation that can take place in the absence of sunlight. Organisms that grow by fixing carbon are called autotrophs. Autotrophs include photoautotrophs, which synthesise organic compounds using the energy of sunlight, and lithoautotrophs, which synthesise organic compounds using the energy of inorganic oxidation. Heterotrophs are organisms that grow using the carbon fixed by autotrophs. The organic compounds are used by heterotrophs to produce energy and to build body structures. "Fixed carbon", "reduced carbon" and "organic carbon" are equivalent terms for various organic compounds. A carbon sink is a natural or artificial reservoir that accumulates and stores some carbon containing chemical compound for an indefinite period. The process by which carbon sinks remove CO_2 from the atmosphere is known as carbon sequestration. Public awareness of the significance of CO_2 sinks has grown substantially. There are different strategies used to enhance this process. Sufficient research activities are directed in this direction.

Hof et al. (1990) developed two alternative economic-analogue models of carbon fixation in trees. The first is a relatively straight-forward economic analogue where carbon serves as the currency and the plant is modelled as a maximiser of net carbon gain (a profit analogue). The second models carbon "revenue" as the minimum of two functions that relate carbon gain to leaf and root biomass, respectively. The plant is then modelled with a MAXMIN operator, such that the minimum net revenue is maximised. Both models reveal that leaves and roots are simultaneously limiting factors, but the second (MAXMIN) model implies that they are not "equally" limiting.

Carbon sequestration is the process in capturing (CO_2) from the atmosphere or capturing anthropogenic (human) CO_2 from large-scale stationary sources like power plants before it is released to the atmosphere. Once captured, the CO_2 gas (or the carbon portion of the CO_2) is compressed and put into long-term storage. This has great potential to reduce significantly the level of carbon that occurs in the atmosphere as CO_2 and to reduce the release of CO_2 to the atmosphere from major stationary human sources, including power plants and refineries. There are two major types of CO_2 sequestration: terrestrial and geologic.

Terrestrial (or biologic) sequestration involves the process to capture CO_2 by plants from the atmosphere and then storing it as carbon in the stems and roots of the plants as well as in the soil. During photosynthesis, plants take in CO_2 and give off oxygen (O_2) to the atmosphere. This oxygen is released, available for our respiration. The plants retain and use the stored carbon for growth. Terrestrial sequestration is the process of land management practices that maximises the amount of carbon that remains stored in the soil and plant material for the long term. This includes non-till farming, wetland management, rangeland management and reforestation. If the soil is disturbed and the soil carbon comes in contact with oxygen in the air, the exposed soil carbon can combine with O_2 to form CO_2 gas and re-enters the atmosphere, reducing the amount of carbon in storage. The US Department of

Energy estimates that anywhere from 1800 to 20 000 billion tonnes of CO_2 could be stored underground in the United States.

Geologic sequestration involves a process in the carbon sequestration process (CCS/0 Process). Unlike terrestrial, or biologic, geologic sequestration, where carbon is stored via agricultural and forestry practices. Geologic sequestration involves injecting carbon dioxide deep underground where it stays permanently [e.g. the US Environmental Protection Agency's (EPA)'s proposed Carbon Pollution Standards for New Power Plants]. On 20 September 2013, the US EPA under President Obama's Climate Action Plan was tasked with reducing carbon pollution from power plants. EPA is proposing carbon pollution standards for new power plants built in the future. CCS is one of the technologies new power plants can employ to meet the standard.

EPA's Greenhouse Gas Reporting Program (GHGRP) collects information from facilities in many industry types that directly emit large quantities of GHGs, suppliers of certain fossil fuels and facilities that inject CO_2 underground. CCS could play an important role in reducing greenhouse gas emissions.

CCS involves a set of high technologies that can greatly reduce CO_2 emissions from new and existing coal and gas-fired power plants and large industrial sources. CCS is a three-step process that consists of three processes:

1 Capture of CO_2 from power plants or industrial processes.
2 Transport of the captured and compressed CO_2 usually through.
3 Injection of CO_2 deep in impermeable, non-porous layers of rock that trap the CO_2 and prevent it from migrating upward.

See: http://prod-mmedia.netl.doe.gov /Video/carbon_sequestration_animation .wmv, and http://prod-mmedia.netl .doe.gov/Video/carbon_sequestration _sept.wmv.

Carbon sequestration means capturing CO_2 from the atmosphere or capturing anthropogenic (human) CO_2 from large-scale stationary sources like power plants before it is released to the atmosphere. Once captured, the CO_2 gas (or the carbon portion of the CO_2) is put into long-term storage.

CCS has the great potential to reduce significantly the level of carbon that occurs in the atmosphere as CO_2 and to reduce the release of CO_2 to the atmosphere from major stationary human sources, including power plants and refineries.

Braakman and Smith (2012) explain the emergence and early evolution of biological carbon fixation. The fixation of CO_2 into living matter sustains all life on Earth, and embeds the biosphere within geochemistry. The six known chemical pathways are used by extant organisms for this function, but their evolution is incompletely known. They construct the complete early evolutionary history of biological carbon fixation, relating all modern pathways to a single ancestral form. They observe that innovations in carbon fixation built the foundation for most major early divergences in the tree of life. Their findings are based on a novel method that fully integrates metabolic and phylogenetic constraints. Comparing gene profiles across the metabolic cores of deep-branching organisms and requiring that they are capable of synthesising all their biomass components, leading to the surprising conclusion that the most common form for deep-branching autotrophic carbon fixation combines two disconnected sub-networks, each supplying carbon to distinct biomass components. Using metabolic constraints they then reconstructed a "phylometabolic" tree with a high degree of parsimony that traces the evolution of complete carbon fixation pathways, and has a clear structure down to the root. Energy optimisation and oxygen toxicity are the two strongest forces of selection. The root of this tree combines the reductive citric acid cycle and the Wood–Ljungdahl pathway into a single connected network.

Okimoto et al. (2013) studied net carbon fixation of a South-East Asian representative mangrove tree, *Rhizophora apiculata*. It was calculated with two different procedures: (i) the gas exchange analysis and (ii) the growth curve analysis methods. The gas exchange analysis method was based on calculated carbon values from the difference between photosynthetic absorption and respiratory emission. These two parameters were calculated by using photosynthetic rates of single-leaves in response to light and temperature and respiratory rates of trunk and branch in response to temperature. These monthly values were adjusted with monthly average measurements of light intensity and temperature to improve estimation accuracy. The value of annual net carbon fixation for three, four, five and nine year old forests was estimated to be 2.5–30.5 Mg C ha^{-1} year^{-1}. The values with the temperature modification increased by 9.3–21.3%, compared to those of 2.2–25.2 Mg C ha^{-1} year^{-1}

without the temperature modification. However, it was found that these estimated values were significantly higher than the results produced by the growth curve analysis method, which produced 1.1–35.2 Mg C ha^{-1} year^{-1}. Results of this study show that further work is required to improve the estimation accuracy for both the gas exchange analysis and growth curve analysis methods. There is a particular need to take into account the respiratory carbon emissions from the plant root for the former method and determination of the maximum biomass at the mature tree phase for the latter method.

Reforestation is a novel technique adopted by EPA to reduce carbon load emitted by factories coal mine areas from the atmosphere. This report of EPA describes the opportunities available to mine land owners, companies and other interested parties to utilise reforestation to clean up and restore former and abandoned mine lands (AMLs) and to generate carbon sequestration credits. The efforts made to increase terrestrial carbon sequestration are based on the premise that reforestation adds to the planet's net carbon storage. This helps to moderate global warming by slowing the growth of carbon emissions in the atmosphere. In a carbon market, each tonne of carbon sequestered is called a carbon credit. Using sequestration, companies can buy or generate these credits, which are then sold or hired by companies to offset their own CO_2 emissions. Storing carbon in forests is cheaper than paying carbon tax; and they have implemented this in coal mining areas in Eastern United States and the South East.

Keller et al. (2003) analysed CO_2 sequestration as a strategy to manage future climate change in an optimal economic growth framework. They approach the problem in two ways: first, by using a simple analytical model and, second, by using a numerical optimisation model which allows us to explore the problem in a more realistic manner. They consider that CO_2 sequestration is not a perfect substitute for avoiding CO_2 production because CO_2 leaks back to the atmosphere and hence imposes future costs. The "efficiency factor" of CO_2 sequestration is expressed as the ratio of the avoided emissions to the economically equivalent amount of sequestered CO_2 emissions. A simple analytical model in terms of a net-present value criterion suggests that short-term sequestration methods such as afforestation can be somewhat (60%) efficient, while long-term sequestration (such as deep aquifer or deep ocean sequestration) can be very (90%) efficient. Their study reveals that CO_2 sequestration methods at a cost within the range of present estimates reduce the economically optimal CO_2 concentrations and climate related damages. The potential savings associated with CO_2 sequestration is equivalent in their utilitarian model to a one-time investment of a small percentage of present gross world product.

The Ecological Society of America reported that, through the process of photosynthesis, plants assimilate carbon and return some of them through respiration. Then the carbon stored in plant tissue is stored in the soil as soil organic matter (SOM). This SOM is a complex mixture of carbon compounds, composed of decomposing plant and animal tissue,

microbes (protozoa, nematodes, fungi and bacteria) and carbon associated with soil minerals. Carbon can be stored in the soil for millennia or quickly released to the atmosphere; further, climatic conditions, natural vegetation, soil texture and drainage affect the amount of the carbon.

With respect to carbon sequestration, the Soil Science Society of America (SSSA) reported that increased long term (20–50 year) sequestration of carbon in soils, plants and plant products will benefit the environment and agriculture. Therefore, crops, grazing and forest lands can be managed efficiently for both economic productivity and carbon sequestration. In many cases, the effective management approach can be achieved by applying the recognised best management practices, such as conservation tillage, efficient nutrient management, erosion control, use of cover crops and restoration of degraded soils. Besides, conversion of marginal arable land to forest or grassland can rapidly increase soil carbon sequestration. Therefore, research is needed to estimate better carbon sequestration obtained by these practices; this research should effectively help a soil carbon sequestration accounting system, suitable to the business sector, utilising soil carbon as a marketable commodity. Implementation of these practices will involve a wide range of disciplines in the basic, agricultural, silvicultural and environmental sciences as well as in the social, economic and political sciences. SSSA recommends that a global increase in soil organic matter will have a timely benefit globally, by reducing the rate of increase in atmospheric CO_2 and increasing the productivity of soil, especially in many areas with degraded soils.

Various programmes such as the Conservation Reserve Program, Wetland Reserve Program, Forestry Incentive Program and Conservation Tillage, lead to both increased above ground carbon sequestration and to increased SOC. Therefore, soils gaining SOC increase plant productivity and environmental quality. Increases in SOC generally improve also soil structure, increase soil porosity and water holding capacity, as well as improve biological health for a vast array of life forms in soil.

With respect to the role of plants in capturing CO_2, Jiménez Pérez et al. (2013) analysed the per unit, dry weight-based biomass in the components of the above-ground biomass of the representative species of the pine–oak forest ecosystem of the Sierra Madre Oriental. The components of the above-ground biomass considered were stem, branches, bark and leaves of the species *Pinus pseudostrobus*, *Juniperus flaccida*, *Quercus laceyi*, *Q. rysophylla*, *Q. canbyi* and *Arbutus xalapensis*. The carbon concentration was determined using a Solids TOC Analyzer, which analyses the carbon concentrations in solid samples by complete combustion. The species with the highest carbon concentration was *J. flaccida* (51.18%), while *Q. rysophylla* had the lowest (47.98%); the component with the highest carbon concentration was the leaves of *A. xalapensis* (55.05%) while the bark of *Q. laceyi* had the lowest (43.65%). Highly significant differences were found for the average carbon concentration by group of species; the group conifers showed an average of 50.76%, while that of the broadleaf species was 48.85%. There were highly significant differences between the various components by species group; the highest concentration was found in the bark of conifers (51.91%), compared to the bark of the broadleaf species, which had the lowest (45.75%). Apart from big tree species various shrubs and herbs have capacity to capture carbon. Utilising a standardised combustion protocol, we estimated the carbon and nitrogen content as well as the C:N ratio of 40 plant species including few medicinal plant species, the results of which are shown in Table 16.1.

It is observed that large variation exists in carbon, nitrogen and carbon/nitrogen content among the studied species. Carbon concentration varies from 25 to 51%. The species containing high carbon concentration are *Litsea glaucensens* (51%), *Syzygium aromaticum* (51%), *Rhus virens* (50%), *Pinus arizonica Engelm* (49%), *Gochanatia hypoleuca* (49%) and *Arbutus xalapensis* (49%). Several species contain more than 45% carbon.

16.3 Conclusions and future research

In the context of the above literature on carbon sequestration, significant research activities have been directed to reduce CO_2 load and pollution, but the success is negligible. In advanced countries millions tonnes of CO_2 are captured and injected into a deeper layer of impervious soil; still they fail to reduce it satisfactorily. The factories which emit excessive CO_2 have to pay carbon credit/heavy tax. In order to pay less tax, they grow plants around the factories thereby attempting to reduce CO_2. We have hypothesised and identified tree species with open canopy and capture more than 50% CO_2 which may be planted in contaminated areas, cities and during

Table 16.1 Carbon (C) and nitrogen (N) content (%) and C:N ratio in different plant species, northeastern Mexico.

Common name	Scientific name	C (%)	N (%)	C:N ratio
Mastranto	*Mentha rotundifolia* (L.)	45.85	3.74	12.26
Parraleña	*Dyssodia setifolia* (Lag) Robins.	39.68	2.34	16.91
Pino blanco	*Pinus arizonica* Engelm	49.32	2.39	20.58
Ocotillo	*Gochanatia hypoleuca*	49.85	3.58	13.90
Alpistle	*Phalaris canariensis* L.	40.73	2.83	14.36
Paistle	*Tillandsia usenoides* L.	44.10	1.56	28.26
Hierba del sapo	*Eryngium heterophyllum* Engelm	40.89	1.75	23.37
Chia	*Salvia hispanica*	44.67	5.24	8.52
Ortiguilla	*Tragia ramosa* Torr	42.68	3.89	10.97
Laurel	*Litsea glauscensens* Kunth	51.34	3.36	15.28
Hierba del pajarito	*Lepidium virginicumun* L.	43.80	4.45	9.82
Gordolobo	*Gnaphalium canascens* DC.	37.73	2.55	14.75
Hierba S. Nicolás	*Chrysactinia mexicana* A. Gray	45.04	3.39	13.27
Neem	*Azadirachta indica*	45.11	5.84	7.71
Colesia	*Eruca sativa* Mill	41.13	5.47	7.51
Gigante	*Nicotiana glauca* Graham	37.93	4.78	7.92
Moringa	*Moringa oleifera*	45.96	6.24	7.35
Canela	*Cinnamom umzeylanicum*	49.33	2.49	19.78
Clavo de olor	*Syzygiumaromaticum = Eugenia caryophyllata*	51.66	2.90	17.80
Charrasquilla	*Mimosa malacophylla* A. Gray	45.14	8.45	5.33
Tatalencho	*Gymnosperma glutinosum* Spreng. Less	46.18	5.89	7.84
Romero	*Rosamrinus officinalis* L.	47.77	4.54	10.51
Yerbabuena	*Mentha piperita* L.	44.13	5.39	8.17
Chile piquín	*Capsicum annum* L. var. glabriusculum	42.93	6.84	6.27
Albahaca	*Ocimum basilicum* L.	38.31	4.65	8.22
Oregano	*Poliomintha longiflora* A. Gray	42.89	4.88	8.77
Betónica	*Hedeoma palmeri* Hemsl	46.37	2.83	16.38
Injerto	*Phoradendron villosum* (Nutt)	40.40	4.92	8.20
Lantrisco	*Rhus virens* A. Gray	50.34	2.27	22.17
Madroño	*Arbutus xalapensis* Kunth	49.09	1.85	26.45
Maguey	*Agave macroculmis* Todaro	41.32	1.35	30.43
Manrrubio	*Marrubium vulgare* L.	40.47	4.55	8.88
Níspero	*Eryobotria japonica* Lindl.	47.97	2.93	16.36
Nogal	*Carya illinoiensis* (Wang)K. Koch.	44.27	3.76	11.77
Nopal de t. año	*Opuntia ficus-indica* L.	25.53	2.35	10.82
Palo blanco	*Celtis laevigata* Willd	39.45	3.01	13.09
Salvia	*Croton suaveolens* Presl	45.16	2.33	19.36
Tronadora	*Tecoma stans* L. Kunth	48.78	3.27	14.89
Tepozan	*Buddleja cordata* Humb	45.70	3.26	14.01

town planning to reduce the CO_2 load, although this should be verified. Concerted research inputs need to be directed to screen forest trees and select species in a particular forest ecosystem in a country with a high capacity of CO_2 fixation with a good landscaping brief.

16.4 Factors affecting the productivity of forest

The productivity of forest is influenced by several physiological and environmental factors prevailing in a forest ecosystem. Here we briefly narrate a few of these parameters.

16.4.1 Transpiration

Transpiration plays a pivotal role in guiding the absorption of water and nutrients by the roots from the soil horizons and its transport and distribution in different organs for metabolic functions and to maintain water balance and the cooling effect of the leaves.

A tree is a woody upright plant has three main sections: roots, trunk (stem, bole) and crown (branches). Transpiration in plants allows for the capillary action that draws water up the plant's vascular system. This is how water gets to the leaves in a tree. Without transpiration, water could not travel against gravity up a tree. Transpiration occurs so as to cool the plant and enable the mass flow of mineral nutrients and water up from roots to shoots.

Transpiration plays a vital role to maintain a supply of water to the cells – essential for their metabolic functions. Transpiration involves complex physico-chemical processes. A tree absorbs more than sufficient water from the soil with the help of roots and the absorbed water carries dissolved minerals from the roots to the leaves through capillary xylem vessels in a watery solution, called the sap. Once the minerals are removed from the sap, the remaining excess water is removed by transpiration. The removal of excess water through transpiration creates a gradient of water potential from the roots upwards through the xylem capillary tubes to the leaves. The accumulation of water in the root cortex exerts a pressure called root pressure which pumps water in the xylem vessels upwards to the leaves at the top of the tree, called the ascent of sap. Water absorbed by roots moves up through capillary xylem vessels in a continuous stream; several factors influence the rate of transpiration, such as light intensity, temperature, relative humidity. The rate of transpiration decreases with decreasing light intensity, temperature or increasing humidity. Besides a few leaf characteristics such as stomatal intensity, size, trichome intensity, waxy coating on leaf surface, and trichrome density, the compact palisade cells and other traits reduce transpiration. Several factors influence transpiration, among which temperature, wind speed, light intensity, humidity and vapour pressure deficit are important. We will mention a few studies on some aspects of transpiration.

Kozlowski (1943) studied the transpiration rates of some forest tree species during the dormant season, showing variability between the species.

Schulze et al. (1985) determined the canopy transpiration and water fluxes in the xylem of the trunk of *Larix* and

Picea trees – a comparison of xylem flow, promoter and cuvette measurement. They concluded that plant water status greatly influences transpiration. The role of stomata and their control by environment and plant factors is discussed.

Kelliher et al. (1992) studied evaporation, xylem sap flow and tree transpiration in a New Zealand broad-leaved forest. They measured total evaporation (E), forest floor evaporation (E_f), tree xylem sap flow (F) and environmental parameters on six consecutive late-summer days under different weather conditions in a well-watered, temperate broad-leaved forest. Two tree species, *Nothofagus fusca* (Hook. f.) Oerst. (red beech) and *N. menziesii* (Hook. f.) Oerst. (silver beech), represented a vertically structured, complex canopy with a one-sided leaf area index of seven. The amount of E (determined by eddy covariance) and the difference between available energy and sensible heat flux densities was generally within 10% on half-hourly and daily bases. On clear days, the Bowen ratio gave a broad plateau of about 1–2 for most of the time with much lower and even negative values around sunrise and sunset. Variable cloudiness induced substantial variation in available energy and Bowen ratio. After rain, daytime Bowen ratios were somewhat lower and relatively constant at about 0.8 when the tree canopy was partially wet. Lysimeter measurements revealed that E_f showed a significant evaporation component and accounted for 10–20% of E, with rates up to 0.5 mm day^{-1}. Agreement between F (measured by a xylem sap flow method in a representative 337 m^2 plot of 14 trees) and tree canopy transpiration ($E–E_f$) was reasonable, with an average disparity of order

10–20% or 0.3 ± 0.1 mm day^{-1} (standard deviation) when the tree canopy was dry. Within the plot, F typically showed a difference of more than an order of magnitude. Tree position, assessed by crown emergence above the general canopy level, strongly affected an individual's contribution to the plot sap flux density. About 50% of the daily plot F emanated from only three emergent trees. Variability in the tree canopy to its aerial environment caused changes in humidity and wind speed leading to changes in stomatal and aerodynamic conductance. Thereby, the varying proportions of radiative and advective energy influenced the tree transpiration rate. Wet canopy evaporation rate was also calculated.

Nagler et al. (2003) made a comparison of transpiration rates among saltcedar, cottonwood and willow trees by sap flow and canopy temperature methods. Transpiration (E_t), measured by stem sap flow gauges, and canopy and air temperature differential ($T_c – T_a$) of *Populus fremontii* (cottonwood), *Salix gooddingii* (willow) and *Tamarix ramosissima* (saltcedar) were compared. Remotely sensed canopy temperature was used to estimate E_t or water stress in these trees in desert riparian zones of the United States and Mexico. Controlled experiments were conducted in which containerised plants were placed closely together and permitted to grow into a single, dense canopy over a summer in a desert climate. During the non-stress part of the experiment, the canopies of all three species showed similar rates of E_t, but saltcedar maintained higher E_t rates and lower r_{sv} than the native trees during stress treatments. For each species, models were developed, using both meteorological data and a canopy, energy-balance equation, to predict daily E_t and stomatal resistance (r_{sv}); these models had standard errors of 15–22% when compared with measured E_t over the unstressed portion of the experiment.

Shirke (2001) studied leaf photosynthesis, dark respiration and fluorescence as influenced by leaf age in an evergreen tree, *Prosopis juliflora*. It was observed that 92% of CO_2 fixed per day was respired by the young leaves. The fluorescence response characteristics revealed that in all types of leaves there was a complete recovery of the photochemical efficiency at sunset. The young and old leaves were photosynthetically less efficient than the mature leaves. They were well adapted to harsh environment.

Tyree (2003) studied hydraulic limits on tree performance transpiration, carbon gain and tree growth. All these are related to the influence of tree size, xylem distribution, growth conditions both within species and between species. It is concluded that high hydraulic condition is necessary both for the productivity of forest tress to have a great implication for commercial forestry.

In view of the importance of transpiration in maintaining water balance, supply of nutrients and the productivity of trees, various studies have been undertaken on the factors influencing transpiration and plant growth.

16.4.2 Photosynthesis and plant productivity

Advanced forest succession and associated accumulations of forest biomass in the Blue Mountains of Oregon and Washington and the Intermountain area have led to increased vulnerability of these forests to insects, diseases and wildfire. Tiedemann et al. (2000) adopted a strategy for the solution of forest health problems with prescribed fire; but would forest productivity and wildlife be at risk. One proposed solution considers a large-scale conversion of these forests to seral conditions that emulate those assumed to exist before European settlement: open-spaced stands (ca. 50 trees ha^{-1}), consisting primarily of ponderosa pine (*Pinus ponderosa* Laws.) and western larch (*Larix occidentalis* Nutt.). They consider here the potential effects of prescribed fire on two aspects of forest management – productivity and wildlife. They suggest a more conservative approach until prescribed fire effects are better understood.

Day et al. (2001) studied age-related changes in foliar morphology and physiology in red spruce and their influence on declining photosynthetic rates and productivity with tree age. They studied age-related trends in needle morphology and gas exchange in a population of red spruce (*Picea rubens* Sarg.) growing in a multi-cohort stand where trees ranged from first-year germinants to trees over 150 years old, as well as in grafted scions from these trees. The field study showed significant age-related trends in foliar morphology, including decreasing specific leaf area and increasing needle width, projected area and width/length ratio. Similar trends were observed in foliage from the grafted scions. Both in situ foliage and shoots of grafted scions from the oldest cohort showed significantly lower photosynthetic rates than their counter parts from younger trees; however, differences in stomatal conductance and internal CO_2 concentrations were not significant. These results suggest that: (a) the foliage of red spruce exhibits age-related trends in both morphology and physiology, (b) age-related decreases in photosynthetic rates contribute to declining productivity in old red spruce, (c) declines in photosynthetic rates result from non-stomatal limitations and (d) age-related changes in morphology and physiology are inherent in meristems and persist for at least three years in scions grafted to juvenile rootstock.

A study was undertaken by Rodrigue et al. (2002) on forest productivity and the commercial value of pre-law reclaimed mined land in the Eastern United States. They investigated the effects of mining practices used prior to the passage of the 1977 Surface Mining Control and Reclamation Act (SMCRA) on forest productivity and commercial value of reclaimed forest sites. They compared the productivity and value of 14 mined and eight non-mined sites throughout the eastern and mid-western coalfield regions. Forest productivity of pre-SMCRA mined sites was equal to or greater than that of non-mined forests, ranging between 3.3 and 12.1 $m^3 ha^{-1} year^{-1}$. Management activities such as planting pine and valuable hardwood species increased the stumpage value of forests on reclaimed mine sites. The results should serve as a benchmark for reforestation success, potential forest productivity and timber value for current reclamation activities.

Tang et al. (2005) demonstrate that tree photosynthesis modulates soil respiration on a diurnal time scale. To estimate this, they measured soil respiration and canopy photosynthesis over an oak–grass savanna during the summer, when the annual grass between trees was dead. Soil respiration measured under a tree crown showed the sum of rhizosphere respiration and heterotrophic respiration; soil respiration measured in an open area represented heterotrophic respiration. Soil respiration was estimated using solid-state CO_2 sensors buried in soils and the flux-gradient method. Canopy photosynthesis was measured from overstorey and flux measurements using the eddy covariance method. They observed that the diurnal pattern of soil respiration in the open was driven by soil temperature, while soil respiration under the tree was coupled with soil temperature. Although soil moisture controlled the seasonal pattern of soil respiration, it had no influence on the diurnal pattern of soil respiration. Soil respiration under the tree controlled by the root component was strongly correlated with tree photosynthesis, but with a time lag of 7–12 h. These results reveal that photosynthesis drives soil respiration in addition to soil temperature and moisture.

McMurtrie and Wang (2006) developed mathematical models of the photosynthetic response of tree stands to rising CO_2 concentrations and temperatures. Two models of canopy photosynthesis, MAESTRO and BIOMASS, are simulated to determine the response of tree stands to increasing ambient concentrations of CO_2 (C_a) and temperatures. The models utilise the same equations to described leaf gas exchange, but differ considerably in the level of detail employed to represent canopy structure and radiation environment. Daily rates of canopy photosynthesis simulated by the two models agree to within 10% across a range of CO_2 concentrations and temperatures. A doubling of C_a leads to modest increases of simulated daily canopy photosynthesis at low temperatures (10% increase at 10 °C), but larger increases at higher temperatures (60% increase at 30 °C). The temperature and CO_2 dependencies of canopy photosynthesis are interpreted in terms of simulated contributions by quantum-saturated and non-saturated foliage. Simulations are developed for periods ranging from a diurnal cycle to several years. Annual canopy photosynthesis simulated by BIOMASS for trees experiencing no water stress is linearly related to simulated annual absorbed photosynthetically active radiation, with light utilisation coefficients for carbon of $\varepsilon = 1.66$ and 2.07 g MJ^{-1} derived for C_a of 350 and 700 μmol mol^{-1}, respectively.

Waring et al. (2006) made an assessment of site index and forest growth capacity across the Pacific and Inland Northwest USA with a MODIS satellite-derived vegetation index. It is difficult to map forest growth potential across regions with different environmental conditions from limited field measurements of productivity. They combined both approaches in this research. They calibrated the midsummer value of NASA's MODIS instrument's enhanced vegetation index (EVI) against site indices (SI) mapped at 10 widely dispersed locations for Douglas-fir or ponderosa pine, ranging in height from 16 to 48 m at 50 years (age at breast height). Median values of EVI derived from a 3×3 km grid centered on commercial forest lands of known productive capacities produced a linear regression with site indices ($R^2 = 0.83$). They matched stand growth properties

generated by a physiologically based stand growth model (3-PG) with site-specific yield tables and inferred, both from model predictions and from the literature, that a close relation exists between maximum leaf area index (max L), maximum periodic annual increment (max PAI) and SI. They tested the ability of median EVI to predict SI values derived from height and tree age measurements made at 5263 federal inventory and analysis (FIA) survey plots in Oregon with comparable success ($R^2 = 0.53$) to that derived from previous application of 3-PG using 1 km resolution of climate and soil data. Based on the general agreement between the two approaches, they used mid-summer EVI valves to generate a 1 km resolution map predicting spatial variation in SI of Douglas-fir over 630 000 km^2 in the Pacific and Inland Northwest USA.

Norby et al. (1992) stated that increased forest growth in response to globally rising CO_2 concentrations could serve an additional sink for the response of trees to increased CO_2; however, it may be modified by the interactions of other environmental resources and stresses, higher-order ecological interactions and internal factors inherent in the growth of large, perennial organisms. In order to verify this hypothesis, they grew yellow-poplar (*Liriodendron tulipifera* L.) saplings for most of three growing seasons with continuous exposure to ambient or elevated concentrations of atmospheric CO_2. Although there was a sustained increase in leaf-level photosynthesis and lower rates of foliar respiration in CO_2-enriched trees, whole-plant carbon storage did not increase. The absence of a significant growth response is explained by changes in carbon allocation patterns, a relative decrease in leaf production and an increase in fine root production. Although these compensatory responses reduced the potential increase in carbon storage in increased CO_2 concentrations, they also stimulated the efficient use of resources over the longer term.

Chasovskykh et al. (2011) studied the biological productivity of forests. A system of spatial analysis of carbon deposition on forest cover using ADABAS and NATURAL software is suggested. The system gives a possibility for automatic actualisation of data of forest biomass plots and of data of National Forest Inventory System (NFIS) that is synchronised with the interactive map-scheme of territorial arrangement of forest cover carbon. The value of carbon capture or sink from atmosphere is determined as difference between the value of deposited carbon change and the value of its atmospheric concentration change in some time interval. This gives a possibility for monitoring the level of air pollution by carbon and other greenhouse gases.

Usoltsev et al. (2012) studied the biological productivity of boreal forests under the influence of man-made pollution. The boreal forests of Russia received indicators of biomass and primary production of spruce, fir, pine and birch stands in gradients of contamination. The results revealed that the biomass and primary productivity of the studied stands depends on the index of toxicity, but the dependence is not linear. There are two stable states (without pollution and under the influence of pollution). The transition between these states is steep.

Lakso (2013) studied aspects of canopy photosynthesis and productivity of the apple tree. He mentioned that the fundamental to the production of apple fruits from sunlight are: (a) light interception by the canopy leaves, (b) potential photosynthetic capability of the leaves, (c) factors (external and internal) that determine actual photosynthesis and (d) distribution of the photosynthetically fixed carbon to the developing organs of the tree. Whole-canopy photosynthesis is a function of the total light interception and the availability of light within the canopy. Structural aspects that affect the interception and distribution of the light to the leaves include leaf amount, distribution, inclination and folding. Uniform leaf distributions will increase interception/leaf area, required for young tree growth. Clustered leaf distributions decrease interception/leaf area, necessary for light penetration into bearing trees. Photosynthetic potentials are genetically limited but generally determined by exposure during leaf development, while tree photosynthesis is primarily light-limited.

Usoltsev (2013) investigated biomass forests of Eurasia and has an extensive database of forests and NPP for Eurasian basic forest-forming species. This currently involves the data from more than 8000 sample plots for forest biomass and more than 2600 sample plots for NPP and biomass (stems, bark, branches, foliage, roots and understorey), as well as traditional taxation indices of forests. This database covers the territory of 43 states of Eurasia.

Ivanova (2014) studied the joint growth of two tree species in Russian forests. Two options of drained habitats were studied: steep slopes of the southern exposition with small stony soils and the lower parts of gentle slopes with thick soils. To describe the biomass dynamics of pine (*Pinus sylvestris* L.) and birch (*Betula pendula* Roth, *B. pubescens* Ehrh.), their interference in the course of formation of the forest stand was constructed on the basis of the connected logistic equations of the model of their coexistence. Ivanova (2014) gives a mathematical description of two alternative ecodynamic series of forest vegetation formation in clear-cutting: regeneration of the initial forests (after cuttings in cowberry shrub pine forests) and the formation of long-derivative grass-reed grass birch forests (after cuttings in grass pine forests).

In view of the pivotal role of photosynthesis on forest productivity, significant progress has been attained on various factors influencing photosynthesis of trees species, capture of CO_2, storage of carbon and storage of potential energy in the timbers of high commercial importance. Clustered leaves do not permit uniform intereception of light and canopy photosynthesis. Uniform interception of light by evenly distributed leaves and canopy photosynthesis contribute to the productivity of tree.

16.4.3 Respiration

Respiration is a process which involves the oxidation of food with the release of energy stored in it. Several enzymes help in guiding various metabolic process in the cells.

Trees need energy for growth. This energy stored in food by photosynthesis is released from the food by the process of respiration. The respiration process occurs in the mitochondria. During the process of respiration, oxygen combines with glucose/carbohydrate and liberates stored energy for growth. Respiration uses between one-quarter and one-half of the food produced in photosynthesis. Respiration uses oxygen and releases energy, CO_2 and water. This is the reverse of photosynthesis.

The chemical reaction involved in respiration is shown below:

$$C_6H_{12}O_6 + 6O_2 \rightarrow 6H_2O + 6CO_2 + Energy$$

The oxidation of glucose occurs in two steps: glycolysis leads to the splitting of glucose, in which process glucose is converted to puruvic acid, leading to the liberation of CO_2, water and energy. During the process of glycolysis, sugar combines with organic phosphosphate to produce glucose-1-phosphate. After this, ATP liberates phosphate and combines with glucose-1-phosphate to form glucose-1,6-diphosphate. Thereafter, a series of phosphorylations leads to the formation of pyruvic acid and ATP, the energy-rich compound which helps in the transfer of energy from one compound to another. This liberaly produces various phosphorylated compounds. Thereby, the six-carbon compound, glucose, is converted to a three-carbon compound, pyruvic acid. Then, the pyruvic acid in the presence of oxygen enters the Krebs cycle, finally liberating CO_2, water and energy, called aerobic respiration. In the absence of oxygen, the pyruvic acid enters an anaerobic process, producing alcohol and less energy. The energy liberated is utilised for cell functions and tree growth maintenance. A high rate of respiration occurs in the meristematic tissue at the shoot and root tips, developing flowers, fruits and so on.

A few research advances on tree respiration are stated herein.

Temperature determines the respiration rate. Janssens et al. (2001) reported that productivity overshadows temperature in determining soil and ecosystem respiration across European forests. They presented CO_2 flux data from 18 forest ecosystems, studied in the European Union, funded by the EUROFLUX project. Overall, mean annual gross primary productivity [GPP, the total amount of carbon (C) fixed during photosynthesis] of these forests was 1380 ± 330 g C m^{-2} year^{-1} (mean $\pm$ SD). On average, 80% of GPP was respired by autotrophs and heterotrophs and released back into the atmosphere (total ecosystem respiration, TER = 1100 ± 260 g C m^{-2} year^{-1}). Mean annual soil respiration (SR) was 760 ± 340 g C m^{-2} year^{-1} (55% of GPP and 69% of TER). Among the investigated forests, large differences were observed in annual SR and TER that were not correlated with mean annual temperature. However, there was a significant correlation between annual SR and TER and GPP among the relatively undisturbed forests. They concluded that: (a) root respiration is limited by the allocation of photosynthates to the roots, related to productivity, and (b) the largest amount of heterotrophic soil respiration is derived from decomposition of young organic matter (leaves, fine roots). It is assumed that differences in SR among forests are likely to depend more on productivity than on temperature. At sites where soil disturbance has occurred (e.g. ploughing, drainage), soil respiration was a larger component of the ecosystem C budget and deviated from the relationship between annual SR (and TER) and GPP observed among the less-disturbed forests. Plant respiratory regulation is considered as too complex for a mechanistic representation in current terrestrial productivity models for carbon load and global change research.

Gifford (2003) adopted simple approaches in an attempt to capture the function of essence of several approaches used in the literature. Respiration may be calculated in terms of growth and maintenance components; and conservatism in the ratio of respiration to photosynthesis (R:P). It is suggested that the assumption of a conservative R:P ratio is an effective and practical approach in the context of C-cycle modelling for global change research and documentation, requiring minimal ecosystem-specific data on respiration. There exist several controversies thought it is to be involved in protecting the plant from "reactive oxygen species". It is reported that short-term respiratory response coefficients of plants (e.g. the Q_{10}) do not predict their long-term temperature response. It is suggested that leaf respiration is not suppressed by light as previously thought. Careful experimental results indicate that plant respiration is not suppressed by elevated CO_2 concentration in a short-term reversible way.

Ryan (1995) studied foliar maintenance respiration of subalpine and boreal trees and shrubs in relation to nitrogen content. A nitrogen (N)-based model of maintenance respiration (R_m) would link R_m with N-based photosynthesis models and enable a simpler estimation of dark respiration flux from forest canopies. In order to test whether an N-based model of R_m would apply generally to foliage of boreal and subalpine woody plants, the author measured R_m (CO_2 efflux at night from fully expanded foliage) for the foliage of seven species of trees and shrubs in the northern boreal forest (near Thompson, Manitoba, Canada) and seven species in the subalpine montane forest (near Fraser, Colorado, USA). CO_2 efflux at 10 °C for the samples was only weakly correlated with sample weight ($r = 0.11$) and leaf area ($r = 0.58$). However, CO_2 efflux per unit foliage weight was highly correlated with foliage N concentration [$r = 0.83$, CO_2 flux at 10 °C (mol kg^{-1} s^{-1}) = 2.62 × foliage N (mol kg^{-1})], and slopes were statistically similar for the boreal and subalpine sites ($P = 0.28$). CO_2 efflux per unit of foliar N was 1.8 times that reported for a variety of crop and wildland species growing in warmer climates.

Metcalfe et al. (2010) studied shifts in plant respiration and carbon use efficiency at a large-scale drought experiment in the eastern Amazon, where the effects of drought on the Amazon rainforest are potentially large and poorly understood. Here, carbon (C) cycling after five years of a large-scale through-fall exclusion (TFE) experiment, excluding about 50% of incident rainfall from an eastern Amazon rainforest, was compared with a nearby control plot. Principal C stocks and fluxes were estimated. Total ecosystem respiration (Reco) and total plant C expenditure (PCE, the sum of net primary productivity (NPP) and autotrophic respiration (Rauto), were elevated on the TFE plot relative to the control. The increase in PCE and Reco was mainly caused by a rise in Rauto from foliage and roots. Heterotrophic respiration did not have remarkable differences between plots. NPP was 2.4 ± 1.4 t C ha^{-1} year^{-1} lower on the TFE than the control. Ecosystem carbon use efficiency, the proportion of PCE invested in NPP, was found lower in the TFE plot (0.24 ± 0.04) than in the control (0.32 ± 0.04). Drought caused by the TFE treatment seemed to drive fundamental shifts in ecosystem C cycling with potentially important consequences for long-term forest C storage.

In conclusion, respiration plays a vital role in the liberation of energy stored as potential energy in plant biomass to guide various metabolic processes in the trees. Although a few studies have been undertaken on respiration of trees and ecosystem, very little study has concentrated to determine genotypic variability in respiration among species, as done in the case of photosynthesis and transpiration. Therefore, more studies need to be directed in this direction.

16.4.4   Tree responses to abiotic stresses and its growth

Karfakis and Andrade (2013) studied the dynamics of functional composition of a Brazilian tropical forest in response to drought stress, with an objective to examine the dynamics of functional composition of a non-flooded Amazonian forest, in response to drought stress in terms of diameter growth, recruitment and mortality. The survey was carried out in the continuous forest of the biological Dynamics of Forest Fragments project 90 km outside the city of Manaus, state of Amazonas, Brazil. All stems >10 cm dbh were identified to species level and monitored in 18 1-ha permanent sample plots from 1981 to 2004. For statistical analysis, all species were aggregated in three ecological guilds.

Two distinct drought events occurred in 1983 and 1997. Results revealed that more early successional species performed better than later successional ones. Response was significant for both events, but was more pronounced for the 1997 event, possibly because of the fact that the event was in the middle of the dry rather than the wet period, as was the 1983 one.

Karfakis and Volkmer-Castilho (2014) studied the growth of forest stands and tree species ecological guilds in undisturbed and selectively logged Amazonian forests, Northern Brazil with the objective of identifying patterns in individual tree diameter growth, both for the tree population overall and as a function of ecological guild in specific for undisturbed and selectively logged forest; and they compared between the two. They collected from two areas of non-flooded forest of the Brazilian Amazon in the states of Amazonas and Para for trees with dbh >10 cm. For purposes of statistical analysis, tree species were clustered in three ecological guilds based on bole wood specific gravity and successional status. Tree growth irrespective of ecological species in the vicinity of logging gaps for the three-year period following the harvest event was significantly more elevated in relation to undisturbed forest and this positive difference was proportionately greater for Amazonas than for Para. Ecological guilds showed a pattern of increasing growth rate with decreasing bole wood specific gravity and successional status when area and site disturbance status were not taken under account. In response to logging, all ecological guilds showed increased growth rates but all were proportional to levels found in the undisturbed forest. Tree species life

history characteristics appear to be a more significant factor affecting tree growth than site disturbance status in response to logging at least for the first three years after the logging event.

Karfakis (2014) studied the regeneration of commercial tree species and long-term compositional and structural changes in a logged and silviculturally treated Brazilian rainforest, 1955–1993. The study consisted of an extensive literature investigation on the long-term silvicultural experiments of Curua Una research station in the state of Para, Brazil. The objective of the work was to assess the long-term ecological and financial sustainability of a set of silvicultural treatments by using numerical figures derived from the literature related to the financial and ecological impacts of the different treatments in the scale of calendar decades. There were five treatments, including undisturbed forest (control). The remaining four were in two categories. The first included measures to increase natural regeneration of commercial species prior to exploitation. The second included measures aimed at increasing stocking of commercials after exploitation. Adequate data was taken to assess the short-term success rate of all five treatments in terms of achieving adequate stocking of commercial tree species in sizes of trees <5 cm dbh in the first few years after initiation. It was also possible to assess tree species community composition, structure and timber financial value for trees >5 cm dbh for a period of 15 years following initial treatment for one and for 36 years after initial treatment for two others. The measures used where the stocking index (S.I.) and the establishment factor (E.F) for the early results and the Shannon–Weinner species diversity index. This included total percentage volume of commercial tree species (TPCTV), total volume of potentially commercial tree species (TPTV), harvestable volume in the year 1993 (HV), number of stems ha^{-1} STD) and stand basal area (in m^2 ha^{-1}) These results suggest that a drastic revision of current legislation must take place in nations with Amazonian forest because these currently include only post-exploitation silvicultural treatments. It is clear that the Brazilian tropical shelterwood system (BTSS) was the most successful in achieving natural regeneration of commercial tree species and better performance of these overall, but only marginally so in relation to the regeneration group felling system (RGFS). However, due to the deliberate planting at high densities of *Vochysia maxima* in the MOAQ system, this ranked first in relation to all other systems, with finally the PEXTSS system ranking last in relation to all of them with a significant difference. This was true for both the stocking index and the establishment factor, meaning that there were not simply more trees of the desired species in these size classes but they were also growing and performing faster overall. A pattern of correlation between stocking index and establishment factor was noted with levels being proportional throughout the treatments. This latter pattern is a logical one as better conditions for establishment are most likely to be correlated with better subsequent survival and early growth of trees for all species. These results are in agreement for similar practices, like the PEXTSS where initial logging of most or some of the harvestable stems

takes place from the Honduras (Karfakis, 2014).

Devi et al. (2008) studied the growth of *Larix sibirica* Ledeb in extreme conditions. The objective of this study was to reconstruct stand structure and growth forms of *Larix sibirica* in undisturbed forest–tundra ecotones of the remote Polar Urals on a centennial time scale. The authors found that the forests had significantly expanded their area. The analysis of forest age structure based on more than 300 trees indicated that more than 70% of the currently upright-growing trees are <80 years old. Tree remnants of the fifteenth to nineteenth centuries were lacking almost entirely. In this way the forest had expanded northwards into the formerly tree-free tundra during the last century by about 20–60 m in altitude. This northward shift of forests was accompanied by significant changes in tree growth forms: while 36% of the few trees that were more than 100 years old were multi-stem tree clusters, 90% of the trees emerging after 1950 were single-stemmed. Tree-ring analysis of horizontal and vertical stems of multi-stemmed larch trees revealed that these trees had been growing in a creeping form since the fifteenth century. In the early twentieth century, they started to grow upright with 5–20 stems per tree individual. The incipient vertical growth led to an abrupt tripling in radial growth and, thus, in biomass production. The results show that the climatic changes are already leaving a fingerprint on the appearance and productivity in the Polar Urals.

Yermakova (2014) investigated the growth of natural and artificial young stands under biotic and abiotic stress. She revealed that the abiotic and biotic stress leads to morphological disorders of tree trunk. The incidence of violations is closely related to the intensity of stress.

In conclusion the prevalence and climatic changes and abiotic stresses in different regions such as tropical, alpine, tundra regions affect the distribution and the productivity of trees.

Bibliography

Braakman R., Smith E. **2012**. The emergence and early evolution of biological carbon-fixation. *PLOS Computational Biology* **8**(4):e1002455.

Chasovskykh V., Usoltsev V., Voronov M. **2011**. Carbon dioxide deposition and air pollution monitoring system by adabas and Natural software. *International Journal of Applied and Basic Research* **6**:69–73.

Day M.E., Greenwood M.S., White A.S. **2001**. Age-related changes in foliar morphology and physiology in red spruce and their influence on declining photosynthetic rates and productivity with tree age. *Tree Physiology* **16**:1195–1204.

Devi N., Hagedorn F., Moiseev P., Bugmann Z.H., Shiyatov S., Mazepa V., Rigling A. **2008**. Expanding forests and changing growth forms of siberian larch at the polar urals treeline during the 20th century. *Global Change Biology* **14**(7):1581–1591.

Gifford R.M. **2003**. Plant respiration in productivity models: conceptualisation, representation and issues for global terrestrial carbon-cycle research. *Functional Plant Biology* **30**(2):171–186.

INEGI. **1986**. *Síntesis Geográfica del Estado de Nuevo León.* Instituto Nacional de Estadística y Geografía, México. 184 pp.

Ivanova N.S. **2014**. Recovery of tree stand after clear-cutting in the Ural Mountains. *International Journal of Bio-resource and Stress Management* **5**(1):90–92.

Janssens I.A., Lankreijer H., Matteucci G., Kowalski A.S., Buchmann N., Epron D., Pilegaard K., Kutsch W., Longdoz B., Grünwald T., Montagnani L., Dore S., Rebmann C., Moors E.J., Grelle A., Rannik Ü., Morgenstern K., Oltchev S., Clement R., Guðmundsson J., Minerbi S., Berbigier P., Ibrom A., Moncrieff J., Aubinet M., Bernhofer C., Jensen N.O., Vesala T., Granier A., Schulze E.-D., Lindroth A., Dolman A.J., Jarvis P.G., Ceulemans R., Valentini R. **2001**. Productivity overshadows temperature in determining soil and ecosystem respiration across European forests. *Global Change Biology* **7**(3): 269–278.

Jiménez Pérez J., Treviño Garza E.J., Yerena Yamallel J.I. **2013**. Carbon concentration in pine–oak forest species of the Sierra Madre Oriental. *Revista Mexicana de Ciencias Forestales* **4**(17):50–61.

Karfakis T.N.S. **2014**. Regeneration of commercial tree species and long term compositional and structural changes in a logged and silviculturally treated Brazilian rainforest 1955–1993. *International Journal of Bio-resource and Stress Management* **5**(2):175–180.

Karfakis T.N.S., Andrade A. **2013**. Dynamics of functional composition of a Brazilian tropical forest in response to drought stress. *International Journal of Environmental, Ecological, Geological and Marine Engineering* **7**(3): 114–118.

Karfakis T.S., Volkmer-Castilho C. **2014**. Growth of forest stands and tree species ecological guilds in undisturbed and selectively logged amazonian forests, northern Brazil. *International Journal of Bio-resource and Stress Management* **5**(3):312–318.

Keller K., Yang Z., Hall M., Bradford D.F. **2003**. Carbon dioxide sequestration: when and how much. *Center for Economic Policy Studies Working Paper No. 94.* Princeton University. 23 pp.

Kelliher F.M., Köstnerb B.M.M., Hollingera D.Y., Byersa J.N., Hunta J.E., McSevenya

T.M., Meserthb R., Weira P.L., Schulze E.-D. **1992**. Evaporation, xylem sap flow, and tree transpiration in a New Zealand broad-leaved forest. *Agricultural and Forest Meteorology* **62**(1/2):53–73.

Kozlowski T.T. **1943**. Transpiration rates of some forest tree species during the dormant season. *Plant Physiology* **18**(2):252–260.

Lakso A.N. **2013**. Aspects of canopy photosynthesis and productivity of the apple tree. In: Symposium on research and development on orchard and plantation systems. *ISHS Acta Horticulturae* **114**:100–109.

McMurtrie R.E., Wang Y.P. **2006**. Mathematical models of the photosynthetic response of tree stands to rising CO_2 concentrations and temperatures. *Plant, Cell and Environment* **16**(1):1–13.

Metcalfe D.B., Meir P., Aragão L.E., Lobo-do-Vale R., Galbraith D., Fisher R.A., Chaves M.M., Maroco J.P., da Costa A.C., de Almeida S.S., Braga A.P., Goncalves P.H., de Athaydes J., da Costa M., Portela T.T., de Oliveira A.A., Malhi Y., Williams M. **2010**. Shifts in plant respiration and carbon use efficiency at a large-scale drought experiment in the eastern Amazon. *New Phytology* **187**(3):608–621.

Nagler P.M., Glenn E.P., Thompson T.L. **2003**. Comparison of transpiration rates among saltcedar, cottonwood and willow trees by sap flow and canopy temperature methods. *Agricultural and Forest Meteorology* **116**(1/2):73–89.

Norby R.J., Gunderson C.A., Wullschleger S.D., O'Neíll E.G., McCracken M.K. **1992**. Productivity and compensatory responses of yellow-poplar trees in elevated CO_2. *Nature* **357**:322–324.

Okimoto Y., Nose A., Oshima K., Tateda Y., Ishii T. **2013**. A case study for an estimation of carbon fixation capacity in the mangrove plantation of *Rhizophora apiculata* trees in Trat, Thailand. *Forest Ecology and Management* **310**:1016–1026.

Rodrigue J.A., Burger J.A., Oderwald R.G. **2002**. Forest productivity and commercial value of pre-law reclaimed mined land in the eastern United States. *Northern Journal of Applied Forestry* **19**(3):106–114.

Ryan M.G. **1995**. Foliar maintenance respiration of subalpine and boreal trees and shrubs in relation to nitrogen content. *Plant, Cell and Environment* **18**(7):765–772.

Saxton K.E., Rawls W.J., Romberger J.S., Papendick R.I. **1986**. Estimating generalized soil–water characteristics from texture. *Soil Science Society of America Journal* **50**(4):1031–1036.

Schulze E.-D., Cermak J., Matyssek R., Penka M., Zimmermann R., Vasicek F., Gries W., Kucera J. **1985**. Canopy transpiration and water fluxes in the xylem of the trunk of Larix and Picea trees – a comparison of xylem flow, porometer and cuvette measurements. *Oceologia* **66**:475–483.

Shirke P.A. **2001**. Leaf photosynthesis, dark respiration and fluorescence as influenced by leaf age in an evergreen tree, *Prosopis juliflora*. *Photosynthetica* **39**(2):305–311.

Tang J., Baldocchi D.D., Liukang X. **2005**. Tree photosynthesis modulates soil respiration on a diurnal time scale. *Global Change Biology* **11**(8):1298–1304.

Tiedemann A.R., Klemmedson J.O., Bull E.L. **2000**. Solution of forest health problems with prescribed fire: are forest productivity and wildlife at risk? *Forest Ecology and Management* **127**(1/3):1–18.

Tyree M.T. **2003**. Hydraulic limits on tree performance: transpiration, carbon gain and growth of trees. *Tree Physiology* **17**:95–100.

Usoltsev V.A. **2013**. *Forest Biomass and Primary Production Database for Eurasia*. CD-version, 2nd edn, enlarged and re-harmonized. Ural State Forest Engineering University, Yekaterinburg. ISBN 978-5-94984-438-0. (http://elar.usfeu.ru/handle/123456789/3059).

Usoltsev V.A., Vorobeichik E.L., Bergman I.E. **2012**. *Biological Productivity of Ural Forests under Condition of Air Pollutions: an Investigation of a System of Regularities*. Ural Stata Forest Engineering University, Yekaterinburg. 366 pp.

Waring R.H., Milner K.S., Jolly W.M., Phillips L., McWethy D. **2006**. Assessment of site index and forest growth capacity across the Pacific and Inland Northwest USA with a MODIS satellite-derived vegetation index. *Forest Ecology and Management* **228**(1/3):285–291.

Yermakova M.V. **2014**. Peculiarities of structure of Scotch pine young growths in the conditions of virgin and lightly disturbed forest communities in middle urals. Bulletin PGTU, a series on forest ecology. *Nature Management* **2**(22):36–45.

CHAPTER 17

Plant nutrients

17.1 Background

Plants require several macro- and micronutrients for their growth and development, called essential elements. The nutrients which are required in a larger proportion are called major nutrients, for example carbon (C), hydrogen (H), nitrogen (N), calcium (Ca), phosphorus (P), magnesium (Mg), potassium (K), oxygen (O) and sulfur (S). On the other hand, some elements are required in a smaller amount and are called minor nutrients, for example iron (Fe), zinc (Zn), chlorine (Cl), copper (Cu), molybdenum (Mo), boron (B) and manganese (Mn). Deficiencies in these micronutrients affect plant growth. The deficiency of any of these causes anomalies in plant growth and metabolism. Various studies have been undertaken in the estimation of these elements in plants.

17.2 Micro- and macronutrient contents and carbon sequestration in ten native shrubs and trees in north-eastern Mexico

The shrubs and trees of the Tamaulipan thornscrub in the semiarid region of Mexico are of great economic importance

for various uses such as timber for furniture, fences and firewood. They are also sources of forage for wild grazing animals because they possess macro- and micronutrients required by the animals.

Leaves contribute greatly to plant growth and productivity through photosynthesis and nutrient content. There is a great diversity among plant species in growth form, leaf size, leaf shape and canopy management. In addition, there exist some general relationships across a wide range of species in leaf traits which determine the carbon fixation strategy between species. The outer canopy leaves and specific leaf area (SLA, leaf area per unit mass) tend to be correlated with leaf nitrogen per unit dry mass, photosynthesis and dark respiration (Wright et al., 2001).

A large variation among species with traits favouring nutrient conservation allows short-term rapid growth. Species having nutrient conservation have a long life span, high leaf mass per area, low nutrient concentration and low photosynthetic capacity (Reich et al., 1997). The availability of nutrients in leaves is essential for efficient plant function. Chapin (1980) reviewed the nature of crop responses to nutrient stress and compared these responses to those of species that evolved under more natural conditions.

Autoecology and Ecophysiology of Woody Shrubs and Trees: Concepts and Applications, First Edition.
Edited by Ratikanta Maiti, Humbero Gonzalez Rodriguez and Natalya Sergeevna Ivanova.
© 2016 John Wiley & Sons, Ltd. Published 2016 by John Wiley & Sons, Ltd.

He emphasised nutritional studies of nitrogen and phosphorus because these elements most commonly limit plant growth (Chapin, 1980). The nutrients present in leaves contribute to plant growth and metabolism. Sufficient research activities have been undertaken on nutrient content and metabolism in leaves. Leaf nutrient content depends on the availability of nutrients present in the soil habitat. Nutrient-poor habitats tend to be dominated by nutrient-conserving species, while fertile habitats tend to be dominated by species with higher short-term productivity per leaf mass (Chapman, 1990).

As leaves age, nutrient resorption occurs and nutrients are withdrawn from leaves prior to abscission and reemployed in the developing tissues (leaves, fruits, seeds). Resorption occurs throughout a leaf life, particularly when the leaves are shaded (Ackerly and Bazzaz, 1995). A major phase of resorption occurs shortly before leaf abscission, which is a highly ordered process of leaf senescence occurring in most species (Nooden, 1988). Around 50% of leaf N and P is recycled via resorption (Aerts, 1996). It is emphasized that the presence of active nutrient sinks has control over resorption (Negi and Singh, 1993).

Rengel and Marschner (2005) studied nutrient availability and management in the rhizosphere and observed that genotypic differences in plants exposed to nutrient deficiency activate a range of mechanisms that lead to increased nutrient availability in the rhizosphere compared with bulk soil. In this respect, plants may change their root morphology, increase the affinity of nutrient transporters in the plasma membrane and exude organic compounds (carboxylates, phenolics, carbohydrates, enzymes etc.). Chemical changes in the rhizosphere lead to an altered abundance and composition of microbial communities. Nutrient-efficient genotypes are adapted to environments with low nutrient. Therefore, understanding the role of plant–microbe–soil interaction governing nutrient availability will enhance environmental sustainability.

Leaf physiology contributes to the growth and development of trees. Wright et al. (2001) developed a strategy that shifted leaf physiology, structure and nutrient content between species in high and low rainfall and high and low nutrient habitats. Most plants absorb nutrients back from their leaves with advancing age. The proportions of nutrients resorbed and the residual nutrient concentration in senesced leaves are different. A major spectrum of strategic variations occurs in plant species with a long life span, high leaf mass per unit area, low leaf nutrient concentrations and low photosynthetic capacity. Quantification of green leaf and senesced leaf N and P concentrations revealed that leaf nutrient concentrations in green and senesced leaves were positively correlated with leaf length across all species and at most sites, excluding nitrogen fixing species. Proportional resorption did not differ with soil nutrients. The results suggest that nutrient losses have affected the residual nutrient concentration rather than proportional resorption per se.

Research has been undertaken on plants which contain nutrients useful for ruminants and wild animals. Lukhele and Van Ryssen (2003) carried out a study on the chemical composition and potential value of subtropical tree species of *Combretum* in southern Africa for ruminants. It was concluded that the foliage tested would not provide suitable resources of N to supplement protein deficiencies in low quality herbage.

The consequences of global climatic change will generate an eventual increase in the aerial temperature and an increase of green house gases, particularly CO_2. Scientists have come to the conclusion that the concentration of CO_2 in the atmosphere, the product of burning of fossil fuels for the generation of energy and deforestation world-wise, has increased to 25%. Forest plantation is done to capture and retain carbon. Forests play an important role in the global C cycle (Brown, 1999).Global warming is attributed to the increased concentration of various gases (known as the greenhouse gas effect): carbon dioxide, methane, nitrous oxide, sulfur dioxide, chlorfluorocarbon, ozone and water vapour. These gases absorb 90% of the infrared radiation, resulting in an average surface temperature of 15 °C (Martínez et al., 2004).

The excess atmospheric carbon released into the atmosphere can be absorbed by photosynthesis by trees and ecosystems (Rodríguez Laguna et al., 2008). This system is called carbon sequestration and can make a significant contribution to reducing global warming (Pimienta de la Torre et al., 2007).

In addition to the estimation of plant nutrient, studies were undertaken to evaluate the nutritive values of plants in the forest; and a few of them are narrated here.

In a study carried out by Alvarado et al. (2013) the contents of Cu, Fe, Mn and Zn were seasonally determined in the leaves of the native shrubs *Castela erecta, Celtis pallida, Forestiera angustifolia,* *Lantana macropoda* and *Zanthoxylum fagara* that are browsed by the white-tailed deer (*Odocoileus virginianus*). The study was conducted in the Tamaulipan thornscrub region, north-eastern Mexico, between summer 2004 and spring 2006. Leaf tissues were collected from three county sites (China, Linares, Los Ramones). Mineral contents were measured using an atomic absorption spectrophotometer. All minerals, in all plants, were significantly different between sites, seasons and years, as well as between double and triple interactions. In general, plants at the Linares site, which had the highest rainfall, had a higher trace element content. Moreover, during the second year, all plant species were higher in micromineral content. Furthermore, during summer these were also higher. Regardless of spatiotemporal variations, all plants had suitable levels of Fe and Mn to satisfy the adult range of white-tailed deer requirements. The Cu and Zn presented a marginal deficiency in some plants especially during the dry seasons (winter and spring). Seasonal variations in minerals could be associated to soil water deficits, excessive irradiance during summer and extreme low temperatures during winter and spring that could have affected leaf development and micromineral concentrations.

The presence of nutrients in plants add nutritive value for the grazing animals in the forest. Several studies have been undertaken to estimate the feed values of various plant species.

Ramírez et al. (2005) estimated the nutritive value and rumen digestion of total plant, leaves and stems of *Dichanthium annulatum*. Male rumen fistulated Pelibuey x Rambouillet sheep, fed alfalfa hay, were used to incubate nylon bags. The crude protein (CP) content was higher in spring and lower in winter ($P < 0.05$). It was higher ($P < 0.05$) in leaves than stems. The cell wall (CW) and its components (hemicellulose, cellulose), insoluble ash and lignin were lower ($P < 0.05$) in spring and higher in winter and were higher ($P < 0.05$) in stems than in leaves. Dry matter, CP and CW of *D. annulatum* were digested by rumen microbes in sheep to a greater extent ($P < 0.05$) during spring than in winter. Higher ($P < 0.05$) nutrient digestion was observed in leaves than in stems. In general, the mineral contents were higher during spring and lower in winter. During spring when precipitation was high (417 mm out of 613 mm), the nutrient content was higher than that in other seasons. Results suggest that, except in winter, *D. annulatum* can be used in all seasons as a good source of nutrients for grazing beef cattle.

Ramirez et al. (1999) evaluated the feed value of foliage from *Acacia rigidula*, *A. berlandieri* and *A. farnesiana*. The seasonal dynamics of the nutritive value and digestion of dry matter (DM) and crude protein (CP) in leaves of *A. rigidula*, *A. berlandieri* and *A. farnesiana*, native shrubs from north-eastern Mexico, was evaluated. An in situ technique was used to estimate the effective degradability of DM (EDDM) and CP (EDCP), using rumen fistulated sheep. In general, the in situ digestibility and EDDM and EDCP were higher ($P < 0.001$) in *A. farnesiana* than in other species. Moreover, the DM and CP in *A. farnesiana* was more highly degraded in the rumen of sheep during summer and fall than in winter. There were little differences among seasons in DM or CP

degradability between *A. berlandieri* and *A. rigidula*. It seems that the high level of condensed tannins in leaves of *A. rigidula* and *A. berlandieri* had detrimental effects on the DM or CP digestion in the rumen of sheep. Thus, leaves from *A. farnesiana* may be considered as a good supplement for grazing ruminants.

Alvarado et al. (2012) evaluated the nutritional value of the native shrubs: *Acacia amentacea*, *Castela erecta*, *Celtis pallida*, *Croton cortesianus*, *Forestiera angustifolia*, *Karwinskia humboldtiana*, *Lantana macropoda*, *Leucophyllum frutescens*, *Prosopis laevigata*, *Sideroxylon celastrinum* and *Zanthoxylum fagara*. Leaf samples were collected seasonally during two years from Summer 2004 to Spring 2006 in three county sites: China, Linares and Los Ramones, in the state of Nuevo Leon, Mexico, and evaluated for the content of their crude protein (CP), neutral detergent fiber (NDF) and acid detergent lignin (ADL). Dry matter digestibility (DMD) was also estimated. The CP content (range of total means = 13–22% dry matter) in most plants significantly varied between sites and seasons and between years. The same pattern occurred for NDF (40–55), ADL (8–22) and DMD (48–73). *Celtis pallida* had the highest nutritional value. However, due to their high CP and DMD and low NDF and ADL, all studied plants in all sites, most seasons and years may be considered as good food sources for white-tailed deer.

Domínguez-Gómez et al. (2014) determined macro- (Ca, K, Mg, Na, P) and micronutrient (Cu, Fe, Mn, Zn) foliar content in *Acacia amentacea* (DC.), *Celtis pallida* (Torr.), *Forestiera angustifolia* (Torr.) and *Parkinsonia texana* (A. Gray). Leaf samples were collected monthly from January throughout December 2009 in China, Linares and Los Ramones counties in the state of Nuevo Leon, Mexico. All nutrients were significantly different between sites, months and species; double and triple interactions were also significant. Among the sites, samples from Los Ramones county showed the higher macronutrient content, followed by China and Linares. As for species, *Celtis pallida* showed the highest values, while *A. amentacea* had the lowest macronutrient content. Micronutrient content showed the following decreasing order: China > Los Ramones > Linares. *Parkinsonia texana* was characteristically the most abundant in micronutrient content, while *A. amentacea* was the poorest in this respect. Regardless of species, site or timing, Ca (total mean = $30\,g\,kg^{-1}$), K (15; except *A. amentacea*), Mg (5.2; except *A. amentacea*), Na (1.9; only *F. angustifolia*), Fe ($100\,mg\,kg^{-1}$), Mn (51; only *F. angustifolia*) and Zn (35; except *A. amentacea*) were determined to be present in suitable amounts to meet nutritional requirements of range ruminants, while Na (1.9; except *F. angustifolia*), P ($1.3\,mg\,kg^{-1}$) and Cu ($6\,mg\,kg^{-1}$) were marginally deficient throughout the year.

Ramírez-Lozano et al. (2010) determined spatio-temporal variations of macro and trace elements in six native plants consumed by ruminants in north-eastern Mexico, that is the Ca, K, Mg, Na, P, Cu, Fe, Mn and Zn contents seasonally, in foliar tissue of native trees (T) and shrubs (S) from north-eastern Mexico, such as *Acacia rigidula* (S), *Bumelia celastrina* (T), *Croton cortesianus* (S), *Karwinskia humboldtiana* (S), *Leucophyllum frutescens* (S) and *Prosopis laevigata*. All minerals in all plants were significantly different between years, sites and seasons; however, some interactions were not significant. In general, plants in Linares site had higher mineral content, followed by Los Ramores and China; in addition, during year two, all plants had higher mineral content; moreover, during summer all plants had higher mineral content, followed by fall, spring and winter. Yearly and seasonal variations in plant minerals might have been related to seasonal water deficits, excessive irradiance levels during summer and extreme low temperatures in winter that could have affected leaf development and senescence. In spite of these differences, all plant species had suitable levels of Ca, Mg, K, Fe and Mn to satisfy grazing ruminant requirements.

In the context of the above, the present study is aimed to estimate the contents of nine micro- and macronutrients including carbon sequestration of 10 native woody plants of the semiarid region, north-eastern Mexico.

17.3 Methodology

This study was carried out at the experimental station of Facultad de Ciencias Forestales, Universidad Autonoma de Nuevo Leon, located in the municipality of Linares (24°47'N 99°32'W), at an elevation of 350 m. The climate is subtropical or semiarid with a warm summer; monthly mean air temperature varies from 14.7 °C in January to 23 °C in August, although during summer the temperature goes up to 45 °C. Average annual precipitation is around 805 mm with a bimodal distribution. The dominant type of vegetation is the Tamaulipan thornscrub or subtropical woodland (SPP-INEGI, 1986). The dominant soil is deep, dark grey,

Table 17.1 The native woody species included in this study.

Common name	Scientific name	Family
Cenizo	*Leucophyllum frutescens* (Berland.) I.M. Johnst.	Scrophulariaceae
Chaparro Prieto	*Acacia rigidula* Benth	Fabaceae
Coma	*Bumelia celastrina* Kunth.	Sapotaceae
Huajillo	*Acacia berlandieri* Benth.	Fabaceae
Anacahuita	*Cordia boissieri* A. DC	Boraginaceae
Granjeno	*Celtis pallida* Torr	Cannabaceae
Panalero	*Forestiera angustifolia* Torr	Oleaceae
Barretilla	*Amyris texana* S. Watson.	Rutaceae
Oreja de rata	*Bernardia myricifolia* (Sheele) S.Wats.	Euphrobiaceae
Lantana	*Lantana macropoda* Torr	Verbenaceae

lime-clay, vertisol with montmorrillonite, which shrinks and swells remarkably in response to changes in moisture content.

Table 17.1 shows the native woody species included in this study.

17.3.1   Chemical analysis

Mature leaf samples (1.0 g dry weight) obtained from each plant and shrub species was used for determining the contents of minerals (Cu, Fe, Mn, Zn, Ca, Mg, K and P). Mineral content was estimated by incinerating samples in a muffle oven at 550 °C during 5 h. Ashes were digested in a solution containing HCl and HNO_3, using the wet digestion technique (Cherney, 2000). Concentrations of Ca (nitrous oxide/acetylene flame), Cu, Fe, Mn, Zn, K and Mg (air/acetylene flame) were determined by atomic absorption spectrophotometry (Varian, model SpectrAA-200), whereas P was quantified spectrophotometrically at 880 nm using a Perkin–Elmer spectrophotometer (Model Lamda 1A). Carbon and nitrogen foliar contents (% dry mass basis) were carried out using 0.020 g of milled and dried leaf tissue in a CHN analyser (Perkin–Elemer, model 2400). Data collected were analysed by means of one-way analysis of variance and means were separated by the Duncan New Multiple Range test at 5% level of probability (Steel and Torrie, 1980).

17.4   Results

The results on micro- (Cu, Fe, Mn, Zn) and macronutrients (Ca, K, Mg, P) are shown in Figures 17.1 and 17.2, respectively. The C, N and the ratio C:N content in leaf tissue are shown in Figure 17.3.

Cu content values ranged from 3.0 to 13.0 $\mu g\,g^{-1}$ dw. *Lantana* contained high value (13 $\mu g\,g^{-1}$ dw) followed by *A. texana* (9.18 $\mu g\,g^{-1}$ dw), *Bernardia myricifolia* (8.03 $\mu g\,g^{-1}$ dw), while other species had minimum values, *Acacia berlandieri* (3.52 $\mu g\,g^{-1}$ dw) and *Cordia boissieri* (3.0 $\mu g\,g^{-1}$ dw).

With respect to Fe, the mean value ranged from 118.12 to 145.0 $\mu g\,g^{-1}$ dw. The species which contained high values were *Lantana macropoda* (145.81 $\mu g\,g^{-1}$ dw),

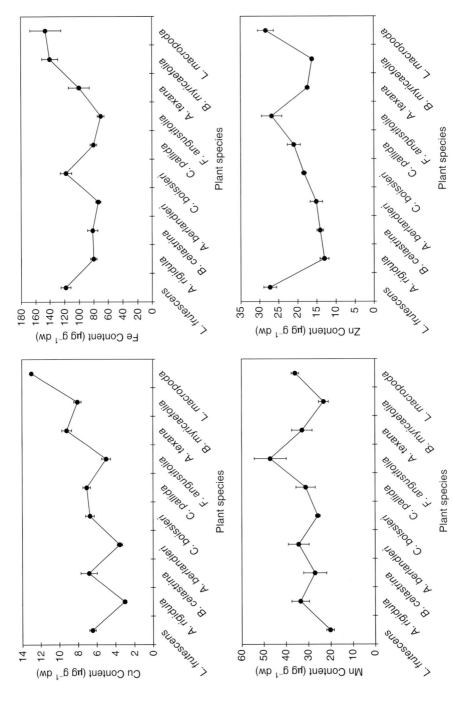

Figure 17.1 Cu, Fe, Mn and Zn content in leaf tissue of different native plant species, north-eastern Mexico.

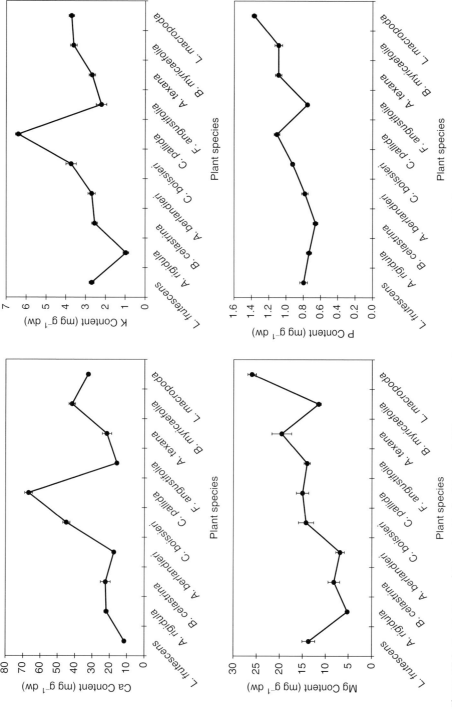

Figure 17.2 Ca, K, Mg and P content in leaf tissue of different native plant species, north-eastern Mexico.

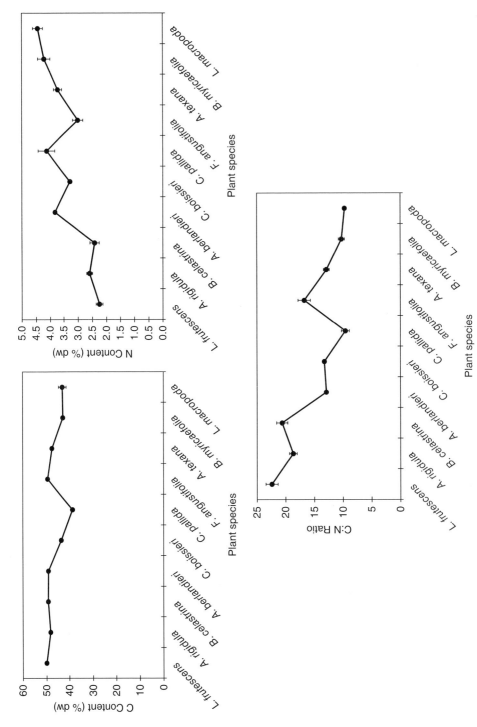

Figure 17.3 Carbon (C), Nitrogen (N) and C:N content in leaf tissue of different native plant species, north-eastern Mexico.

Bernardia myricifolia (133.73 µg g^{-1} dw), *Leucophyllum frutescens* (118.12 µg g^{-1} dw), followed by *Bumelia celastrina* (81 µg g^{-1} dw), *Acacia rigidula* (80 µg g^{-1} dw), *Acacia berlandieri* (78.46 µg g^{-1} dw) and *Forestiera angustifolia* (69.95 µg g^{-1} dw).

Mn ranged from 21.15 to 47.0 µg g^{-1} dw. The species which contained high values were *Forestiera angustifolia* (47.0 µg g^{-1} dw), *Lantana macropoda* (35.81 µg g^{-1} dw), *A. texana* (32.74 µg g^{-1} dw). Minimum values were found in *Leucophyllum frutescens* (20.22 µg g^{-1} dw) and *Bumelia celastrina* (25.67 µg g^{-1} dw).

The mean value of Zn ranged from 12.01 to 32.46 µg g^{-1} dw. The species which contained the highest value was *Lantana macropoda* (32.46 µg g^{-1} dw), followed by *Leucophyllum frutescens* (27.23 µg g^{-1} dw), *Forestiera angustifolia* (26.73 µg g^{-1} dw), *Celtis pallida* (20.97 µg g^{-1} dw), *Cordia boissieri* (18.33 µg g^{-1} dw), *Amyris texana* (17.40 µg g^{-1} dw), *Bernardia myricifolia* (16.17 µg g^{-1} dw), *Acacia berlandieri* (15.08 µg g^{-1} dw), *Bumelia celastrina* (14.08 µg g^{-1} dw), *Acacia rigidula* (13.01 µg g^{-1} dw).

The mean value of P ranged from 0.6 to 1.37 µg g^{-1} dw. *Lantana macropoda* had a high value (1.37 µg g^{-1} dw), followed by *Acacia rigidula* (0.79 µg g^{-1} dw), *Bumelia celastrina* (0.6 µg g^{-1} dw).

The mean value of Mg ranged from 5.21 to 26.04 µg g^{-1} dw. *Lantana macropoda* had a high value (26.04 µg g^{-1} dw) followed by *.A. texana* (19.56 µg g^{-1} dw), *Celtis pallida* (15.03 µg g^{-1} dw), *Cordia boissieri* (14.26 µg g^{-1} dw), *Forestiera angustifolia* (14.03), *Leucophyllum frutescens* (13.69 µg g^{-1} dw), *Bernardia myricifolia* (11.54 µg g^{-1} dw), *Bumelia celastrina* (8.15 µg g^{-1} dw), *Acacia berlandieri* (6.80 µg g^{-1} dw) and *Acacia rigidula* (5.21 µg g^{-1} dw).

There existed large variability in Ca content among species. The mean value of Ca ranged from 11.50 to 66.78 µg g^{-1} dw. The species *Celtis pallida* contained the maximum content (66.78 µg g^{-1} dw) followed by *Cordia boissieri* (44.95 µg g^{-1} dw), *Bernardia myricifolia* (41.79 µg g^{-1} dw), *Lantana macropada* (32.4 µg g^{-1} dw), *Acacia rigidula* (21.91 µg g^{-1} dw), *Amyris texana* (21.69 µg g^{-1} dw), *Forestiera angustifolia* (15.8 µg g^{-1} dw), *Acacia berlandieri* (17.55 µg g^{-1} dw), *Leucophyllum frutescens* (11.5 µg g^{-1} dw).

With respect to carbon fixation, there existed a large variability in carbon fixation among the species studied. Among them, the species that fixed a high carbon content were: *Leucophyllum frutescens* (49.97%), *Forestiera angustifolia* (49.47%), *Bumelia celastrina* (49.35%), *Acacia berlandieri* (49.18%), followed by *Acacia rigidula* (48.23%), *Cordia boissieri* (43.43%), *Lantana macropoda* (42.91%), *Bernardia myricifolia* (42.69%) and the minimum by *Celtis pallida* (38.66%). The mean value of N content ranged from 2.25 to 4.43%. Among the species studied, *Bernardia myricifolia* and *Lantana macropoda* accounted for 4.43%, *Celtis pallida* 4.12%, and they contained the higher values. *Acacia berlandieri* achieved 3.82%, *Acacia texana* 3.72%, *Cordia boissieri* 3.28%, *Forestiera angustifolia* 3.0%, *Acacia rigidula* 2.6%, *Bumelia celastrina* 2.42%, *Leucophyllum frutescens* 2.25%.

In regard to carbon/nitrogen, there were large variations in C:N ratio. The highest C:N ratio was obtained in *Leucophyllum frutescens* (22.39) followed by *Bumelia celastrina* (20.6), *Acacia rigidula* (18.87), *Acacia berlandieri* (17.3), *Forestiera angustifolia* (16.73), *Cordia boissieri* (13.24), *Amyris texana* (12.66), *Lantana macropoda* (9.69) and *Celtis pallida* (9.69).

17.5 Discussion

The leaves of native plants studied are good sources of nutrients fpr grazing ruminants and wild animals like white-tailed deer in the Tamaulipan thornscrub, north-eastern Mexico. The chemical analysis reveals that all the species contained higher nutrients required for the ruminants as has been found in other studies (Khan et al. 2007a, b). Ruminants are tolerant of excess Mn in dietary components. A study was undertaken in Sooney valley in Pakistan to assess the concentrations of mineral elements such as Mn, Fe and Zn. Micronutrient concentrations for forage varied in the leaves and pods among species: Mn with ranges of 3.92–5.09 and 5.9–6.83, Zn 0.027–0.076 and 0.028–0.064, Fe 20.72–25.43 and 25.35–32.94, Cu 0.38–054 and 0.34–0.51 $mg\,g^{-1}$, respectively. The plants showed significant differences for Zn and Mn contents of leaves and non-significant differences for pods. It is concluded the forage species were found to possess the required amounts of nutrients necessary for grazing anaimals (Ahmad et al., 2008). Cu concentration found in the present study was much higher than the critical level for the requirement of livestock as suggested by Anon 1964 (cited by Ahmad et al., 2008). Ruminants are more sensitive to Cu toxicity than non-ruminants, the tolerance range of Cu is 100 $mg\,kg^{-1}$ cattle and 25 $mg\,kg^{-1}$ for sheep (Anon, 1964; cited by Ahmad et al., 2008).

The Zn content found in all the species was much higher than the critical level suggested by Mayland et al. (1980) and Khan et al. (2007a). Lukhele and Van Ryssen (2003) undertook a study on chemical composition and potential value of subtropical tree species of *Combretum* in southern Africa for ruminants. It was concluded that the foliage tested would not be a suitable source of N to supplement protein deficiencies in low-quality herbage. In another study, concentrations of K, Fe, Zn, Cu, Cr, Ni and Co were analysed in several taxa of *Quercus*. No specific differences in section level were observed but differences at specific and interspecific levels were significant. There were variations in different elements in different taxa (Özcan and Baycu, 2005). The chemical composition of 80 species of trees with forage value was undertaken. The crude protein of trees starting at 7–8% varied among species (González et al., 2007). Studies were undertaken on the macromineral status (Ca, K, Mg, Na and contents) at three phenological stages of some range shrubs of the Gadon hills, district Swabi, Keybeer Pukhtunkhw, Pakistan. Variations in macromineral contents varied among species at different phenological stages (Sher et al., 1912). A study was undertaken on litter phosphorus (P) and nitrogen (N) dynamics in shifting cultivation of maize in the Yucatan Peninsula, Mexico (Read and Lawrence, 2003). Litter P concentration declined with forest age, while litter N did not differ between age classes. Average P concentration from the southern-west region was 0.87 $mg\,g^{-1}$ almost double the litter P concentration in the drier central and northern regions (0.44 and 0.45 $mg\,g^{-1}$, respectively). Average N concentration range was 1.1–1.2%. It was concluded that the litter nutrient dynamics in the secondary dry tropical forests are strongly influenced by water and nutrient availability.

The nutrient analysis of 10 native woody species revealed that all the species contained high amount of both macro- and micronutrients required for plant growth and metabolisn as well as adaptation to semiarid environments. In this respect, sufficient research inputs were undertaken on leaf traits (Wright et al., 2001), nutrient conservation (Reich et al., 1997), nutrient availability in the habitat (Chaplin, 1980; Grime, 1997; Rangel and Marschner, 2005) and nutrient resorption from senescence leaves before abscission (Negi and Singh, 1993; Ackerly and Bazzaz, 1995).

The present study showed a large variability in carbon fixation (carbon sequestration) among the species studied, ranging from 38.66 to 49.97%. Among them, the species fixing high carbon were: *Leucophyllum frutescens* (49.97%), *Forestiera angustifolia* (49.47%), *Bumelia celastrina* (49.35%) and *Acacia berlandieri* (49.18%). Interestingly, all these species have an open canopy indicating a high capacity for photosynthesis and carbon sequestration; all these four species also contain high macro- and micronutrients associated with carbon fixation. A similar study was undertaken by Yerena-Yamallel et al. (2011) in the Tamaulipan thornscrub. The concentration ranged from 44.25 to 47.08%. *Cordia boissieri* (44.25%), *Acacia farnesiana* (44.52%), *Forestiera angustifolia* (47.08%). A study was undertaken to quantify carbon accumulation in the aerial biomass of a temperate community of northern Durango, Mexico. The concentration of carbon for *Pinus* spp. and *Quercus* was 0.56 and 0.54, respectively (Silva-Arredondo and Návar-Cháidez, 2009). The expansion factors are parameters which transform the volume of trees or unit of surface area in carbon density in each vegetation (Garcia et al., 2004). The constant increase of CO_2 and emission of greenhouse gas in the atmosphere has a great impact on global warming, associated with fossil fuel combustion. The increase of CO_2 emissions in the atmosphere is leading to an increase in global warming; but gases utilised by plants should mitigate the increase of CO_2 through the process of photosynthesis. The plants of the forests convert this gas through their accumulation of carbon during photosynthesis; thereby the carbon is stored in trees. Therefore, the sustainable management of habitats in Mexico where large extensions of forests have the capacity to capture carbon is essential (Navar et al., 2005).

The carbon concentrations of several biomass components of 25 trees were quantified with the help of an automatic equation to estimate biomass and carbon concentration. The carbon concentration factors were modelled as a function of the dasometric factors of the trees. The results revealed that the carbon expansion factors are dependent on environmental height (Silva-Arredondo and Návar-Cháidez, 2009). It is estimated that around 50% of dry weight is deposited as carbon (Husch et al., 2003). A study carried out by Yerena Yamall et al (2012) in 21 species of conifers found that the carbon content ranged from 45.67% in *Pinus remota* to 51.18% in *Juniperus flaccida*.

17.6 Conclusions and future studies

The present study was undertaken to estimate nine macro- and micronutrient contents (Cu, Fe, Mn, Zn, Ca, K, Mg, P, C and N) of ten native woody species in Linares, north-eastern region of Mexico, that is *Leucophyllum frutescens, Acacia rigidula, Bumelia celastrina, Acacia berlandieri, Cordia boissieri, Celtis pallida, Forestiera angustifolia, Amyris texana, Bernardia myricifolia* and *Lantana macropoda*. Large variations were observed between these species in their nutrient contents.

The values of mineral contents were much higher than those required by grazing ruminants. The mean value of C fixation (sequestration) ranged from 36.66 to 49.97%. The species capturing high carbon (about 50%) were: *Leucophyllum frutescens, Forestiera angustifolia, Bumelia celestrina* and *Acacia berlandieri*. These species may be recommended for plantations in highly contaminated areas to reduce pollution.

The concentration of nutrients in leaves is influenced by environmental factors, both climatic and edaphic. Therefore, there is a necessity to estimate the nutrient contents of the same species grown in different localities and also to study its relation to growth in different environments. Besides, there is a necessity to select more species with high carbon fixation which have the capacity to fix high carbon from an atmosphere contaminated with a high CO_2 load. There is also a necessity to establish the relationship, if any, with leaf canopy and crown architecture.

17.7 Research needs for the conservation of native trees and carbon sequestration

Increasing global warming owing to an increase of greenhouse gases associated with drought and other abiotic stresses is causing great concern and endangering native plant species/trees. This in turn is affecting crop productivity under sustainable agriculture. This is also associated with a tremendous increase in the human population, leading to increased hunger and poverty. In this situation there is a great necessity for research on the conservation of native plant species, with special reference to trees.

With respect to native plant species, there is a great necessity to develop efficient techniques for breaking seed dormancy and the propagation of plants and trees. No efficient techniques are available except traditional methods. In this regard, in north-east Mexico a few native species such as *Agave lecheguilla, Yucca* spp, *Euphorbia antisyphilitica* and wild chilli, *Capsicum annuum*, are overexploited by the semiarid farmers for their economic importance, which may lead to the extinction of these species in the long term. Many cacti are also endangered and disappearing. Though simple techniques are available, no attempts are being made by scientists to enhance plant propagation. We have developed simple techniques for inducing seed germination in about 40 *Cactus* spp. and also *Capsicum annuum*.

With respect to forest, climate change due to increase of CO_2 and other greenhouse gases have a direct impact on the growth and survival of tree species. The climatic factors prevailing in different semiarid, tropical and temperate regions affect the growth of trees. The changes in climate factors related to atmospheric conditions are in the solar radiation, light, air, temperature, precipitation, relative humidity and intensity of light.

Climate determines the distribution of vegetation in a forest ecosystem. There exists a good relation of climate with the conservation and development of forestry. It is essential for foresters to have a good understanding of climate change and its impact on forest productivity and to take necessary measures to protect the forest. It is difficult to determine the effects of climate change on plant growth and human activity and to adopt effective measures to mitigate them.

We are very much concerned about how the human activities such as the burning of fossil fuels, conversion of forests to agricultural lands and other illegal activities cause a significant increase of CO_2 and other greenhouse gases in the atmosphere. On the other hand, both forests and human use of forest products contribute to a gradual increase of greenhouse gases in the atmosphere. Fortunately, the trees and forests with their ability to absorb CO_2 and carbon have an opportunity to mitigate climate change. Greenhouse gases cause the retention of heat in the lower atmosphere due to absorption and re-radiation by clouds and other gases.

The earth receives radiant energy from solar radiation for its utilisation by plants for photosynthesis and other human activities. The short-wave solar energy (visible) received from the sun passes through the atmosphere, thereby warming the earth's surface. Long-wave thermal radiation is absorbed by a number of greenhouse gases. These greenhouse gases accumulate in small amounts in the atmosphere and reflect long-wave thermal radiation in all directions. Some of the radiation is directed towards the earth's surface. The amount of greenhouse gases in the atmosphere influences global temperature. The greenhouse gases affecting climate change at the earth's surface are water vapour (H_2O), carbon dioxide (CO_2), nitrous oxide (NO_2), ozone (O_3), carbon monoxide (CO) and chlorofluorocarbon (CFC). With increasing agriculture, animal husbandry, over-grazing and an increase in human population indirectly increasing the level of these greenhouse gases, increased global warming thereby endangers the security of life for humans, animals and so on. Over and above, an incessant logging of trees for timber has a direct impact in the increase of greenhouse gases. An increase in the accumulation of these gases has a direct effect on forest growth and productivity.

Tree species have the capacity to capture CO_2 from the atmosphere, thereby reducing this gas noxious for human health and saving our lifetime carbon concentration. This in turn leads to the accumulation of carbon, the source of energy for fuel after combustion; this is of high commercial importance for the wood industry. Tree species having a high capacity for carbon sequestration (carbon fixation) could have a high potential for reducing the CO_2 load through the process of photosynthesis.

We reported that trees with an open canopy are expected to be more efficient in photosynthesis due to their capacity to capture solar radiation and greater carbon fixation. We selected a few tree species with an open canopy with about 50% carbon sequestration. This could be a potential line of research for forest scientists. In this respect, we raised a question in the natural resource management (NRM) group in Linkedin: Tree species with high carbon fixation could be effectively used in planting in highly contaminated areas, factory sites and cities with a high carbon load in their atmosphere.

In response to my discussion on the importance of tree species with high carbon fixation for reducing the CO_2 load in polluted areas, various arguments were raised among experts in the NRM group in Linkedin. Several of them support this concept, while others have reservations. We want to mention the comments of a few of them. Franklin Oppong-Obiri commented that the natural systems in the world have their own mechanisms to tackle the threats from external sources, such as human activities. The capacity of forests to absorb carbon emissions is a vital aspect in managing climate change impact. In this respect, incessant logging has caused the depletion of forest resources, with catastrophic consequences. It causes great changes in the environment, particularly weather patterns and its effects on agricultural productivity. Therefore, to counter this trend is the RED+ (Reducing Emissions and Deforestation) where tree plantations are being encouraged so that carbon offsets can be calculated as dividends for their owners. He recommends that the planting of trees with a high carbon fixation within highly contaminated areas could be considered a practical step in redevelopment efforts in town planning.

Marie-Paule Nougaret mentioned in her book that, in C3 photosynthesis, trees have two kinds of highly efficient carbon fixation: with the addition of extra CO_2 juvenile trees grow faster, but may also have a shorter life. She gave a few examples. She added that leaves can absorb some licuens and soil aerobic bacteria and may break petroleum-originated gaseous components much faster when some plant is growing on top. I guess C4 grasses might be better for quick results. Her book has 400 scientific references, with about 300 plants, 90 cities, 90 pollutants quoted in the index (http://www.actes-sud.fr/catalogue/ecologie-developpement-durable/la-cite-des-plantes).

Franklin Oppong-Obiri commented that a lot of plantation projects under the RED+ do not seem to realise this initiative. The trees being planted are basically based on their economic values. He expects that a lot of research could be conducted in this area to improve the penetration of such tree species in combating atmospheric pollutants.

Abere Sodienye, an Owner, suggests that there is great necessity to know the rate and capacity of carbon fixation by such plants to be able to imagine/determine how long it will take to rid the area of excess carbon. In Nigeria, the issue of Ogoni land has lingered for years now and the planned clean-up exercise might benefit from this discussion. Also the Bonny marshlands and mudflats are heavily contaminated with crude oil and nobody is attempting a solution yet.

Alen Berta mentioned that, as for southeast Europe, the carbon ratio in wood is around 50%; therefore they should give an advantage to fast-growing hardwood species, since they have a greater wood density than softwood deciduous and coniferous species. Under this condition, hardwood species have more weight and consequently more carbon fixed and stored.

Ratikanta mentioned that they selected the species with 50% carbon fixation in Mexico, *Leucophyllum frutescens*, *Forestiera angustifolia*, *Bumelia celestrina* and *Acacia berlandieri*.

Chris Hermansen suggested that fast-growing trees with high rates of carbon sequestration are often heavy users of water; such as *Eucalyptus* spp. He mentioned that eucalypts have been banned in some parts of India due to its water use and much risk of ground water depletion. Some put forward other ideas.

Fast-growing species plantations can be used to offset a dependency on native forests for firewood and building materials. Therefore, these species can provide reasonable conditions for agroforestry (e.g. establishing shade-grown coffee or cocoa below a fast-growing overstorey).

Some fast-growing species, such as various *Populus*, have been shown to absorb heavy metals, which could be useful in some industrial contexts. The grown trees must be harvested and buried somewhere; they cannot be used for firewood, construction and so on. Of course, growth rates can be further enhanced by combining a forest plantation with certain kinds of sewage treatment programmes, where the trees benefit from the nutrients contained in sewage.

Yves Nathan Mekembom mentioned that most of the species listed belong to the legume family. The species in this family are very fast-growing and are used in agroforestry systems to improve the nitrogen concentration in the soil, thereby urging the necessity to explore more fast-growing species.

Tony Freeman suggested that there is no necessity to look for trees around the world; instead, one can use the local trees that are compatible with the soil types you are planting in. In turn, this will support local animal, bird and insect communities.

In view of the above discussion, we may suggest the following options to reduce carbon load from the atmosphere:

1 Selection of native tree species with a high capacity of carbon sequestration, with promotion for their planting in polluted areas and in development planning in townships.
2 Selection of legumes, *Acacia* and C4 plants.
3 Preferably the selection of fast-growing species in this respect and heart wood species.

We feel these are potential lines of research to reduce the CO_2 load in the atmosphere; and concerted research inputs needs to be directed in this direction. At the same time incessant logging and other human activities and expansion of agriculture in logged areas need to be rigorously controlled by the forest authority. Rigorous training needs to be given to forest rangers to control this menace. Conservation of forest health and its growth can save our lives from the menace of contamination in the atmosphere. There is a necessity to develop efficient techniques to propagate native plant species and trees through seed germination.

Bibliography

Ackerly D.D., Bazzaz F.A. **1995**. Leaf dynamics, self-shading and carbon gain in seedlings of a tropical pioneer tree. *Oecologia* **101**(3): 289–298.

Aerts R. **1996**. Nutrient resorption from senescing leaves of perennials: are ther general principles. *Journal of Ecology* **84**:507–606.

Ahmad K., Khan Z.I., Ashraf M.Y., Ashraf M., Valeem E.E. **2008**. Forage evaluation for some some trace elements: A case study in the Soone Valley, Pakistan. *Pakistan Journal of Botany* **40**(3):999–1004.

Alvarado M., González Rodríguez H., Ramírez Lozano R.G., Cantú Silva I., Gómez Meza M.V., Cotera Correa M., Jurado Ybarra E., Domínguez-Gómez T.G. **2013**. Trace elements in native shrubs consumed by white-tailed deer (*Odocoileus virginianus*) in northeastern Mexico. *Journal of Applied Animal Research* **41**(3):277–284.

Alvarado M., Gonzalez-Rodriguez H., Ramirez-Lozano R.G., Cantu-Silva I., Gomez-Meza M.V., Cotera-Correa M., Jurado Ybarra E., Dominguez-Gomez T.G. **2012**. Chemical composition and digestion of shurbs browsed by white-tailed deer (*Odocoileus virginianus* Texanus). *Journal of Animal and Veternary Advances* **11**(23):4428–4434.

Brown S. **1999**. *Guidelines for Inventorying and Monitoring Carbon Effects in Forest-base Projects.* Winrock International for the World Bank, Arlington.

Chapin F.S. **1980**. The mineral nutrition of wild plants. *Annual Review of Ecology and Systematics* **11**:233–260.

Chaplin F.S. **1980**.The mineral nutrition of wild plants. *Annual Review of Ecology and Systematics* **11**:233–260.

Chapman F.S., Schultz E.-D., Money H.A. **1990**. The ecology and economics of storage in plants. *Annual Review of Ecology and Systematics* **21**:423–447.

Cherney D.J.R. **2000**. Characterization of forages by chemical analysis. In: Givens D.I., Owen E., Axford R.F.E., Omed H.M. (eds). *Forage Evaluation in Ruminant Nutrition*. CAB International, Wallingford. Pp. 281–300.

Domínguez-Gómez T.G., Ramíirez-Lozano R.G., González-Rodríguez H., Cantúu-Silva I., Gomez-Meza M.V., Alvarado M del S. **2014**. Mineral content in four browse species from Northeastern Mexico. *Pakistan Journal of Botany* **46**(4):1421–1429.

Garcia C., Vayreda J., Sabaté S., Ibáññez J. **2004**. *Main Components of the Aboveground Biomass: Expansion Factors*. Centre de Recerca Ecological i Applications Forestals, University of Barcelona. 23 pp.

González J.C., Ayala A., Gutierrez E. **2007**. Composición química de especies arbóreas con potencial forrajero de la región de tierra caliente, Michoacán, México. *Revista Cubana de Ciencia Agrícola* **41**(1):87–93.

Grime P. **1997**. Evidence for the existence of three primary strategies in plants and its relevance to ecological and evolutionary theory. *American Naturalist* **111**:1169–1194.

Husch B., Beers T.W. Kershaw J.A. **2003**. *Forest Mensuration*. John Wiley & Sons, Inc., New Jersey. 433 pp.

Khan Z.I., Ashraf M., Javed I., Ermidou-Pollet S. **2007b**. Transfer sodium from soil and forage to sheep and goats grazing in a semiarid region of Pakistan: Influence of the seasons. *Trace Elements and Electrolytes* **24**:49–54

Khan Z.I., Hussain A., Ashraf M., McDowell M.Y., Huchzermeyer B. **2007a**. Copper nutrition of goats grazing native and improved pasture with seasonal variation in semiarid regions of Pakistan. *Small Ruminants Research* **67**(2/3):138–148.

Lukhele M.S., Van Ryssen J.B.J. **2003**. Chemical composition and potential nutritive value of subtropical tree species in southern Africa for ruminants. *South African Journal of Animal Science* **33**(2):132–141.

Martínez J., Fernández Bremauntz A., Osnaya P. **2004**. *Cambio Climático: una Visión Desde México*. SEMARNAT, México, D.F. 523 pp.

Mayland H.F., Rosenau R.C., Florence A.R. **1980**. Grazing cow and calf responses to zinc supplementention. *Journal of Animal Science* **51**:966–974.

Navar J., Gonzalez N., Graciano J. **2005**. Carbon sequestration by forest plantations of Durango, Mexico. *Madera y Bosques* **1**:15–34.

Negi G.C.S., Singh S.P. **1993**. Leaf nitrogen dynamics with particular reference to retranslocation in evergreen and deciduous tree species in Kumaun Himalaya. *Canadian Journal of Forest Research* **23**:349–357.

Nooden L.D. **1988**. The phenomena of senescence and ageing. In: *Senescence and Ageing in*

Plants. L.D. Nooden and A.C. Leopold (eds). Academic Press, Michigan. 526 pp.

Özcan T., Baycu G. **2005**. Some elemental concentrations in the acorns of Turkish *Quercus* L. (Fagacea) Taxa. *Pakistan Journal of Botany* **37**(2):361–371.

Pimienta de la Torre D., Domínguez Cabrera G., Aguirre Calderón O., Javier Hernández F., Jiménez Pérez J. **2007**. Estimación de biomasa y contenido de carbono de *Pinus cooperi* Blanco, en Pueblo Nuevo, Durango. *Madera y Bosques* **13**(1):35–46.

Ramírez R.G., González-Rodríguez H., García-Dessommes G., Morales-Rodríguez R. **2005**. Seasonal trends in chemical composition and digestion of Dichanthium annulatum (Forssk.) Stapf. *Journal of Applied Animal Research* **28**(1):35–40.

Ramirez R.G., Gonzalez-Rodriguez H., Gomez-Meza M.V., Perez-Rodriguez M.A. **1999**. Feed Value of Foliage from *Acacia rigidula*, *Acacia berlandieri* and *Acacia farnesiana*. *Journal of Applied Animal Research* **16**(1):23–32.

Ramírez-Lozano R.G., González-Rodríguez H., Gómez-Meza M.V., Cantú-Silva I., Uvalle-Sauceda J.I. **2010**. Spatio temporal variations of macro and trace mineral contents in six native plants consumed by ruminants at northeastern Mexico. *Tropical and Subtropical Agroecosystems* **12**(2):267–281.

Read L., Lawrence D. **2003**. Litter nutrient dynamics during succession in dry tropical forests of the Yucatan: Regional and seasonal effects. *Ecosystems* **6**:747–761.

Reich P.B., Walters M.B., Ellsworth D.S. **1997**. From tropics to tundra: Global convergence in plant functioning. *Proceedings of the National Academy of Sciences of the United States of America* **94**(25):13730–13734.

Rengel Z., Marschner P. **2005**. Nutrient availability and management in the rhizosphere: exploiting genotypic differences. *New Phytologist* **168**(2):305–312.

Rodríguez Laguna R., Jiménez Pérez J., Meza Rangel J., Aguirre Calderón O., Razo Zárate R. **2008**. Contenido de carbono en un bosque tropical subcaducifolio en la reserva de la biosfera el cielo, Tamaulipas, México. *Revista Latinoamericana de Recursos Naturales* **4**(2):215–222.

Sher Z., Hussain F., Saleem M. **1912**. Macro-mineral status at three phenological stages of some range shrubs of Gadoon hills, district Swabi, Khyber Pakhtunkhwa, Pakistan. *Pakistan Journal of Botany* **44**(2):711–716.

Silva-Arredondo F.M., Návar-Cháidez J. de J. **2009**. Estimating carbon expansion factors in temperate forest communities of northern Durango, Mexico. *Revista Chapingo, Serie Ciencias Forestales y del Ambiente* **15**(2):155–160.

SPP-INEGI. **1986**. *Síntesis Geografía del Estado del Nuevo León*, Instituto Nacional de Geografía Estadística e Información, México, D.F.

Steel R.G.D., Torrie J.H. **1980**. *Principles and Procedures of Statistics: a Biometrical Approach, 2nd edn.* McGraw-Hill, New York. 632 pp.

Wright L.J., Reich P.B., Westoby M. **2001**. Strategy shifts in leaf physiology, structure and nutrient content between species of high- and low-rainfall and high- and low-nutrient habitats. *Functional Ecology* **15**(4):423–434.

Yerena Yamallel J.I., Jiménez Pérez J., Aguirre Calderón O.A., Treviño Garza E.J., Alanís Rodríquez E. **2012**. Concentración de carbono en el fuste de 21 especies de coníferas del noreste de México. *Revista Mexicana de Ciencias Forestales* **3**(13):49–56.

Yerena-Yamallel J.I., Jiménez-Pécrez J., Aguirre-Calderón O.A., Treviño-Garza E.J. **2011**. Concentración de carbono en la biomasa aérea del matorral espinoso tamaulipeco. *Revista Chapingo, Serie Ciencias Forestales y del Ambiente* **17**(2):282–291.

CHAPTER 18

Litterfall and forest productivity

18.1 Litterfall studies in north-eastern Mexico

Litterfall and litter decomposition are key processes in nutrient cycling for forest ecosystems (Baker et al., 2001). In addition to these processes, throughfall and stemflow are the main sources to maintain soil fertility at the forest floor (Vasconcelos and Luizão, 2004). Litterfall plays a fundamental role in nutrient turnover and in the transfer of energy between plants and soil, the main source of organic material and nutrients being accumulated in the uppermost layer of the soil. Nutrient release from decomposing litter is an important internal pathway for nutrient flux in forest ecosystems (Santa-Regina et al., 2005).

Evaluation of litterfall production is important for understanding nutrient cycling, forest growth, successional pathways, carbon fluxes, disturbance ecology and interactions with environmental variables in forest ecosystems (Vasconcelos and Luizão, 2004). Despite the great number of well-documented floristic studies carried out at the Tamaulipan thornscrub or subtropical thornscrub woodlands in north-eastern Mexico, few studies have been carried out to address the spatial and temporal patterns of litterfall and nutrient deposition at different altitude gradients where plant species composition is different.

Nutrient cycling is of key importance in forest communities where sustainable litterfall production is an essential part of the above-ground net primary production and depends upon the nutritional status of soil. In the north-eastern region of Mexico, the vegetation depends on the bio-geochemical cycles of plant nutrients contained in plant detritus. Therefore, this region provides an opportunity to investigate the litterfall production and nutrient returns of forest and native vegetation in order to gain a better understanding of how to sustain and improve productivity in response to changes in resource availability. González Rodríguez et al. (2008) studied the spatial and seasonal litterfall deposition pattern in the Tamaulipan thornscrub, north-eastern Mexico. Litterfall and litter decomposition are key fundamental processes in nutrient cycling for the woodland ecosystems at the Tamaulipan thornscrub of north-eastern Mexico, which is characterised by a wide range of taxonomic groups exhibiting differences in growth patterns, leaf life spans, textures, growth dynamics and phenological development. During two consecutive years (November 2004 to October 2006), monthly litterfalls and their respective constituents

Autoecology and Ecophysiology of Woody Shrubs and Trees: Concepts and Applications, First Edition.
Edited by Ratikanta Maiti, Humbero Gonzalez Rodriguez and Natalya Sergeevna Ivanova.
© 2016 John Wiley & Sons, Ltd. Published 2016 by John Wiley & Sons, Ltd.

were quantified at three county sites (Los Ramones, China and Linares) located in the state of Nuevo Leon, Mexico. At each site, litterfall deposition was quantified in an undisturbed thornscrub experimental plot (20×20 m). At each plot, seven (replications) litter traps were scattered over the entire area. Each trap covered an area of 0.16 m^2 (0.4×0.4 m) and was placed approximately 0.3 m above the soil level to intercept litterfall. At each sampling date, the collected litter was sorted manually into the following categories: leaves, branches (< 2 cm in diameter), reproductive structures (flowers, fruits and seeds) and fine residues (unidentified, fine plant residues such as bark, pieces of insect bodies or faeces). The samples were then dried to a constant weight at $65\,°C$ for 72 h. Total mean annual litterfall deposition was 4472, 6743 and 4788 kg ha^{-1} year^{-1} for Los Ramones, China and Linares, respectively. Of total annual litter production, leaves averaged about 67%, followed by branches (ranging from 11% at Los Ramones to 22% at Linares), reproductive structures (7% Linares to 15% Los Ramones) and fine residues litter (4% Linares to 8% China). Differences in spatial and temporal litterfall deposition between sites might be related to plan phenology, community plan structure and environmental variables such as extreme temperatures and heavy rainfall events. This has implications not only for nutrient cycling in the forest soil of different stand communities, but also for maintaining fundamental ecological and ecosystems processes which in turn support and sustain life for invertebrate fauna, thereby enhancing organic matter mineralisation and improving the soil physical structures of the forest ecosystem.

18.2 Methodology

An experimental plot (2500 m^2) was delimited at each studied site. At each site, 10 canisters of 1.0×1.0 m were randomly scattered inside the plot. The canisters were made of a wooden bevelled frame (10 cm) with a plastic mesh (1.0 mm) on the bottom. The canisters were placed at 0.30 m above soil surface; in this way throughfall is allowed to pass through. In this study, the sampling area, number of canisters and height above soil level are within the range of previous studies (Fang et al., 2007; Zhou et al., 2007). The frequency of litter sampling was at intervals of 30 days between 21 December 2006 and 20 December 2007.

18.2.1 Chemical analyses

Since both macro- (Ca, K, Mg, P and N) and micro-nutrients (Cu, Fe, Mn and Zn) play essential roles in plant cell metabolism and plant life cycle, most of the total litterfall is constituted mainly by leaves (from 74 to 86%, on a yearly basis), and because the contribution of each litterfall component deposition and macronutrient deposition is through leaves, the authors documented the content and deposition pattern of micronutrients on a monthly as well as on a yearly basis, since very little information of this kind is available.

Quintuplicate 2.0 g of leaf litter samples collected from each canister were subjected to micronutrients analysis. First, leaf samples were incinerated during 5 h in a furnace at $550\,°C$. Ashes were digested by the wet digestion technique using a solution containing HCl and HNO$_3$ (Cherney, 2000). The Cu, Fe, Mn and Zn contents

($\mu g\,g^{-1}$ dry weight; air flame/acetylene) were determined by atomic absorption spectrophotometry using a Varian spectrophotometer (model SpectrAA-200). Micronutrient deposition ($mg\,m^{-2}$) at each site was calculated by multiplying the leaf-fall production of each sampling date by the mineral content of the corresponding date of sampling, site and replication, and then adding them over the entire year. The cumulative monthly values at each site were used.

18.3 Results and discussion

López-Hernández et al. (2013) quntified the production of litterfall and potential of nutrients in three sites of Nuevo Leon, north-eastern Mexico, with an objective to estimate and quantify the monthly dynamics (January to December 2009) of litterfall deposition and macro (Ca, K, Mg, N and P) and microminerals (Cu, Fe, Mn and Zn) of litterfall leaves in the counties of Los Ramones, China and Linares in Nuevo León State, Mexico. Annual deposition was 321.5 (Los Ramones), 431.6 (China) and 462.9 (Linares) $g\,m^{-2}\,year^{-1}$. Leaves represented the highest component ($154.4–304.5\,g\,m^{-2}\,year^{-1}$) of total litterfall production. Branches varied from 48.25 to $75.16\,g\,m^{-2}\,year^{-1}$. Reproductive structures varied from 58.4 to $98.9\,g\,m^{-2}\,year^{-1}$ and other components (unidentified material, bodies and insect feces) ranked from 19.9 to $29.2\,g\,m^{-2}\,year^{-1}$. Annual deposition of Ca in the three sites varied from 3.4 to 8.1, K from 1.4 to 3.2, Mg from 0.76 to 1.6, N from 3.0 to 5.5 and P from 0.05 to $0.11\,g\,m^{-2}\,year^{-1}$. Cu ranked from 0.84 to 2.8, Fe from 14.7 to 21.7, Mn from 5.5 to 11.8 and Zn from 3.3 to $5.2\,g\,m^{-2}\,year^{-1}$. In winter mineral deposition was higher. Moreover, the highest mineral deposition occurred in Linares; and Los Ramones was the lowest. There were spatio-temporal variations in the quantity of literfall collected and the return of minerals, but not in the leaf quality between study sites.

Domínguez-Gómez et al. (2014) estimated the micro-elements through leafall in different types of vegetation, north-eastern Mexico. The objective of this study was to determine the content and deposition of Cu, Fe, Mn and Zn in leaf litterfall samples collected in four sites, north-eastern Mexico. Site 1 was located at 1600 m of altitude in a pine forest, mixed with deciduous trees; site 2 was at 550 m in the ecotone of a *Quercus* spp. forest and submontane scrub; sites 3 and 4 (at 350 and 300 m, respectively) were located in the Tamaulipan thornscrub vegetation. Leaf litterfall samples at each site ($2500\,m^2$) were obtained from 10 canisters of $1.0\,m^2$ that were randomly situated at each site. The Cu annual deposition ($g\,ha^{-1}\,year^{-1}$) was significantly different between sites, being highest in site 4 (23.2) and lowest in site 1 (4.1). Fe deposition was also significantly higher in site 4 (522.2) and lower in site 1 (120.0). Mn was higher in site 2 (479.4) and lower in site 1 (64.6). Zn was significantly higher in site 1 (62.8) and lower in site 1 (24.3). Micronutrient annual order deposition was as follows: Fe > Mn > Zn > Cu. Differences in deposition may be attributable to environmental conditions and plant species composition at each studied plant community. Site 1 was located in pine (*Pinus pseudostrobus* Lindl.) forest mixed with deciduous trees

(*Quercus* spp) located at the Experimental Forest Research Station of the Universidad Autonoma de Nuevo Leon in the Sierra Madre Oriental Mountain Range, Iturbide county (24°43′N, 99°52′W, 1600 m of altitude). Annual mean air temperature is about 14°C, average annual rainfall is approximately 639 mm (INEGI, 2002). The soils are mainly rocky and comprise upper cretaceous lutite or siltstone. Site 2 was located in the ecotone of a *Quercus* spp. forest and the Piedmont scrub vegetation (24° 46′N; 99° 41′W; 550 m of altitude) in the "Ejido Crucitas" in Linares County. Average total annual rainfall is about 755 mm (INEGI, 2002). Site 3 was located at the Experimental Research Station of the Faculty of Forest Sciences of the Universidad Autonoma de Nuevo Leon (24°47′N, 99°32′W, 350 m of altitude) in Linares county. Average total annual rainfall is about 805 mm (INEGI, 2002). Peak rainfall months are May, June and September (González-Rodríguez et al., 2004). Site 4 was located in the "Ejido Hacienda de Guadalupe" in Linares county (24°54′N, 99°32′W, 300 m of altitude). Total cumulative annual rainfall is about 672 mm. The vegetation at sites 1 and 3 is known as the Tamaulipan thornscrub or subtropical thornscrub woodlands. In the last three sites, the climate is subtropical and semi-arid with a warm summer and monthly mean air temperature ranges from 14.7°C in January to 22.3°C in August, although daily high temperatures of 45°C are common during summer. The dominant soils are deep, dark-gray, lime-clay vertisols, with montmorillonite, which shrink and swell noticeably in response to changes in soil moisture content (González-Rodríguez et al., 2008).

González-Rodríguez et al. (2011) studied literfall deposition and leaf litter nutrient return in different locations at north-eastern Mexico. The aim of this study was to determine the litterfall production and macronutrient (Ca, K, Mg, N and P) deposition through leaf litter in four sites with different types of vegetation. Site one (Bosque Escuela) was located at 1600 m a.s.l. in a pine forest mixed with deciduous trees, site two (Crucitas) was at 550 m a.s.l. in the ecotone of *Quercus* spp. forest and Tamaulipan thornscrub and sites three and four (Campus at 350 m a.s.l. and Cascajoso at 300 m a.s.l., respectively) were in Tamaulipan thornscrub. Litter constituents (leaves, reproductive structures, twigs and miscellaneous residues) were collected at 15-day intervals from 21 December 2006 to 20 December 2007. Collections were carried out in 10 litter traps (1.0 m^2) randomly situated at each site of approximately 2500 m^2. Total annual litterfall deposition was 4407, 7397, 6304 and 6527 kg ha^{-1} year^{-1} for Bosque Escuela, Crucitas, Campus and Cascajoso, respectively. Of the total annual litter production, leaves were highest, varying from 74 (Bosque Escuela) to 86% (Cascajoso) followed by twigs from 4 (Cascajoso) to 14% (Crucitas), reproductive structures from 6 (Bosque Escuela) to 10% (Crucitas) and miscellaneous litterfall from 1 (Campus) to 12% (Bosque Escuela). The Ca annual deposition was significantly higher in Cascajoso (232.7 kg ha^{-1} year^{-1}), followed by Campus (182.3), Crucitas (130.5) and Bosque Escuela (30.3). The K (37.5, 32.5, 24.8, 7.2, respectively) and Mg (22.6, 17.7, 13.7, 4.5, respectively) annual deposition followed the same pattern as Ca. However, N was highest in Campus (85.8) followed by Crucitas (85.1), Cascajoso (68.3) and Bosque Escuela (18.3). P was highest in Campus and Crucitas (4.0) followed by Cascajoso (3.4) and Bosque Escuela (1.4). On an annual basis for all sites, the order of nutrient deposition through leaf litter was Ca > N > K > Mg > P, whereas on a site basis of total nutrient deposition (Ca + N + K + Mg + P), the order was Cascajoso > Campus > Crucitas > Bosque Escuela. Ca, K, Mg, N, and P nutrient use efficiency values in leaf litter were higher in Bosque Escuela, while lower figures were acquired in Cascajoso and Crucitas sites. It seems that the highest litterfall deposition was found in the ecotone of *Quercus* spp. forest and Tamaulipan thornscrub; however, the Tamaulipan thornscrub vegetation alone had better leaf litter nutrient return.

Litterfall being an important source of nutrients for the growth of trees, more concerted research needs to be directed in assessing the quantity of litterfall of the tree species growing in different agro-climatic regions, with an analysis of nutrients in relation to the productivity of trees.

Bibliography

Baker T.T., Lockaby B.G., Conner W.H., Meier C.E., Stanturf J.A., Burke M.K. **2001**. Leaf litter decomposition and nutrient dynamics in four southern forested floodplain communities. *Soil Science Society of America Journal* **65**(4):1334–1347.

Cherney D.J.R. **2000**. Characterization of forages by chemical analysis. In: Givens D.I., Owen E., Axford R.F.E., Omed H.M. (eds) *Forage Evaluation in Ruminant Nutrition*. CAB International, Wallingford, UK, pp. 281–300.

Domínguez-Gómez T.G., González-Rodríguez H., Cantu-Silva I., Ramirez Lozano R.G.,

Gomez-Meza M.V., Alvarado M. **2014**. Deposition of micro-elements through leaf fallen from different types of vegetation, north-eastern Mexico. *International Journal of Bioresource and Stress Management* **5**(1):001–006.

Fang J.Y., Liu G.H., Zhu B., Wang X.K., Liu S.H. **2007**. Carbon budgets of three temperate forest ecosystems in Dongling Mt., Beijing, China. *Science China Series D – Earth Sciences* **50**:92–101.

González R.H., Cantú Silva I., Gómez Meza M.V., Jordan W.R. **2000**. Seasonal plant water relationships in *Acacia berlandieri*. *Arid Soil Research and Rehabilitation* **14**(4):343–357.

González-Rodríguez H., Cantú-Silva I., Gómez-Meza M.V., Ramírez-Lozano R.G. **2004**. Plant water relations of thornscrub shrub species, north-eastern Mexico. *Journal of Arid Environments* **58**(4):483–503.

González-Rodríguez H., Cantú-Silva I., Ramírez-Lozano R.G., Gómez-Meza M.V., Domínguez-Gómez T.G., Bravo-Garza J., Maiti R.K. **2008**. Spatial and seasonal litterfall deposition pattern in the Tamaulipan thorscrub, northeastern Mexico. *International Journal of Agriculture Environment and Biotechnology* **1**(4):177–181.

González-Rodríguez H., Domínguez-Gómez T.G., Cantú-Silva I., Gómez-Meza M.V., Ramírez-Lozano R.G., Pando-Moreno M., Fernández C.J. **2011**. Litterfall deposition and leaf litter nutrient return in different locations at Northeastern Mexico. *Plant Ecology* **212**(10):1747–1757.

Haase R. **1999**. Litterfall and nutrient return in seasonally flooded and non-flooded forest of the Pantanal, Mato Grosso, Brazil. *Forest Ecology and Management* **117**(1/3):129–147.

INEGI **2002**. *Uso Actual del Suelo en los Nucleos Agrarios. Aspectos Geográficos de Nuevo Leon*, Instituto Nacional de Estadistica, Geografia e Informatica, México, D.F., http://nl.inegi.gob.mx/territorio/espanol/cartcat/uso.html (accessed 20 May 2009).

López-Hernández J.M., González-Rodríguez H., Ramírez-Lozano R.G., Cantú-Silva I., Gómez-Meza M.V., Pando-Moreno M.,

Estrada-Castillón A.E. **2013**. Producción de hojarasca y retorno potencial de nutrientes en tres sitios del estado de Nuevo León, México. *Polibotánica* **35**:41–64.

Santa-Regina J., Salazar S., Leonardi S., Rapp M. **2005**. Nutrient pools to the soil through organic matter in several *Castanea sativa* Mill. coppices of mountainous Mediterranean climate areas. *Acta Horticulturae* **693**:341–348.

Vasconcelos H.L., Luizão F.J. **2004**. Litter production and litter nutrient concentrations in a fragmented Amazonian landscape. *Ecological Applications* **14**(3):884–892.

Zhou G., Guan L., Wei X., Zhang D., Zhang Q., Yan J., Wen D., Liu J., Liu S., Huang Z., Kong G., Mo J., Yu A. **2007**. Litterfall production along successional and altitudinal gradients of subtropical monsoon evergreen broadleaved forests in Guangdong, China. *Plant Ecology* **188**:77–89.

CHAPTER 19

Nutrient cycling

19.1 Background

A large number of plant species grow mutually in an forest ecosystem. The trees absorb nutrients and moisture for the growth of the plants depending on the availability of nutrients in the soil. These nutrients are stored in the leaves and branches which fall down onto the ground after maturity as litterfall. This litterfall is decomposed by bacterial actions/microrganisms in the soil and then the nutrients are released, which are again absorbed by the plants to maintain the nutrient cycle.

Agricultural practices and soil management practices are the most significant anthropogenic activities that disturb the physical and chemical characteristics of soil (Buckley and Schmidt, 2001). By being inadequate, they lead to soil degradation (Michelena et al., 2008). Lately and due to the importance of soil as an essential component of the ecosystem's health, interest has increased in determining the consequences of agricultural practices on soil properties (Schoenholtz et al., 2000). Considerable research has focused on the identification of indicators to estimate the current state and trends in the quality of soils through an evaluation of their physical and chemical properties, looking for

sustainable management strategies (Sacchi and De Pauli, 2002; Inzunza-Ibarra and Curtis, 2005; Alejo-Santiago et al., 2012).

19.2 Studies on nutrient cycling

The north-east of Mexico includes a large extension of arid and semiarid areas that have been particularly affected by soil degradation caused by the transformation of natural ecosystems into irrigated agricultural lands (McCready et al., 2005). A clear example of this problem occurs in the grasslands of the southern region of the Chihuahua Desert, which exhibits a great biological diversity so that its preservation is of imperative importance (Arriaga et al., 2000). According to studies conducted in the area Pronatura Noreste (PN) and The Nature Conservancy (TNC) in 2007, Estrada-Castillon et al. (2010) suggested that such diversity is strongly threatened mainly by the conversion of the gypsophila and halophile grasslands into crop lands, particularly dedicated to grow potatoes. In the area, this crop is cultivated using a high amount of fertilisers and other agrochemicals and, even so, peasants claim that potato yields notoriously decrease after one cycle of cultivation. The land is

Autoecology and Ecophysiology of Woody Shrubs and Trees: Concepts and Applications, First Edition.
Edited by Ratikanta Maiti, Humbero Gonzalez Rodriguez and Natalya Sergeevna Ivanova.
© 2016 John Wiley & Sons, Ltd. Published 2016 by John Wiley & Sons, Ltd.

abandoned after just one year of use and new grasslands are cleared. Hence, in order to preserve these grassland ecosystems, there is a need to establish new management strategies for sustainable agricultural practices and an unavoidable requisite for these strategies to be successful is to know the dynamics of the properties of soils subjected to this type of agricultural use. Thus, this research was conducted to compare the chemical characteristics of the soil in areas with different abandonment times and an adjacent area of natural grassland, looking for an explanation for the abandonment of the areas and setting the basis for management practices that encourage the continued use of already-cleared areas. The hypothesis was that, after potato cultivation, the land is depleted of some of the essential nutrients or its chemical characteristics change in a way that prevents the proper development of the crop.

Silva-Arredondo et al. (2013) evaluated the chemicals of a soil impacted by intensive agriculture in north-eastern Mexico. This study was carried out in a semiarid area within the Chihuahuan Plateau. The objective of the study was to evaluate the long-term changes in the soil chemical properties after one cycle of potato cultivation with an intensive application of agrochemicals. Sampling sites were established in areas that had 1, 2, 5 and 10 years of abandonment after potato cultivation; and an area of native grassland was also used as control. Potato cultivation is carried out by a company that rents the land to the peasants, so cultivation is done in the same way all over this area. At each site, the following variables were evaluated at 0–10 and 10–30 cm soil depths: pH, CEC, EC, SOM and $CaCO_2$.

Mireles Posadas (2014) umdertook a study on nutrient deposition via twigs in north-eastern Mexico. She mentioned that a healthy layer of litterfall in the soil is a vital component of any forest ecosystem. Litterfall protects and conserves soil, enables rainwater to be filtered, acts as a seed bank and provides refuges for wildlife. Without litterfall, the forests could not exist and reproduce. The cycle of nutrients is one of the fundamental aspects in the productivity and dynamics of forest ecosystems and is part of its stability. The objective of this research was to determine the deposition of macro- (K, Mg and P) and micronutrients (Cu, Fe and Zn) through litterfall production, via twig components in three sites of the Tamaulipan thornscrub, north-eastern Nuevo Leon, Mexico. Litterfall production was sampled over 12 months (January–December, 2009). The experimental plots were located at three research sites: Los Ramones, China, and Linares. At each research site, an experimental plot of 2500 m^2 (50 × 50 m) was previously established. Ten canisters (1 m^2) were randomly distributed within each experimental plot in order to collect litterfall samples. Twigs were manually separated from litterfall samples. Determinations of K, Mg, Cu, Fe and Zn were determined by atomic absorption spectrophotometry and P by colorimetric analysis using an UV-visible spectrophotometer. Since most of the sampling dates for twig and nutrient deposition data did not follow the assumptions of normal distribution and homogeneity of variances even though data were logarithmically transformed, the Kruskal–Wallis non-parametric test was used to detect significant differences between the research sites at each sampling month. Total litterfall annual deposition ($g\,m^{-2}\,year^{-1}$) was 321.5 (Los Ramones), 431.6 (China) and 462.9 (Linares). Twig annual deposition ($g\,m^{-2}\,year^{-1}$) was 48.2 (Los Ramones), 64.5 (China) and 75.1 (Linares). The annual contribution ($mg\,m^{-2}\,year^{-1}$) of K ranged from 2470.6 (Linares) to 1392.4 (China), Mg from 127.8 (Linares) to 100.7 (Los Ramones) and P from 0.83 (Linares) to 5.76 (China). The annual contribution ($\mu g\,m^{-2}\,year^{-1}$) of Cu varied from 670.3 (Linares) to 301.1 (Los Ramones), Fe from 9835.6 (China) to 8533.4 (Linares) and Zn from 2173.5 (Linares) to 1985.3 (Los Ramones). The order of nutrient deposition was as follows: Linares > China > Los Ramones. The spatial and temporal nutrient deposition was related to prevailing environmental conditions, to plant structure composition at each research site, to mechanisms of nutrient absorption and accumulation among plant tissues and to differences in the physical and chemical soil properties at the research sites.

Bibliography

Aghasi B., Jalalian A., Honarjoo N. **2010**. The comparison of some soil quality indexes in different land uses of Ghareh Aghaj watershed of Semirom, Isfahan, Iran. *International Journal of Environmental and Earth Sciences* **1**(2):76–80.

Alejo-Santiago G., Salazar-Jara F.I., García-Paredes J.D., Arrieta-Ramos B.G., Jiménez-Meza V.M., Sánchez Monteón A.L. **2012**. Degradacion fisico-quimica de suelos agricolas en San Pedro Lagunillas, Nayarit. *Tropical and Subtropical Agroecosystems* **15**(2):323–328.

Arriaga L., Espinoza J.M., Aguilar C., Martinez E., Gomez L., Loa E. **2000**. *Regiones Terrestres Prioritarias de Mexico, Escala 1:1000000.*

Comision Nacional para el Conocimiento y Uso de la Biodiversidad, Mexico, D.F.

Bowman R., Vigil M., Nielsen D., Anderson R. **1999**. Soil organic matter changes in intensively cropped dryland systems. *Soil Science Society of America* **63**:186–191.

Buckley D.H., Schmidt T.M. **2001**. The structure of microbial communities in soil and the lasting impact of cultivation. *Microbial Ecology* **42**(1):11–21.

Estrada-Castillon E., Scott-Morales L., Villarreal-Quintanilla J., Jurado-Ybarra E., Cotera-Correa M., Cantu-Ayala C., Garcia-Perez J. **2010**. Clasificacion de los pastizales halofilos del noreste de Mexico asociados con perrito de las praderas (*Cynomys mexicanus*): diversidad y endemismo de especies. *Revista Mexicana de Biodiversidad* **81**:401–416.

Inzunza-Ibarra M.A., Curtis M.H. **2005**. Variation of soil chemical properties in irrigated and non-irrigated areas of the Laguna Region of Mexico. *TERRA Latinoamericana* **23**(4):429–436.

McCready B., Mehlman D., Kwan D., Abel B. **2005**. *The C Prairie Wings Project: A Conservation Strategy for the Grassland Birds of the Western Great Plains.* USDA Forest Service General Technical Report PSW-GTR-191, pp. 1158–1161.

Michelena R., Morras H., Irurtia C. **2008**. *Degradacion Fisica por Agricultura Continua de Suelos Franco-LLimosos de la Provincia de Cordoba.* INTA CIRN, Instituto de Suelos, Argentina.

Mireles Posadas C.G. **2014**. *Deposit of nutrientes via el componente ramas en hohareska en tres sitios del Norested de Mexico.* BSc Thesis in Forest Engineering, Universidad Autónoma de Nuevo León, Facultad de Ciencias Forestales, Linares, Mexico.

Presley D.R., Ransom M.D., Kluitenberg G.J., Finnell P.R. **2004**. Effects of thirty years of irrigation on the genesis and morphology of two semiarid soils in Kansas. *Soil Science Society of America* **68**:1916–1926.

Pronatura Noreste A.C. **2007**. *Plan de Conservacion de los Pastizales del Altiplano Mexicano*

2006–2010, The Nature Conservancy (compiladores), Coahuila, Nuevo Leon, San Luis Potosi y Zacatecas. Monterrey, N.L., Mexico, 171 pp.

Sacchi G., De Pauli C. **2002**. Evaluación de los cambios en las propiedades físicas y químicas de un Argiustol údico por procesos de degradación. *Agrociencia* **6**(2):37–46.

Schoenholtz S.H., Miegroet H.V., Burgerc J.A. **2000**. A review of chemical and physical properties as indicators of forest soil quality: challenges and opportunities. *Forest Ecology and Management* **138**(1/3):335–356.

Silva-Arredondo I., Pando Moreno M., González Rodríguez H., Scott Morales L. **2013**. Chemical properties of a soil impacted by intensive agriculture, north-eastern Mexico. *International Journal of Bioresource and Stress Management* **4**(2):126–131.

Steel R.G., Torrie J.H. **1980**. *Principles and Procedures of Statistics. A Biometrical Approach*, 2nd Edn. McGraw-Hill International, Tokyo.

Wackerly D.D., Mendenhall W., Scheaffer R.L. **2002**. *Estadistica Matematica con Aplicaciones*, 6th Edn. Thomson, Mexico, D.F.

Wang Y., Zhang X., Huang C. **2009**. Spatial variability of soil total nitrogen and soil total phosphorus under different land uses in a small watershed on the Loess Plateau, China. *Geoderma* **150**:141–149.

CHAPTER 20

Plant water relations and forest productivity

20.1 Background

The growth and development of a plant is highly dependent on the availability of soil moisture, soil nutrients and water content in the plant cell. The plant cell requires 85–89% water to maintain its dynamic vital activity and enzymatic processes. A decrease in the water content in the cell below this level reduces the metabolic activity of plant cells and, thus, plant growth. Therefore, there is a great necessity to maintain the water balance between plant cells, starting from the roots to the leaves. The water content in the cell maintains water potential in the cell and the hydrostatic balance between the cells. From the roots up to the leaves there is a gradual decrease in water content, maintaining a water potential in the cells which forces the movement of water from the root cells up to the leaves, leading to a loss of water through transpiration. Plant cells need to maintain this hydrostatic balance for maintaining plant growth.

In order to understand this dynamic process there is a great necessity to understand the sequential processes occurring in the plant cell such as morpho-anatomical and physiological mechanisms, for example water relations of the plant cell, water in the cell, the water potential of plant cells, water relations and cellular translocation in plant absorption, transpiration and water balance during drought. These parameters vary largely in semiarid, tropical rainforest, temperate forest in different agro-climatic regions of different countries. Very little progress has been achieved on research in these directions. This needs an interdisciplinary team to address this aspect for assessing the productivity of forests and their management.

Water is absorbed by roots and moves from the peripheral root cells to the interior of the cells. It then reaches the endodermis and finally the xylem vessels in the roots owing to the water potential gradients which start from the peripheral root cells to xylem cells in the roots. Water once entering the xylem vessels is retained in the narrow vessels owing to the adhesion force between water molecules. Xylem vessels form capillary tubes connected from the roots upwards to the stems and pump water from the roots upwards by root pressure. Following this process, water moves upwards (called the ascent of sap) owing to the presence of a deficit of vapour pressure in the leaf cells (caused by the continuous loss of water by transpiration). In a humid climate the rate of transpiration is lower

Autoecology and Ecophysiology of Woody Shrubs and Trees: Concepts and Applications, First Edition.
Edited by Ratikanta Maiti, Humbero Gonzalez Rodriguez and Natalya Sergeevna Ivanova.
© 2016 John Wiley & Sons, Ltd. Published 2016 by John Wiley & Sons, Ltd.

than that on hot summer days which controls directly the speed of transpiration and movement of water from the roots upwards to the leaves. This is the driving force for the growth and development of a plant. Optimum moisture is essential for plant growth. Lack of soil moisture cause drought and reduces growth. If a forest species is susceptible to drought, this causes a decrease in growth. Some species are resistant/tolerant to drought due to the presence of morpho-anatomical and physiological mechanisms.

20.2 Plant water relation studies in north-eastern Mexico

The climate in north-eastern Mexico is characterised by alternating favourable and unfavourable periods of soil water content. This affects plant growth and development throughout the year. Shrubs and trees growing in this region under adverse environmental conditions have to seasonally adjust their morpho-physiological traits to cope successfully with changes in soil water availability (Bucci et al., 2008). Plants differ widely in their capacity to cope with drought. Adaptations exist to explain these differences and these can be conveniently related to the capacity to maintain water status (water potential and/or relative water content, RWC). Plants under such conditions regulate their water status using several strategies, for example osmotic adjustment, stomatal aperture, turgor maintenance, root distribution and leaf canopy properties (Rhizopoulou et al., 1997). The main type of vegetation in north-eastern Mexico, known as the Tamaulipan thornscrub, is distinguished by a wide range of taxonomic groups exhibiting differences in growth patterns, leaf life spans, textures, growth dynamics and phenological development (Reid et al., 1990; McMurtry et al., 1996). This semiarid shrubland, which covers about 200 000 km^2 including southern Texas and north-eastern Mexico, is characterised by an average annual precipitation of 805 mm and a yearly potential evapo-transpiration of about 2200 mm. Vegetation is utilised as forage for livestock and wildlife and for fuel-wood, timber for construction, traditional medicine, fencing, charcoal, agroforestry and reforestation practices in disturbed sites (Reid et al., 1990). Since water availability is the most limiting factor controlling tree growth, survival and distribution in dry climates (Newton and Goodin, 1989), the great diversity of native shrubs in this region reflects the plasticity of these species to cope with a harsh environment. These shrub and tree plants have evolved key morphological and physiological traits suited for adaptation to environmental constraints, especially in drought-prone regions. The strategies involve early leaf abscission, limited leaf area, an extensive and deeper root system, epidermal wax accumulation associated with reduction of water loss by stomatal closure and accumulation of organic and inorganic solutes (Newton et al., 1991).

The study of native species in this region provides an opportunity to investigate, from an ecophysiological perspective, the response of shrub species to changes in resource availability, for example soil moisture content, in order to gain a better understanding of how such an ecosystem may sustain biomass productivity.

However, a few studies (Stienen et al., 1989; González Rodríguez et al., 2000, 2004, 2009, 2010) have attempted to directly relate the water status of native shrub species across a summer drought in this region of Mexico. This study was conducted to assess xylem water potentials (Ψ) and to estimate the relationship between plant water potentials and soil water availability in native shrub species.

20.3 Methodology

This study was carried out in 2009 at El Abuelo Ranch (25°40′N, 99°27′W; elevation 200 m asl) in Los Ramones county, state of Nuevo Leon, Mexico. The climate is semiarid with a warm summer. Annual mean air temperature and rainfall is about 22 °C and 700 mm, respectively. Peak rainfall occurs in May, June and September. The main type of vegetation is known as Tamaulipan thornscrub or sub-tropical thornscrub woodlands (SPP-INEGI, 1986). The dominant soils are deep, dark-grey, lime-clay vertisols with montmorillonite, which shrink and swell noticeably as soil moisture content varies.

Five plants from each of the native shrub species *Forestiera angustifolia* Torr. (Oleaceae; evergreen shrub with stiff and dense branches), *Celtis pallida* Torr. (Ulmaceae; evergreen and spiny shrub with oval and smooth-edged leaves), *Acacia amentacea* Benth (Leguminosae; deciduous shrub with mycrophyllous compound leaves) and *Parkinsonia texana* (A. Gray) S. Watson var. *macra* (I. M. Johnst.) Isely (Leguminosae; small thorny deciduous tree with twice compound leaves) were randomly selected within

a 20 × 20 m² previously established (González-Rodríguez et al., 2004) and an undisturbed experimental thornscrub plot for determination of xylem water potential (Ψ, MPa). Since pre-dawn and mid-day Ψ measurements are influenced by environmental conditions, and the purpose was to detect changes in the plant water relation status, measurements were conducted, when possible, at 15-day intervals between 14 January and 29 September 2009. The Ψ measurements were taken from terminal twigs at 6 a.m. (pre-dawn, Ψ_{pd}) and 2 p.m. (mid-day, Ψ_{md}), local time. Water potential was estimated using a Scholander pressure bomb (Model 3005, Soil Moisture Equipment Corp, Santa Barbara, CA, USA; Ritchie and Hinckley, 1975). One terminal shoot, with fully expanded leaves, was excised and sampled from the middle and shaded side of each plant. Measurements were performed within 10–25 s after collecting the samples. Pressure was applied to the chamber at 0.05 MPa s⁻¹. For safety reasons and as per operating instructions, the lowest limit of the pressure chamber was 7.3 MPa.

Air temperature (°C) and relative humidity (%) were registered on a daily basis using a Pro data logger (HOBO Pro Temp/RH Series, Forestry Suppliers, Inc., Jackson, MS, USA). Daily precipitation (mm) was obtained from a tipping bucket rain gauge (Forestry Suppliers, Inc.). Air temperature and relative humidity were used to calculate vapour pressure deficit (VPD, kPa; Rosenberg et al., 1983). Gravimetric soil water content on each sampling date was determined in soil cores at depths of 0–10, 10–20, 20–30, 30–40 and 40–50 cm, respectively using a soil sampling tube (Soil Moisture Equipment

Corp.). Gravimetric soil water content was determined by drying soil samples in an oven at 105 °C for 72 h and was expressed on a dry weight basis (kg kg^{-1}).

20.4   Results and discussion

López-Hernández et al. (2010) studied the adaptation of native shrubs to drought stress in north-eastern Mexico. Native shrubs that grow in the semiarid regions of north-eastern Mexico are important feed resources for range ruminants and white-tailed deer. They also provide high quality fuel-wood and timber for fencing and construction. Since water stress is the most limiting factor in this region, the present research focused on studying how seasonal xylem water potentials of native shrubs such as *Forestiera angustifolia* (Oleaceae), *Celtis pallida* (Ulmaceae), *Acacia amentacea* (Leguminosae) and *Parkinsonia texana* (Leguminosae) are related to soil water availability. During the wettest period, Ψ at pre-dawn ranged from −0.62 MPa (*A. amentacea* and *F. angustifolia*) to −1.10 MPa (*P. texana*). In contrast, during the driest period, pre-dawn Ψ varied from −1.78 MPa (*P. texana*) to −3.94 MPa (*C. pallida*). With respect to mid-day Ψ data, on the wettest sampling date, Ψ values ranged from −0.62 MPa (*A. amentacea*) to −1.57 MPa (*C. pallida, F. angustifolia* and *P. texana*). In contrast, on the driest sampling date, *C. pallida, A. amentacea* and *P. texana* tended to achieve higher (−2.21 MPa) Ψ than *F. angustifolia* which acquired a Ψ value around −3.50 MPa. Since the shrub species *A. amentacea* and *P. texana* achieved higher pre-dawn and mid-day Ψ values under water stress conditions, these species could be considered as drought adapted species, while *F. angustifolia* which acquired lower water potentials, may not be suitable to drought and, thus, may be in a physiological disadvantage under limited water conditions. The study recommends that the first two species may serve as a pertinent model to study the strategies of adaptation to drought at high tissue water potential while the later may serve as an adequate model to study plant adaptation to drought at low tissue water potential.

González-Rodríguez et al. (2010) estimated the xylem water potential in ten native plants of north-eastern Mexico. Since water stress is the most limiting factor in north-eastern Mexico, the present study focused to characterise the xylem water potentials (MPa) of ten native tree and shrub species such as *Acacia rigidula* (Leguminosae; shrub), *Bumelia celastrina* (Sapotaceae; tree), *Castela texana* (Verbenaceae; shrub), *Celtis pallida* (Ulmaceae; shrub), *Forestiera angustifolia* (Oleaceae; tree), *Karwinskia humboldtiana* (Rhamnaceae; shrub), *Lantana macropoda* (Simaroubaceae; shrub), *Leucophyllum frutescens* (Scrophulariaceae; shrub), *Prosopis laevigata* (Leguminosae; tree) and *Zanthoxylum fagara* (Rutaceae; tree) under drought and high soil water content. Under drought conditions, *P. laevigata, A. rigidula* and *C. texana* achieved higher Ψ at pre-dawn with values of −2.72, −2.78 and −3.42 MPa, respectively, while a minimum value of −6.82 MPa was observed in *Z. fagara*. Similarly, higher Ψ at mid-day was registered in *C. texana, celastrina* and *P. laevigata* with values around −4.15 MPa, while lower values (< −7.0 MPa) were acquired by *L. macropoda, K. humboldtiana* and *Z. fagara*. In contrast, under high soil water content, Ψ at pre-dawn varied from −0.52 MPa (*K. humboldtiana*) to −1.63 MPa (*C. texana*). With respect to mid-day data, values ranged from −1.43 MPa (*L. macropoda*) to −2.28 MPa (*C. texana*). Since the plant species, *A. rigidula*, *B. celastrina*, *C. texana* and *P. laevigata* achieved higher pre-dawn and mid-day values under drought conditions, the results indicated that these species could be considered as drought-adapted species, while *L. macropoda*, *K. humboldtiana* and *Z. fagara* which acquired lower water potentials may not be suitable to drought and, thus, may be in a physiological disadvantage under limited water conditions. The study suggests that the first four species may serve as a pertinent model to study the strategies of adaptation to drought at high tissue water potential while the later may serve as an adequate model to study plant adaptation to drought at low tissue water potential. The implications of this study suggest that the species respond differently to drought through the employment of different strategies and there is scope for forest and range management practices in the selection of drought tolerant species for planting and reforestation of drought-prone areas.

As an approach to understanding how seasonal plant water potentials are related to soil water availability and to describe if osmotic adjustment occurs as a mechanism of adaptation to drought stress, the adaptive responses of ten native shrubs species (*Acacia berlandieri*, *Pithecellobium ebano*, *Cordia boissieri*, *Helietta parvifolia*, *Pithecellobium pallens*, *Acacia rigidula*, *Eysenhardtia polystachya*, *Diospyros texana*, *Randia rhagocarpa* and *Bernardia myricaefolia*) of the north-eastern region of Mexico were investigated. Pre-dawn water potential values between the wettest and driest period ranged from −0.5 to −7.3 MPa for the shrub species studied. The capacity for osmoregulation was observed among six shrub species. This value varied from −1.11 to −2.65 MPa. Seasonal patterns in water potentials could be explained by the soil water availability in a range from 65 to 87%. The results have indicated that the response of a shrub species to evade drought stress is related to their water and osmotic potentials and to the response of interacting to environmental variables, specifically soil water availability.

Native trees and shrubs that grow in the semiarid regions of northeastern Mexico are important feed resources for range ruminants and white-tailed deer. They also provide high-quality fuelwood and timber for fencing and construction. Since water stress is the most limiting factor in this region, the present work focused on studying how the seasonal leaf water potentials (Ψ) of native tree species are related to soil water availability and evaporative demand components. The studied tree species were: *Cordia boissieri* (Boraginaceae), *Condalia hookeri* (Rhamnaceae), *Diospyros texana* (Ebenaceae) and *Bumelia celastrina* (Sapotaceae). Determinations of Ψ in the four native tree species were at 10-day intervals between 10 July and 30 November 2007. Ψ was monitored in five different plants per species at 6 a.m. (Ψ_{pd}, pre-dawn) and 2 p.m. (Ψ_{md}, mid-day). Air temperature, relative humidity, vapour pressure deficit, precipitation and soil water content were registered throughout. Ψ data were subjected to one-way ANOVA and correlation analysis. During the wettest period (10 September) Ψ_{pd} ranged from −0.72 (*C. boissieri*) to

−1.30 MPa (*B. celastrina*), in contrast, during the driest period (30 November), Ψ_{md} varied from −2.90 (*B. celastrina*) to −6.10 MPa (*D. texana*). Diurnal Ψ values were negatively correlated with air temperature and vapour pressure deficit; in contrast, a positive relationship was found with relative humidity. The ability of tree species to cope with drought stress depends on the pattern of water uptake and the extent to which water loss is controlled through the transpirational flux.

Mixed pine–oak forests are exposed to extreme environmental conditions where water availability is a limiting factor. In order to determine the adaptation of four tree species (*Arbutus xalapensis, Juniperus flaccida, Pinus pseudostrobus, Quercus canbyi*) to water stress, diurnal leaf water potentials (Ψ_w) were measured under natural drought and non-drought conditions in the Sierra Madre Oriental (Himmelsbach et al., 2010). Furthermore, the relation between leaf Ψ_w and environmental variables was analysed. The ANOVA revealed significant differences in Ψ_w between sampling dates (block variable), sampling hours and species (treatment variables) with no significant (p-value > 0.05) interaction between the treatment variables. In general, all species showed high pre-dawn and low mid-day values that declined progressively with increasing drought and soil-water loss. During the dry period, *J. flaccida* had the lowest Ψ_w followed by *Q. canbyi* and *A. xalapensis*, but all species recovered with rapid higher potentials after the onset of the rainy season at the end of May. In comparison, *P. pseudostrobus* showed less seasonal fluctuations. Differences in leaf Ψ_w were significant (p-value < 0.01) between all species, except for

the two conifers (p-value > 0.05). Correlations between Ψ_w and environmental variables were highly significant for soil moisture content in the morning hours (6 a.m. and 8 a.m.) and evaporative demand components in the afternoon (12 p.m. to 4 p.m.), depending on the species. *A. xalapensis, J. flaccida* and *Q. canbyi* showed strong correlations with climatic variables; hence these variables are better indicators of measured site conditions than *P. pseudostrobus*. In contrast, *P. pseudostrobus* showed weak correlations with climatic variables. This indicates that the species employed different strategies to overcome periods of drought. The latter species seems to tolerate drought (water stress) using its deeper rooting system, while the former species use other physiological strategies to overcome water stress. In conclusion, all species are considered as suitable candidates for reforestation programs in the Sierra Madre Oriental. Nevertheless, their suitability depends on the environmental site conditions, especially with respect to soil characteristics in the area of improvement.

Soil water potential (Ψ_s) is often estimated by measuring leaf water potential before dawn (Ψ_{pd}), based on the assumption that the plant water status has come into equilibrium with that of the soil (Kavanagh et al., 2007). However, it has been documented for a number of plant species that stomata do not close completely at night, allowing for nocturnal transpiration and thus preventing nocturnal soil–plant water potential equilibration. The potential for nighttime transpiration necessitates testing the assumption of nocturnal equilibration before accepting it as a valid estimate. They determined the magnitude of disequilibrium between four temperate conifer species across three height classes through a replicated study in northern Idaho. Based on both stomatal conductance and sap flux measurements, Kavanagh et al. (2007) confirmed that the combination of open stomata and high nocturnal atmospheric vapour pressure deficit (D) resulted in nocturnal transpiration in all four species. Nocturnal stomatal conductance ($g_{s\text{-noc}}$) averaged about 33% of mid-morning conductance values. They used species-specific estimates of $g_{s\text{-noc}}$ and leaf specific conductance to correct values for nocturnal transpiration at the time the samples were collected. Compared with the unadjusted values, corrected values reflected a significantly higher (when $D > 0.12$ kPa). These results demonstrate that comparisons between species, canopy height classes and sites, and across growing seasons can be influenced by differential amounts of nocturnal transpiration, leading to flawed results. Consequently, it is important to account for the presence of nocturnal transpiration, either through a properly parameterised model or by making measurements when D is sufficiently low that it cannot drive nocturnal transpiration.

20.5   Conclusions and research needs

A series of studies undertaken by Dr. Humberto González and his team indicates clearly that native woody species adapted to semiarid arid regions show a large variability in seasonal water potential. Seasonal variations in water potentials can be explained by the variability of soil water and root growth and development. Research needs to be directed to select plant species tolerant to drought and study other parameters, such as stomatal conductivity, morpho-anatomical traits and physiological and biochemical mechanisms of resistance, including rooting patterns.

20.6   Soil water potential

Soil moisture content is related to soil water potential, which contributes greatly the growth and the productivity of trees. Various studies have been undertaken on the effects of soil water content on tree growth and its productivity.

McCutchan and Shackel (1992) studied stem-water potential as a sensitive indicator of water stress in prune trees (*Prunus domestica* L. cv. French). The relative sensitivity of plant- and soil-based measures of water availability were compared for prune trees subjected to a range of irrigation regimes under field conditions. Over the growing season, leaf- and stem-water potentials (Ψ) measured at mid-day showed large differences between frequently irrigated trees and unirrigated trees that were growing on stored soil moisture. Stem Ψ was less variable than leaf Ψ, and the daily variability in stem Ψ was closely related to daily variability in evaporative demands, as measured by vapour pressure deficit (VPD). Owing to lower variability, stem Ψ showed the small stress effect of a moderate, 50% soil moisture depletion irrigation interval, whereas leaf Ψ did not. The relation between soil water content and estimated orchard evapotranspiration (ET) was influenced by local differences in soil texture within the experimental plot. The relation between stem Ψ and ET, however, was not influenced by soil texture and, in addition, was very similar to the relation between stem Ψ and leaf stomatal conductance. Both relationships indicated that a 50% reduction in leaf and canopy level water loss characteristics was associated with relatively small reductions (0.5–0.6 MPa) in stem Ψ. Stem Ψ appears to be a sensitive and reliable plant-based measure of water stress in prune and may be a useful tool for experimental work and irrigation scheduling.

Saxton et al. (1986) investigated the soil water characteristics from texture. Soil water potential and hydraulic conductivity relationships with soil water content are necessary for many plant and soil water studies, but the measurement of these relationships is costly, difficult and often impractical. Some studies used a large data base to establish statistical correlations between soil texture and selected soil potentials and also between selected soil textures and hydraulic conductivity. This study was undertaken to extend these results using mathematical equations for continuous estimates over broad ranges of soil texture, water potentials and hydraulic conductivities. Results from the recent statistical analyses were used to calculate water potentials for a wide range of soil textures, then these were fitted by multivariate analyses to provide continuous potential estimates for all inclusive textures. Similarly, equations were developed for unsaturated hydraulic conductivities for all inclusive textures. While the developed equations only reveal a statistical estimate and only a textural influence, they provide quite useful estimates for many usual soil water cases. The equations give an excellent computational efficiency for model applications and the textures can be used as calibration parameters where field or laboratory soil water characteristic data are available. Predicted values were successfully compared with several independent measurements of soil water potential.

Thinning improves leaf water potential in a tree. Bréda et al. (1995) studied the effects of thinning on soil and tree water relations, transpiration and growth in an oak forest (*Quercus petraea* (Matt.) Liebl.) with the objective of quantifying the effects of crown thinning on the water balance and growth of a stand, and to analyse the ecophysiological modifications induced by canopy opening on individual tree water relations. They undertook a thinning experiment in a 43 year old *Q. petraea* stand by removing trees from the upper canopy level. Soil water content, rainfall interception, sap flow, leaf water potential and stomatal conductance were monitored for two seasons following thinning. Seasonal time courses of leaf area index (LAI) and girth increment were also determined. The results revealed that pre-dawn leaf water potential was significantly higher in trees in the thinned stand than in a closed stand, as a consequence of higher relative extractable water in the soil. The higher water availability in the thinned stand was due to decreases in both interception and transpiration. In two-year studies, an increase in transpiration was observed in the thinned stand without any modification in LAI, whereas changes in transpiration in the closed stand were accompanied by variations in LAI. Thinning increased inter-tree variability in sap flow density, which was closely related to a leaf area competition index. Stomatal conductance showed little variation inside the crown; and differences in stomatal conductance between the treatments appeared only during a water shortage and affected mainly the closed stand. Thinning enhanced tree growth as a result of a longer growing period due to the absence of summer drought and higher rates of growth.

Dye (1996) studied the response of *Eucalyptus grandis* trees to soil water deficits subjected to soil drying at two field sites in the Mpumalanga province of South Africa in order to determine the relation between transpiration rate and soil water availability. It was hypothesised that, with this relationship defined, simple modelling of the soil water balance could be used to predict what fraction of potential transpiration was taking place at a given time. He grew a stand of three year old *E. grandis* trees, while nine year old trees were growing on Site 2, situated 2 km away. At each site, plastic sheeting was laid over the ground to prevent soil water recharge, thereby allowing the roots in the soil to induce a continuous progressive depletion of soil water. Measurements of pre-dawn xylem pressure potential, leaf area index, growth and sap flow rates indicated that prevention of soil water recharge caused only moderate drought stress. He observed that modelling the water balance of deep rooting zones is impractical for the purpose of simulating non-potential transpiration rates because of uncertainties about the depth of the root system, the soil water recharge mechanism and the water retention characteristics of the deep subsoil strata. It is concluded that predicting the occurrence and severity of soil water deficits from the soil water balance is not feasible at these sites.

Jackson et al. (1999) studied partitioning of soil water among tree species in a Brazilian Cerrado ecosystem. They determined source water used by woody perennials in a Brazilian savanna (Cerrado) by comparing the stable hydrogen isotope composition (δD) of xylem sap and soil water at different depths during two consecutive dry seasons (1995 and 1996). Plant water status and rates of water use were also determined and compared with xylem water δD values. It was assessed that soil water δD declined with increasing depth in the soil profile. Mean δD values were $-35°$ for the upper 170 cm of soil and $-55°$ between 230 and 400 cm depth at the end of the 1995 dry season. Soil water content increased with depth, from 18% near the surface to about 28% at 400 cm. A similar pattern of decreasing soil water δD with increasing depth was observed at the end of the 1996 dry season. Concurrent analyses of xylem and soil water δD values revealed a distinct partitioning of water resources between 10 representative woody species (five deciduous and five evergreen). Among these species, four evergreen and one deciduous species contained water primarily in the upper soil layers (above 200 cm), whereas three deciduous and one evergreen species tapped deep sources of soil water (below 200 cm). Among evergreen species, minimum leaf water potentials were also negatively correlated with xylem water δD values, suggesting that access to more readily available water at greater depth permitted maintenance of a more favourable plant water status. No significant relationship between xylem water δD and plant size was observed in two evergreen species, suggesting a strong selective pressure for small plants to rapidly develop a deep root system. The degree of variation in soil water partitioning, leaf phenology and leaf longevity was consistent with the high diversity of woody species in the Cerrado.

Goldhamer et al. (1999) studied the sensitivity of continuous and discrete plant and soil water status monitoring in peach trees subjected to deficit irrigation. To characterise tree responses to water deficits in shallow- and deep-rooted conditions, parameters developed using daily oscillations from continuously measured soil water contest and trunk diameter were compared with traditional discrete monitoring of soil and plant water status in lysimeter and field-grown peach trees [*Prunus persica* (L.) Batsch "O'Henry"]. Evaluation was undertaken during the imposition of deficit irrigation for 21 days, followed by full irrigation for 17 days. The maximum daily available soil water content fluctuations (MXAWCF) taken at any of the four monitored root zone depths responded most rapidly to the deficit irrigation. The depth of the MXAWCF increased with time during the deficit irrigation. Parameters based on trunk diameter monitoring, including maximum daily trunk diameter (MXTD), correlated well with established physiological parameters of tree water status. Statistical analysis of the differences in the measured parameters relative to fully irrigated trees during the first 10 days of deficit irrigation ranked the sensitivity of the parameters in the lysimeter as MXAWCF > MNTD > MDS > MXTD > stem Ψ = A = pre-dawn leaf Ψ = leaf Ψ. Equivalent analysis with the field-grown trees ranked the sensitivity of the parameters as MXAWCF > MNTD > MDS > stem Ψ = leaf Ψ = MXTD = pre-dawn leaf Ψ >

A. Following a return to full irrigation in the lysimeter, MDS and all the discrete measurements except A quickly returned to predeficit irrigation levels. Tree recovery in the field-grown trees was slower and incomplete due to inadequate filling of the root zone. Fruit size was significantly reduced in the lysimeter, but minimally affected in the field-grown trees.

Donovan et al. (2003) measured pre-dawn plant water potential (Ψ_w, measured with leaf psychrometers); and surrogate measurements with a pressure chamber (termed Ψ_{pc} here) were used to infer comparative ecological performance, based on the expectation that these plant potentials reflect the wettest soil Ψ_w accessed by roots. It is known that some species show substantial pre-dawn disequilibrium (PDD), defined as plant Ψ_w or Ψ_{pc} at pre-dawn, which is substantially more negative than the Ψ_w of soil accessed by roots. In the western Great Basin desert, the magnitude of PDD calculated as soil Ψ_w minus pre-dawn leaf Ψ_w was as large as 1.4 and 2.7 MPa for two co-dominant shrub species, *Chrysothamnus nauseosus* and *Sarcobatus vermiculatus*, respectively. The magnitude of PDD calculated as soil Ψ_w minus pre-dawn Ψ_{pc} was smaller, up to 0.6 and 2.1 MPa for *C. nauseosus* and *S. vermiculatus*, respectively. For both species, mechanisms contributing to PDD included night-time transpiration and putative leaf apoplastic solutes, but not hydraulic conductance limitations. Hydraulic lift also occurred in both species and likely contributed to PDD for *Sarcobatus*. Mechanisms contributing to PDD affect the relationships between plants and their soil resources and also the potential for plant–plant interactions

Himmelsbach et al. (2011) studied the acclimation of three co-occurring tree species to water stress and their role as site indicators in mixed pine–oak forests in the Sierra Madre Oriental, Mexico. Water availability and salt excess are limiting factors in Mexican mixed pine–oak forest. In order to characterise the acclimatation of native species to these stresses, the leaf water (Ψ_w) and osmotic potentials (Ψ_s) of *Juniperus flaccida*, *Pinus pseudostrobus* and *Quercus canbyi* were measured under natural drought and non-drought conditions under two different aspects in the Sierra Madre Oriental. Factorial ANOVA revealed significant differences in Ψ_w and Ψ_s between the two aspects, species and sampling dates. In general, all species showed high pre-dawn and low mid-day values that declined progressively with increasing drought and soil-water loss, seasonal and diurnal.

Diverse studies reveal variation in soil water and soil water potential in relation to tree growth. Its variability owing to variation in soil structure and environment influences the leaf water potential and tree growth. It is essential to estimate soil water potential of a site before studying leaf water potential, which is an indicator of water stress.

Bibliography

Anderson L.J., Brumbaugh M.S., Jackson R.B. **2001**. Water and tree-understory interactions: a natural experiment in a savanna with oak wilt. *Ecology* **82**(1):33–49.

Bréda N., Granier A., Aussenac G. **1995**. Effects of thinning on soil and tree water relations, transpiration and growth in an oak forest [*Quercus petraea* (Matt.) Liebl.]. *Tree Physiology* **15**:295–306.

Brown M.B., Forsythe A.B. **1974**. Robust tests for the equality of variances. *Journal of the American Statistical Association* **69**(346):364–367.

Bucci S.J., Scholz F.G., Goldstein G., Meinzer F.C., Franco A.C., Zhang Y., Hao G-Y. **2008**. Water relations and hydraulic architecture in Cerrado trees: adjustments to seasonal changes in water availability and evaporative demand. *Brazilian Journal of Plant Physiology* **20**(3):233–245.

Bussotti F., Bettini D., Grossoni P., Mansuino S., Nibbi R., Soda C., Tani C. **2002**. Structural and functional traits of *Quercus ilex* in response to water availability. *Environmental and Experimental Botany* **47**(1):11–23.

Castro-Díez P., Puyravaud J.P., Cornelissen J.H.C. **2000**. Leaf structure and anatomy as related to leaf mass per area variation in seedlings of a wide range of woody plants species and types. *Oecologia* **124**(4): 476–486.

De Micco V., Aronne G. **2008**. Twig morphology and anatomy of Mediterranean trees and shrubs related to drought tolerance. *Botanica Helvetica* **118**(2):139–148.

Donovan L.A., Richards J.H., Linton M.J. **2003**. Magnitude and mechanisms of disequilibrium between pre-dawn plant and soil water potentials. *Ecology* **84**(2):463–470.

Dye P.J. **1996**. Response of *Eucalyptus grandis* trees to soil water deficits. *Tree Physiology* **16**(1/2):233–238.

Evans R.D., Black R.A., Loescher W.H., Fellows R.J. **1992**. Osmotic relations of the drought tolerant shrub *Artemesia tridentata* in response to water stress. *Plant Cell and Environment* **15**(1):49–59.

Goldhamer D.A., Fereres E., Mata M., Girona J., Cohen M. **1999**. Sensitivity of continuous and discrete plant and soil water status monitoring in peach trees subjected to deficit irrigation. *Journal of the American Society for Horticultural Science* **124**:437–444.

González-Rodríguez H. Cantú-Silva I., Ramírez-Lozano R.G., Gómez-Meza M.V., Uvalle

Sauceda J.I., Maiti R.K. **2010**. Characterization of xylem water potential in ten native plants of north-eastern Mexico. *International Journal of Bio-resource and Stress Management* **1**(3):219–224.

González-Rodríguez H., Cantú Silva I., Gómez Meza M.V., Jordan W.R. **2000**. Seasonal plant water relationships in *Acacia berlandieri*. *Arid Soil Research and Rehabilitation* **14**(4):343–357.

González-Rodríguez H., Cantú Silva I., Gómez Meza M.V., Ramírez Lozano R.G. **2004**. Plant water relations of thornscrub shrub species, northeastern Mexico. *Journal of Arid Environments* **58**(4):483–503.

González-Rodríguez H., Cantú Silva I., Gómez Meza M.V., Ramírez Lozano R.G., Pando Moreno M., Molina Camarillo I.A., Maiti R.K. **2009**. Water relations in native trees, northeastern Mexico. *International Journal of Agriculture Environment and Biotechnology* **2**(2):133–141.

Himmelsbach W., Garza E.J.T., Rodríguez H.G., Tagle M.A.G., Estrada Castillón A.E., Aguirre Calderón O.A. **2010**. Wasserpotenziale im Tages- und Jahresverlauf als Indikatoren von Umweltfaktoren, gemessen an vier vergesellschafteten Baumarten in der Sierra Madre Oriental, Mexiko (Site-conditions reflected by seasonal and diurnal leaf water potentials of four co-occurring tree species in the Sierra Madre Oriental, Mexico). *Forstarchiv* **81**(3):110–117.

Himmelsbach W., Treviño-Garza E.J., González-Rodríguez H., González-Tagle M.A., Gómez Meza M.V., Aguirre Calderón O.A., Estrada Castillón A.E., Mitlöhner R. **2011**. Acclimation of three co-occurring tree species to water stress and their role as site indicators in mixed pine-oak forests in the Sierra Madre Oriental, Mexico. *European Journal of Forest Research* **131**(2):355–367.

Jackson P.C., Meinzer F.C., Bustamante M., Goldstein G., Franco A., Rundel P.W., Caldas L., Igler E., Causin F. **1999**. Partitioning of soil water among tree species in a Brazilian Cerrado ecosystem. *Tree Physiology* **19**(11):717–724.

Kavanagh K.L., Pangle R., Schotzko A.D. **2007**. Nocturnal transpiration causing disequilibrium between soil and stem pre-dawn water potential in mixed conifer forests of Idaho. *Tree Physiology* **27**:621–629.

Kolb T.E., Stone J.E. **2000**. Differences in leaf gas exchange and water relations among species and tree sizes in an Arizona pine-oak forest. *Tree Physiology* **20**(1):1–12.

Kozlowski T.T., Pallardy S.G. **2002**. Acclimation and adaptive responses of woody plants to environmental stresses. *The Botanical Review* **68**(2):270–334.

Lacher W.L. **1983**. *Physiological Plant Ecology*. Springer, Berlin, 203 pp.

Liu M.Z., Jiang G.M., Li Y.G., Niu S.L., Gao L.M., Ding L., Peng Y. **2003**. Leaf osmotic potentials of 104 plant species in relationship to habitats and plant functional types in Hunshandak Sandland, Inner Mongolia, China. *Trees* **17**(6):554–560.

López-Hernández J.M., González-Rodríguez H., Cantú-Silva I., Ramírez-Lozano R.G., Gómez-Meza M.V., Pando-Moreno M., Sarquís-Ramírez J.I., Coria-Gil N., Maiti R.K., Sarkar N.C. **2010**. Adaptation of native shrubs to drought stress in north-eastern Mexico. *International Journal of Bio-resource and Stress Management* **1**(1):30–37.

McCutchan H., Shackel K.A. **1992**. Stem-water potential as a sensitive indicator of water stress in prune trees (*Prunus domestica* L. cv. French). *Journal of the American Society for Horticultural Science* **117**:607–611.

McMurtry C.R., Barnes P.W., Nelson J.A., Archer S.R. **1996**. *Physiological responses of woody vegetation to irrigation in a Texas subtropical savanna. La Copita Research Area: 1996 Consolidated Progress Report*. Texas Agricultural Experiment Station, College Station, TX. pp. 33–37.

Montagu K.D., Woo K.C. **1999**. Recovery of tree photosynthetic capacity from seasonal drought in the wet-dry tropics: the role of phyllode and canopy processes in *Acacia auriculiformis*. *Australian Journal of Plant Physiology* **26**(2):135–145.

Nagler P.M., Glenn E.P., Thompson T.L. **2003**. Comparison of transpiration rates among saltcedar, cottonwood and willow trees by sap flow and canopy temperature methods. *Agricultural and Forest Meteorology* **116**(1/2):73–89.

Newton R.J., Funkhouser E.A., Fong F., Tauer C.G. **1991**. Molecular and physiological genetics of drought tolerance in forest species. *Forest Ecology and Management* **43**(3/4):225–250.

Newton R.J., Goodin, J.R., **1989**. Moisture stress adaptation in shrubs. In: McKell, C.M. (ed.), *The Biology and Utilization of Shrubs*. Academic Press, San Diego, pp. 365–383.

Otieno D.O., Kurz-Besson C., Liu J., Schmidt M.W.T., Vale-Lobo do R., David T.S., Siegwolf R., Pereira J.S., Tenhunen J.D. **2006**. Seasonal variations in soil and plant water status in a *Quercus suber* L. stand: roots as determinants of tree productivity and survival in the Mediterranean-type ecosystem. *Plant and Soil* **283**(1/2):119–135.

Ott L. **1993**. *An Introduction to Statistical Methods and Data Analysis*, 2nd Edn. Duxbury Press, Boston, Massachusetts, 775 pp.

Reid N., Marroquin J., Beyer-Münzel P. **1990**. Utilization of shrubs and trees for browse, fuelwood and timber in the Tamaulipan thornscrub in northeastern Mexico. *Forest Ecology and Management* **36**(1):61–79.

Rhizopoulou S., Heberlein K., Kassianou A. **1997**. Field water relations of Capparis spinosa L. *Journal of Arid Environments* **36**(2):237–248.

Ritchie G.A., Hinckley T.M. **1975**. The pressure chamber as an Instrument for ecological research. *Advances in Ecological Research* **9**:165–254.

Rosenberg N.J., Blad B.L., Verma S.B. **1983**. *Microclimate; the Biological Environment*, 2nd Edn. John Wiley & Sons, Inc., New York, pp. 170–172.

Saxton K.E., Rawls W.J., Romberger J.S., Papendick R.I. **1986**. Estimating generalized soil-water characteristics from texture. *Soil Science Society of America Journal* **50**(4): 1031–1036.

SPP-INEGI. **1986**. *Síntesis Geografía del Estado del Nuevo León*. Instituto Nacional de Geografía Estadística e Información, México, D.F.

Steel R.G.D., Torrie J.H. **1980**. *Principles and Procedures of Statistics. A Biometrical Approach*, 2nd Edn. McGraw-Hill, New York, 632 pp.

Stienen H., Smits M.P., Reid N., Landa J., Boerboom J.H.A. **1989**. Ecophysiology of 8 woody multipurpose species from semiarid northeastern Mexico. *Annals of Forest Sciences* **46**:454–458.

CHAPTER 21

Cold tolerance of trees

21.1 Studies on the cold tolerance of trees

The cold temperatures prevailing in cold temperate and alpine regions affect tree growth and productivity. One of the most limiting climatic constraints on tree growth is the lowest temperature. Cold injury caused by freezing temperature affects tree productivity in the northern hemisphere.

De Hayes et al. (1989) described winter injury to red spruce (*Picea rubens* Sarg.) trees from 12 provenances in a plantation near Colebrook, NH, USA, and quantified them from spring 1986 throughout 1989. Winter injury was detected exclusively on current-year needles (i.e. buds and twigs were virtually uninjured), and 34–68% of trees were injured each year. Laboratory assessments of cold tolerance of current and one year old needles of trees representing four red spruce and four balsam fir [*Abies balsamea* (L.) Mill.] provenances demonstrated that current-year needles of red spruce, but not balsam fir, were 8–12 °C less cold tolerant than one year old needles in late autumn and early winter. Also, throughout the winter, current-year needles of red spruce were about 10 °C less cold tolerant on average than those of balsam fir, a sympatric species that does not suffer winter injury. It appears that

the rate of freezing injury susceptibility is observed naturally for red spruce in early winter. The remarkable differences in foliar winter injury and percentage of trees injured were highly significant in 1986, 1987 and 1989 and marginally significant in 1988. Trees from Quebec, New York and New Brunswick were consistently the least injured each winter, whereas trees from Massachusetts and New Hampshire were consistently among the most injured. It is expected that repeated winter injury is a major contributor to the decline of red spruce in northern forests.

Canham et al. (1990) compared light regimes beneath closed canopies and tree-fall gaps for five temperate and tropical forests using fish-eye photography of intact forest canopies and a model for calculating light penetration through idealised gaps. Beneath intact canopies, analyses of canopy photographs revealed that sunflecks potentially contribute 37–68% of the seasonal total photosynthetically active radiation. In all of the forests, potential sunfleck duration was brief (4–6 min) but the frequency distributions of potential sunfleck duration varied because of differences in canopy geometry and recent disturbance history. Analysis of the photographs indicated that incidence angles

Autoecology and Ecophysiology of Woody Shrubs and Trees: Concepts and Applications, First Edition.
Edited by Ratikanta Maiti, Humbero Gonzalez Rodriguez and Natalya Sergeevna Ivanova.
© 2016 John Wiley & Sons, Ltd. Published 2016 by John Wiley & Sons, Ltd.

for photosynthetically active radiation beneath closed canopies were not generally vertical for any of the forests, but considerable variation was observed both among and within sites in the contribution of overhead versus low-angle lighting. Measurements of light penetration through idealised single-tree gaps in old growth Douglas-fir–hemlock forests revealed that such gaps had little effect on the understorey light regimes because of the high ratio of canopy height to gap diameter. However, single-tree gaps in the other four forest types showed significant overall increases in understorey light levels. Significant spatial variation was also observed n seasonal total radiation in and around single-tree gaps. The results demonstrated that there could be a significant penetration of light into the understorey adjacent to a gap, particularly at high latitudes. As gap size increased, both the mean and the range of light levels within the gap increased but even in large gaps the potential duration of direct sunlight was generally brief ($< 4\,$h). The major differences in gap light regimes of the five forests were mainly a function of canopy height and latitude. The effects of latitude should also lead to differences in gap light regimes across the geographic range of individual forest types.

Chlorophyll fluorescence is an indicator of a species for cold tolerance. Adam and Perkins (1993) assessed cold tolerance in *Picea* using chlorophyll fluorescence. Chlorophyll fluorescence measurements were undertaken on field-collected red spruce foliage exposed to controlled freezing temperatures to determine the utility of chlorophyll fluorescence kinetics in assessing low temperature tolerance.

No significant decreases in fluorescence relative to unfrozen control foliage were detected at progressively lower temperatures until a critical temperature was reached, whereupon rapid, irreversible decreases in fluorescence occurred. Weekly comparisons between cold tolerance estimates derived using chlorophyll fluorescence inflection point temperatures and foliar visual LT_{20} (the temperature at which 20% of foliage showed necrosis) estimates were determined throughout the winter of 1991–1992. No significant differences were found in weekly cold tolerance estimates between the two methods were evident. The weekly absolute difference in mean cold tolerance using the two methods was $3.0\,°$C. These results demonstrated that chlorophyll fluorescence is a rapid, consistent, and reproducible method of determining the low temperature tolerance of spruce foliage.

Swanson (1993) studied cold tolerance and maximal thermogenic capacity under cold stress in helox gas mixtures (approx. 79% He, 21% O_2) for Dark-eyed Juncos (*Junco hyemalis*) and American Tree Sparrows (*Spizella arborea*) wintering in south-eastern South Dakota. Junco data were compared to previous data on juncos wintering in the milder climate of western Oregon. Dark-eyed Juncos from South Dakota (SDJU) tolerated colder helox temperatures than Oregon birds (ORJU). This improved cold resistance was not associated with increased thermogenic capacity or SMR in SDJU, as SMR and total did not differ significantly and the mass-specific was greater in ORJU. SDJU were significantly larger and showed significantly lower thermal conductance than ORJU. This is important to improved cold tolerance capabilities in the former. SDJU gave significantly higher total than sympatric American Tree Sparrows (ATSP) but the mass-specific did not vary as the juncos were significantly heavier than sparrows. SDJU tolerated slightly colder temperatures than did ATSP. Thermal conductance and SMR did not differ between the two species. Sympatric wintering juncos and sparrows, which showed similar ranges and habits, also demostrated similar metabolic performance in the cold.

Bigras (1997) studied the root cold tolerance of black spruce seedlings, using viability tests in relation to survival and regrowth. Root systems of six-month, cold-hardened, container-grown black spruce seedlings [*Picea mariana* (Mill.) B.S.P.] were exposed to 0, −5, −10, −15, −20 or −22.5 °C. Freezing-induced damage to fine roots, coarse roots and the whole root system was assessed by various viability tests, including leakage of electrolytes, leakage of phenolic compounds, water loss, root and shoot water potentials and live root dry mass. To assess the long-term effects of freezing-induced root damage, seedling survival and regrowth were estimated. Leakage of both electrolytes and phenolic compounds showed difference *d* among fine roots, coarse roots and whole root systems. In coarse roots and the whole root system, leakage of electrolytes, leakage of phenolic compounds, water loss and root and shoot water potentials showed a correlation with the percentage of live root dry mass which, in turn, was highly correlated with seedling survival and regrowth. Compared with the live root dry mass, the electrolyte and phenolic leakage, water loss and root and shoot water potentials showed a low correlation with

seedling survival and regrowth. Among the viability tests, electrolyte leakage of the whole root system correlated most closely with seedling survival and regrowth. Under freezing conditions that destroyed less than 50+ of each seedling's root system, about 70+ of the seedlings survived and subsequent growth was little affected, whereas under freezing conditions that destroyed 70+ of each seedling's root system, only about 30+ of the seedlings survived and subsequent growth was reduced compared with that of undamaged plants.

Ögren (1997) studied the relationship between temperature, respiratory loss of sugar and premature dehardening in dormant Scots pine seedlings. Increased intracellular sugar concentration is an important contributor to the increased cold tolerance of conifers in winter. Two year old seedlings of Scots pine (*Pinus sylvestris* L.) were grown and cold-hardened in the field and then exposed to different temperature regimes for 16 weeks while dormant. For determining short-term carry-over effects, after the temperature treatments, all seedlings were exposed to 5.5 °C and watered before the assessment of non-structural carbohydrates and cold tolerance. Needle sugar concentration was reduced by 54, 32, 21 and 9% following treatment at 5.5, 0, −1.5 and −8.5 °C, respectively. Sugar concentration did not decrease as much in root tissues as in needles because starch was mobilised in roots. Cold tolerance of needles was analysed by controlled freezing; and the temperature causing an initial 10% damage (LT_{10}) was plotted as a function of needle sugar concentration, revealing a strong, linear relationship. When one-third of the initial sugars had been consumed, LT_{10} had increased from −24.5 to −16.5 °C; and when one-half had been consumed, LT_{10} had increased to −10 °C.

Walters and Reich (2000) studied plants in a low-light CO_2 exchange: a component of variation in shade tolerance among cold temperate tree seedlings to determine whether enhanced whole-plant CO_2 exchange in moderately low to high light occurred at the cost of greater CO_2 loss rates at very low light levels. They investigated the tolerance of first-year seedlings of intolerant *Populus tremuloides* and *Betula papyrifera*, intermediate *Betula alleghaniensis* and tolerant *Ostrya virginiana* and *Acer saccharum* grown in moderately low light (7·3% of open sky) and low light (2·8%). They predicted that, compared with shade-tolerant species, intolerant species would have characteristics leading to greater whole-plant CO_2 exchange rates in moderately low to high light levels, with higher CO_2 loss than shade-tolerant rates at very low light levels. *A. saccharum*, a less-tolerant species grown in both light treatments, showed greater mass-based photosynthetic rates, leaf, stem and root respiration rates, leaf mass plant mass ratios and leaf area:leaf mass ratios, similar whole-plant light compensation points and leaf-based quantum yields.

Coder (2011) discussed the cold tolerance of trees. Trees are found all over the globe in a variety of climates. Too low a temperature disrupts a tree's reactions. The tree growth line observed on high mountains is affected by low temperatures, wind stress and water availability problems. Trees are not usually observed in areas or at altitudes where sustained temperatures in the dormant season fall below −40 °C. Cold damage in trees arises from three complex events: (a) rate of temperature change, (b) lowest temperature reached in chilling, but not freezing and (c) lowest temperature reached in freezing. Freezing (frost) begins when liquid water starts to change state into solid ice crystals.

Huey et al. (2010) studied the evolution of the thermal sensitivity of ectotherm performance. Most ectothermal animals have variable body temperatures, because physiological rates are temperature sensitive. An ectotherm's behavioural and ecological performance can be influenced by body temperature.

Very few studies are available on the cold tolerance of trees in high alttudes. Therefore, concerted research activities need to be undertaken on the effect of cold on the physiological functions and genotypic variability of temperate and alpine trees as well as trees of the tropical and semiarid tropics.

Bibliography

Adam G.T., Perkins T.D. **1993**. Assessing cold tolerance in *Picea* using chlorophyll fluorescence. *Environmental and Experimental Botany* **33**(3):377–382.

Bigras F.J. **1997**. Root cold tolerance of black spruce seedlings: viability tests in relation to survival and regrowth. *Tree Physiology* **17**(5):311–318.

Canham C.D., Denslow J.S., Platt W.J., Runkle J.R., Spies T.A., White P.S. **1990**. Light regimes beneath closed canopies and tree-fall gaps in temperate and tropical forests. *Canadian Journal of Forest Research* **20**(5):620–631.

Coder K.D. **2011**. *Trees and Cold Temperatures. Environmental Tolerance Series*, Warnell, London, pp. 1–10.

De Hayes D.H., Waite C.E., Ingle M.A., Williams M.W. **1989**. Winter injury susceptibility and cold tolerance of current and year-old needles of red spruce trees from several provenances. *Forest Science* **36**(4):982–994.

Guy C.L. **1990**. Cold accelimation and freezing stress tolerance: role of protein metabolism. *Annual Review of Plant Physiology and Plant Molecular Biology* **41**:187–223.

Huey R.B., Joel G., Kingsolver J.G. **2010**. Evolution of thermal sensitivity of ectotherm performance. *Trends in Ecology and Evolution* **4**(5):131–135.

Huey R.B., Kingsolver J.G. **1997**. Root cold tolerance of black spruce seedlings: viability tests in relation to survival and regrowth. *Tree Physiology* **17**(5):311–318.

Ögren E. **1997**. Relationship between temperature, respiratory loss of sugar and premature dehardening in dormant Scots pine seedlings. *Tree Physiology* **17**(1):47–51.

Swanson D.L. **1993**. Cold tolerance and thermogenic capacity in dark-eyed Juncos in winter: Geographic variation and comparison with American tree Sparrows. *Journal of Thermal Biology* **18**(4):275–281.

Sykes M.T., Prentice I.C., Cramer W. **1996**. A bioclimatic model for the potential distributions of north European tree species under present and future climates. *Journal of Biogeography* **23**(2):203–233.

Walters M.B., Reich P.B. **2000**. Trade-offs in low-light CO_2 exchange: a component of variation in shade tolerance among cold temperate tree seedlings. *Functional Ecology* **14**(2):155–165.

CHAPTER 22

Heat stress tolerance of trees

22.1 Studies on the heat stress tolerance of trees

Coder (2011) discussed heat stroke in trees. He mention that summer exposes people and trees to a number of hot and dry weeks, thereby damaging many of the old, young and soil-limited trees in parking lots, along streets, on open squares and in surrounding pavement, leading to a number of tree symptoms. The old term "heat stroke" fits trees where heat loads have been extreme and caused problems. Trees require an optimum growing condition across the temperature range 70–85 °F (21–30 °C). Hot temperatures can injure and kill living tree systems. A thermal death threshold occurs at approximately 115 °F (46 °C). The thermal death threshold varies depending upon the duration of hot temperature, the absolute highest temperature and the tissue age, thermal mass, water content of tissue and ability of the tree to make adjustments to temperature changes. Tree temperature is generally just at or slightly above air temperature. Trees dissipate heat by long-wave radiation or convection of heat into the air and by transpiration (water loss from leaves). Transpiration plays an important role in the dissipation of tree heat loads, like heat radiation to the surroundings

and wind cooling. When transpiration is limited by hot temperatures and the tree is surrounded by non-evaporative surfaces (hard surfaces), leaf temperatures may rise above the thermal death threshold.

Turner and Kramer (1980) in their book explained the adaptation of plants to water and high temperature stress. They reviewed the response of plants to water and high temperature stress; and they emphasised the adaptation and acclimation of plants to such stresses, including plant morphological and physiological mechanisms of adaptation to stress. They also discussed the adaptation of plants to hot and dry environments in both natural and agricultural communities, the importance of these adaptations on the yield, survival and productivity of these communities, adaptations to high-temperature stress and the interactions and integration of adaptations to stress.

Smillie and Hetherington (1983) investigated stress tolerance and stress-induced injury in crop plants using chlorophyll fluorescence. In vivo stress responses of leaf tissue were estimated by F_R, the maximal rate of the induced rise in chlorophyll fluorescence. The time taken for F_R to decrease by 50% in leaves at 0 °C was used as a measure of chilling tolerance. Relative heat tolerance was also indicated by the decrease in

Autoecology and Ecophysiology of Woody Shrubs and Trees: Concepts and Applications, First Edition.
Edited by Ratikanta Maiti, Humbero Gonzalez Rodriguez and Natalya Sergeevna Ivanova.
© 2016 John Wiley & Sons, Ltd. Published 2016 by John Wiley & Sons, Ltd.

F_R in heated leaves while changes in vivo resulting from photoinhibition, ultraviolet radiation and photobleaching could also be measured.

Mittler (2006) discussed abiotic stress, the field environment and stress combinations. Farmers and breeders know well that the simultaneous occurrence of several abiotic stresses, rather than one particular stress condition, can be most lethal to crops. It is also known that the co-occurrence of different stresses is rarely addressed by the molecular biologists who study plant acclimation. Studies have revealed that the response of plants to a combination of two different abiotic stresses cannot be directly extrapolated from the response of plants to each of the different stresses applied individually. Tolerance to a combination of different stress conditions, particularly those that mimic the field environment, should direct future research programmes aimed at developing transgenic crops and plants with enhanced tolerance to naturally occurring environmental conditions.

Valladares and Pearcy (1997) studied the interactions between water stress, sun-shade acclimation, heat tolerance and photoinhibition in the sclerophyll shrub *Heteromeles arbutifolia.* They used gas exchange and chlorophyll fluorescence techniques to evaluate the acclimation capacity of *H. arbutifolia* to the multiple co-occurring summer stresses of the California chaparral. They examined the influence of water, heat and high light stresses on the carbon gain and survival of sun and shade seedlings via a factorial experiment involving a slow drying cycle applied to plants grown outdoors during the summer. The photochemical efficiency of PSII showed a diurnal, transient decrease ($\delta F/F_m'$) and a chronic decrease or photoinhibition (F_v/F_m) in plants exposed to full sunlight. Water stress increased both transient decreases of $\delta F/F_m'$ and photoinhibition. Effects of decreased $\delta F/F_m'$ and F_v/F_m on carbon gain were observed only in well-watered plants, since in water-stressed plants they were overidden by stomatal closure. Reductions in photochemical efficiency and stomatal conductance were observed in all plants exposed to full sunlight, even in those that were well-watered. This demonstrated that *H. arbutifolia* sacrificed carbon gain for water conservation and photoprotection (both structurally via shoot architecture and physiologically via down-regulation) and that this response was triggered by a hot and dry atmosphere together with high PFD, before severe water, heat or high PFD stresses occurred. They observed fast adaptive adjustments of the thermal stability of PSII (diurnal changes) and a superimposed long-term acclimation (days to weeks) to high leaf temperatures. Water stress increased the resistance of PSII to high temperatures both in the dark and over a wide range of PFD. Low PFD protected photochemical activity against inactivation by heat while high PFD exacerbated damage of PSII by heat. The greater interception of radiation by horizontally restrained leaves relative to the steep leaves of sun-acclimated plants induced photoinhibition and increased leaf temperature. When transpirational cooling was decreased by water stress, leaf temperature surpassed the limits of chloroplast thermostability. The remarkable acclimation of water-stressed plants to high leaf temperatures was insufficient for the semi-natural environmental conditions of the experiment. Summer stresses characteristic of Mediterranean-type climates (high leaf temperatures in particular) are a potential limiting factor for seedling survival in *H. arbutifolia*, especially for shade seedlings lacking the crucial structural photoprotection provided by steep leaf angles.

Chaves et al. (2002) showed that plants are frequently subjected to periods of soil and atmospheric water deficit during their life cycle. Plant responses to water scarcity are complex, involving deleterious and/or adaptive changes under field conditions. Changes in the root to shoot ratio or the temporary accumulation of reserves in the stem are associated with alterations in nitrogen and carbon metabolism. C-metabolism is an important defence mechanism under conditions of water stress and is accompanied by down-regulation of photochemistry and, in the longer term, of carbon metabolism.

Bita and Gerats (2003) studied plant tolerance to high temperature in a changing environment. Global warming is predicted to have a general negative effect on plant growth due to the damaging effect of high temperatures on plant development. The increasing threat of climatological extremes associated with very high temperatures may cause catastrophic loss of crop productivity and result in wide-spread famine. In this review, they assessed the impact of global climate change on agricultural crop production. High temperature stress has a wide range of effects on plants in terms of physiology, biochemistry and gene regulation pathways. However, strategies need to be adopted for crop improvement for heat stress tolerance. They presented recent advances in research on all these levels of investigation and focussed on potential leads that may help to understand more fully the mechanisms that make plants tolerant or susceptible to heat stress. There is a differential effect of climate change in terms of both geographic location and the crops that will likely show the most extreme reductions in yield as a result of expected extreme fluctuations in temperature and global warming in general. High temperature stress has a wide range of effects on plants in terms of physiology, biochemistry and gene regulation pathways. However, strategies exist to crop improvement for heat stress tolerance. In their review, the authors presented recent advances of research on all these levels of investigation and focussed on potential leads that may help to understand more fully the mechanisms that make plants tolerant or susceptible to heat stress. Finally, they reviewed possible procedures and methods which could lead to the generation of new varieties with sustainable yield production, in a world likely to be challenged by increasing population, higher average temperatures and larger temperature fluctuations.

Feder and Hofmann (1999) investigated heat-shock proteins, molecular chaperones and the stress response of plants. Molecular chaperones, including the heat-shock proteins (Hsps), are a characteristic feature of cells in which these proteins cope with stress-induced denaturation of other proteins. The function of Hsps at the molecular and cellular level is becoming well-explored. The authors' complementary focus concentrated on the Hsps of both model and non-model organisms undergoing stress in nature, on the roles of Hsps in the stress physiology of whole multicellular eukaryotes and the tissues and organs they comprise and on the ecological and evolutionary correlates of variation in Hsps and the genes that encode them. This focus disclosed that: (a) the expression of Hsps can occur in nature, (b) all species have *hsp* genes but they vary in the patterns of their expression, (c) Hsp expression can be correlated with resistance to stress and (d) species' thresholds for Hsp expression are correlated with levels of stress that they naturally undergo.

Bibliography

Bita C.A., Gerats T. **2003**. Plant tolerance to high temperature in a changing environment: scientific fundamentals and production of heat stress-tolerant crops. *Frontiers in Plant Science* **4**:273.

Chaves M.M, Pereira J.S., Marocco J., Rodrigues M.L., Ricardo C.P.P., Osório M.L., Pintero C., Carvalho I., Faria T., Pinheiro C. **2002**. How plants cope with water stress in the field? photosynthesis and growth. *Annals of Botany* **89**(7):907–916.

Coder K.D. **2011**. *Trees and Cold Temperatures*. Environmental Tolerance Series, Warnell, London, pp. 1–10.

Feder M.E., Hofmann G.E. **1999**. Heat shock proteins, molecular chaperones, and the stress respnse: evolutionary and ecological physiology. *Annual Review of Physiology* **61**: 243–282.

Mittler R. **2006**. Abiotic stress, the field environment and stress combination. *Trends in Plant Science* **11**(1):15–19.

Smillie R.M., Hetherington S.E. **1983**. Stress tolerance and stress-induced injury in crop plants measured by chlorophyll fluorescence in vivo: chilling, freezing, ice cover, heat, and high light. *Plant Physiology* **72**(4):1043–1050.

Turner N.C., Kramer P.J. **1980**. *Adaptation of Plants to Water and High Temperature Stress*. In: Proceedings of seminar held from November 6 to 10, 1978, at the Carnegie Institution of Washington, Department of Plant Biology, Stanford, California.

Valladares F., Pearcy R.W **1997**. Interactions between water stress, sun–shade acclimation, heat tolerance and photoinhibition in the sclerophyll *Heteromeles arbutifolia*. *Plant, Cell and Environment* **20**(1):25–36.

CHAPTER 23

Seed characteristics, seed dormancy, germination and plant propagation

There exists large variability in seed characteristics of tree species with respect to seed surface colour, striations, seed size, dimensions and so on. Very little information is available in the literature on the seed characteristics and seed structure of tree species.

23.1 Seed dormancy and germination

The seed dormancy of a species may be defined as its period of incapacity for germination until the advent of favourable conditions, such as suitable temperature and rain. Fruits/seeds of a particular tree species fall onto the ground after seed dispersal and remain there being exposed to changes in the environment until the advent of a favourable season, such as the precipitation in the following season. Seed dormancy functions as a mechanism for adaptation to adverse conditions. With the advent of a favourable season, the seeds start germinating and emerge, depending on the species. This is termed regeneration.

In this context, Finch-Savage and Leubner-Metzger (2006) made a review on seed dormancy and the control of germination. Seed dormancy is considered a barrier to complete germination for an intact viable seed. Seed dormancy is an inherent seed property that defines the environmental conditions in which a seed is able to germinate. It is determined by genetics with a substantial environmental influence which is mediated partly by the plant hormones abscisic acid and gibberellins. Dormancy is present in all the higher plants in all major climatic regions. Forest seeds from different trees fall to the ground after seed dispersal and remain dormant until the advent of favourable conditions. It is a mechanism of adaptation of the species to unfavourable conditions. The adaptation has resulted in divergent responses to the environment.

23.1.1 Techniques for breaking seed dormancy

Two types of dormancy are observed: (a) seed coat or external dormancy and (b) internal (endogenous) dormancy. The following types of dormancy are also reported by various scientists.

Morphological dormancy (MD) occurs in seeds with embryos that are underdeveloped but have differentiated inside the seed into cotyledons and hypocotyl/radical. These embryos are not (physiologically) dormant, but simply need time and

Autoecology and Ecophysiology of Woody Shrubs and Trees: Concepts and Applications, First Edition.
Edited by Ratikanta Maiti, Humbero Gonzalez Rodriguez and Natalya Sergeevna Ivanova.
© 2016 John Wiley & Sons, Ltd. Published 2016 by John Wiley & Sons, Ltd.

favourable conditions to grow and germinate, for example celery (*Apium graveolens*, Apiaceae; Jacobsen and Pressman, 1979).

Morphophysiological dormancy (MPD) is present in seeds with underdeveloped embryos, but in addition they have a physiological component to their dormancy (Baskin and Baskin, 2004). These seeds therefore need a dormancy-breaking treatment, such as a defined combination of warm and/or cold stratification, which in some cases can be replaced by GA application. There are eight known levels of MPD.

Physical dormancy (PY) is caused by the presence of water-impermeable layers of palisade cells in the seed or fruit coat that control water movement. Mechanical or chemical scarification can break PY dormancy (Baskin and Baskin, 2004). Chemical scarification is generally done by treating seeds with different concentrations of sufuric acids or simply boiling water, as in the case of *Prosopis* and wild chilli observed in our study.

Combinational dormancy (PY+PD) is observed in seeds with water-impermeable coats (as in PY) combined with physiological embryo dormancy (Baskin and Baskin, 2004).

Below, we mention different techniques reported in breaking seed dormancy (Evans and Frank, 1999).

23.2   Seed scarification

Commercial growers scarify seeds by soaking them in concentrated sulfuric acid. Seeds are placed in a glass container and covered with sulfuric acid. The seeds are gently stirred and allowed to soak for 10 minutes to several hours, depending on the species for mechanical scarification. Seed coats can also be put in boiling water, filed with a metal file, rubbed with sandpaper, cut a little with a knife or cracked gently with a hammer to weaken the seed coat.

23.3   Seed stratification

Another type of imposed dormancy observed in seeds is internal dormancy regulated by the inner seed tissues. This dormancy inhibits the seeds of many species from germinating when environmental conditions are not favourable for seedling survival and serves as a mechanism for survival against adverse conditions. Besides, there are several different degrees or types of internal dormancy. However, another type of internal dormancy requires special treatments to overcome. Seeds having this type of dormancy will not germinate until subjected to a particular duration of moist pre-chilling and/or moist/warm periods.

Cold stratification (moist pre-chilling) involves mixing seeds with an equal volume of a moist medium (sand or peat, for example) in a closed container and storing them in a refrigerator (approximately 40 °F; 4 °C). Periodic checking should be done to see that the medium is moist but not wet. The length of time it takes to break dormancy varies with particular species. Warm stratification is similar, except temperatures are maintained at 68–86 °F (20–30 °C), depending on the species. In the case of wild chilli, Chile piquin (*Capsicum annuum*), the seeds are kept in in a test tube mixed with cowdung mixture for seven days. Then, the seeds are sown on a soil surface in the presence of light in a growth chamber or natural environment.

In the following we narrate a few research advances in breaking seed dormancy and inducing seed germination of a few tree species.

Zoghi et al. (2011) studied the effect of different treatments on seed dormancy breaking and germination of caspian locust (*Gleditschia caspica*). The tree caspian locust (*Gleditschia caspica*) is a special species in hyrcanian forests. The conservation of this species is essential as a gene source; thus, seedling production and their use is necessary in afforestation programmes. This study was carried out to investigate the effects of different treatments for dormancy breaking and germination of caspian locust seed. The authors adopted different techniques, such as concentrated sulfuric acid at three levels, humidity sand at two levels and seeds soaking in water with two levels. The results revealed that scarification with sulfuric acid treatment for 1.5 h was best.

Raich and Khoon (1990) studied the effects of canopy openings on tree seed germination in a Malaysian dipterocarp forest. They investigated the germination of 43 tree species native to the lowland forests of Malaysia on forest soil in trays placed in: (a) closed-canopy forest, (b) an artificial forest gap and (c) a large clearing. Large variations were observed in germination among the habitats, with only seven species germinating well in all three sites. Seed germination of most species demonstrated clear patterns of shade tolerance or intolerance identical to those long recognised for tree seedlings both in mature and secondary forests. Most forest canopy species germinated well in the gap, but germination in the areas of large clearing declined drastically or were zero. It was also observed that the regeneration of these species in large clearings was delayed even when seeds were present. Canopy-induced inhibition of germination was observed in several pioneer species, but the seeds moved from the forest into the gap or clearing germinated rapidly. Results revealed that natural treefall gaps did not inhibit the germination of most species, but allowed pioneer species to germinate and therefore regenerate. Although germination occurred in all three locations, the level of germination for particular species varied greatly among habitats, depending on species. They concluded that germination requirements play a very important role in controlling the species composition of regeneration in forest understorey, gap and large clearing habitats.

Vázquez-Yanes and Orozco-Segovia (1993) discussed patterns of seed longevity and germination in the tropical rainforest. Tropical rainforest plants produce seeds showing a wide range of shapes, size, structures, chemical composition, water content, dormancy mechanisms and pattern of longevity. Prompt germination is a general characteristic. Most rainforest seeds remain alive for a short time.

Vázquez-Yanes and Smith (1982) studied phytochrome control of seed germination in the tropical rainforest trees *Cecropia obtusifolia* and *Piper auritum* and its ecological significance. The tropical forest pioneer trees *C. obtusifolia* and *P. auritum* germinate and establish in large light gaps of the forest canopy in the rain forest of south-eastern Mexico. Seed germination in both species is under photocontrol and is triggered when the red:far-red ratio (R:FR) of the incident light increases, due to a reduction of the green canopy density. Exposure to simulated light canopies delayed and reduced germination. The light environment inside the forest inhibits germination totally. Experiments with alternate R and FR light treatments revealed the need for long periods of exposure to R light for germination and demonstrated a strong reversibility of R light stimulation by FR light in both species. This property of seeds and germination capacity may be related to the detection of light gap size and its differentiation from the normal sunflecks of the forest.

Eira and Caldas (2000) described seed dormancy and germination as a concurrent process. Dormancy is termed a physiological state in which germination is blocked by a seed-related mechanism, as opposed to a lack of germination due to inadequate environmental conditions. Dormancy can be induced by environmental and/or maternal effects during seed development or after dispersal and can involve several mechanisms which inhibit seed germination (imbibition, activation of metabolism, visible growth). Both the quantitative aspects of dormancy and the cyclic variations in degree of dormancy which are observed in natural seed populations can be explained by quantitative variations in the levels of growth promoters, inhibitors, receptors or other metabolites or physiological processes which respond to low and high temperatures, drying, photoperiod, light quality/intensity or other environmental factors. Breaking dormancy requires that all the necessary elements for germination be in place and functioning; the absence or block of a single essential process is sufficient to cause dormancy.

Dormancy may also be important in distributing the germination of a seed lot over time or space. Tolerance of adverse conditions may be improved by dormancy but does not depend on dormancy

Debeaujon et al. (2000) studied the influence of the testa on seed dormancy, germination and longevity in *Arabidopsis*. The testa of higher plant seeds protects the embryo against adverse environmental conditions mainly by controlling germination through the imposition of dormancy and by limiting the detrimental activity of physical and biological agents during seed storage. To analyse the function of the testa in the model plant *Arabidopsis*, they compared mutants affected in testa pigmentation and/or structure for dormancy, germination and storability. The seeds of most mutants showed reduced dormancy. Moreover, unlike wild-type testas, mutant testas were permeable to tetrazolium salts. These changed dormancy and tetrazolium uptake properties were related to defects in the pigmentation of the endothelium and its neighbouring crushed parenchymatic layers, as determined by vanillin staining and microscopic observation. Structural aberrations such as missing layers or a modified epidermal layer in specific mutants also affected dormancy levels and permeability to tetrazolium. Both structural and pigmentation mutants deteriorated faster than the wild types during natural aging at room temperature, with structural mutants being the most strongly affected.

Kyereh et al. (2001) studied the effect of light on the germination of 11 forest trees in Ghana. They tested seed germination in light and dark and responses to irradiance and light quality in shadehouse experiments for 19 West African tropical forest tree species representing a wide range of ecological types. The results revealed that percentage germination decreased in the dark only for three small-seeded species that are common in forest soil seed banks: *Musanga cecropioides, Nauclea diderrichii* and *Milicia excelsa.* Percentage germination of the other 16 species, including four widely regarded as "pioneers", was unaffected. The effects of different irradiances in shadehouses were significant for some species, but there was no consistent pattern. Irradiance effects in forest gaps, where the seeds received only natural wet season rainfall, were more widespread and substantial and were most commonly shown as a depression of percentage germination at high irradiance. Effects of light quality (neutral vs green shade; red/far-red = 0.43) were insignificant at 5% irradiance in shadehouses for all species except *Nauclea diderrichii*. In growth chamber experiments, the low energy response was only evident at 1.0 μmol m^{-2} s^{-1} (<1% of unshaded forest irradiance) in *Musanga* and *Nauclea*. The speed of germination was affected by irradiance in many species, but the effect was small compared with differences between species, in which the time to complete germination varied between 3 weeks and over 6 months. Seeds of *Ceiba pentandra* and *Pericopsis elata* planted in deep forest shade (2% irradiance) and in a small gap (30% irradiance) germinated well in both sites and showed exponential biomass growth in the gap; but there was a linear decline in mean seedling biomass and subsequent death in deep shade.

A tremendous stimulus was generated in the different kinds of seed dormancy (Baskin, 2004). The book by Baskin (2004) contains eight chapter headings: reproductive strategies in plants, predispersal hazards, dispersal, seed banks, dormancy, germination, seedling establishment, regeneration and diversity. It discusses the significance of various life cycles, the concept of reproductive effort, strategies for reducing seed loss, plant–animal interactions in the pollination of seeds, the role of seed reserves in soil on plant population dynamics and the different regeneration requirements of coexisting species.

González-García et al. (2003) reported negative regulation of abscisic acid signalling in the *Fagus sylvatica* FsPP2C1 plays a role in seed dormancy regulation and the promotion of seed germination. FsPP2C1 was previously isolated from beech (*Fagus sylvatica*) seeds as a functional protein phosphatase type-2C (PP2C) with all the conserved features of these enzymes and high homology to ABI1, ABI2 and PP2CA, PP2Cs identified as negative regulators of ABA signalling. The expression of FsPP2C1 was induced upon abscisic acid (ABA) treatment and was also up-regulated during the early weeks of stratification. Besides, this gene was specifically expressed in ABA-treated seeds and was rarely detectable in vegetative tissues. In this report, to provide genetic evidence on FsPP2C1 function in seed dormancy and germination, they used an overexpression approach in *Arabidopsis* because transgenic work is not feasible in beech. Constitutive expression of FsPP2C1 under the cauliflower mosaic virus 35S promoter confers ABA insensitivity in Arabidopsis seeds, leading to a reduced degree of seed dormancy. In addition, transgenic *35S:FsPP2C1* plants are able to germinate under unfavourable conditions, such as inhibitory concentrations of mannitol, NaCl or paclobutrazol. In vegetative tissues, *Arabidopsis* FsPP2C1 transgenic plants show ABA-resistant early root growth and a diminished induction of the ABA-response genes *RAB18* and *KIN2*, but no effect on stomatal closure regulation. Seed and vegetative phenotypes of Arabidopsis *35S:FsPP2C1* plants suggest that FsPP2C1 negatively regulates ABA signalling. The ABA inducibility of *FsPP2C1* expression, together with transcript accumulation mainly in seeds, could play an important role in modulating ABA signalling in beechnuts through a negative feedback loop. Finally, the authors suggested that negative regulation of ABA signalling by FsPP2C1 is a factor contributing to promote the transition from seed dormancy to germination during the early weeks of stratification.

Finch-Savage and Leubner-Metzger (2006) presented an integrated view of the evolution, molecular genetics, physiology, biochemistry, ecology and modelling of seed dormancy mechanisms and their control of germination. This involved seed structure and seed germination, definition of seed dormancy, evolution of seed structure and seed dormancy, a classification system for seed dormancy and the physiological and molecular basis of dormancy. They argued that adaptation has taken place on a theme rather than via fundamentally different paths and they identified similarities underlying the extensive diversity in the dormancy response to the environment.

Liu et al. (2010) reported that H_2O_2 mediates the regulation of ABA catabolism and GA biosynthesis in *Arabidopsis* seed dormancy and germination. H_2O_2 is known as a signal molecule in plant cells, but its role in the regulation of abscisic acid (ABA) and gibberellic acid (GA) metabolism and hormonal balance is not yet clear. In this study, it was found that H_2O_2 affected the regulation of ABA catabolism and GA biosynthesis during seed imbibition and thus exerted control over seed dormancy and germination. As seen by quantitative RT-PCR (QRT-PCR), H_2O_2 up-regulated ABA catabolism genes (e.g. *CYP707A* genes), resulting in a decreased ABA content during imbibition. This action required the participation of nitric oxide (NO), another signal molecule. At the same time, H_2O_2 also up-regulated GA biosynthesis, as shown by QRT-PCR. When an ABA catabolism mutant, *cyp707a2*, and an overexpressing plant, *CYP707A2-OE*, were tested, their ABA content was negatively correlated with GA biosynthesis. Exogenously applied GA was able to override the inhibition of germination at low concentrations of ABA but had no obvious effect when ABA concentrations were high. It was concluded that H_2O_2 mediates the up-regulation of ABA catabolism, probably through an NO signal, and also promotes GA biosynthesis. High concentrations of ABA inhibit GA biosynthesis but a balance of these two hormones can jointly control the dormancy and germination of *Arabidopsis* seeds.

Oracz et al. (2009) studied the mechanisms involved in seed dormancy alleviation by hydrogen cyanide. They unravelled the role of reactive oxygen species as key factors of cellular signaling during germination. The physiological dormancy of sunflower (*Helianthus annuus*) embryos can be overcome during dry storage (after-ripening) by applying exogenous ethylene or hydrogen cyanide (HCN) during imbibition. The present study was undertaken to provide a comprehensive model, based on oxidative signalling by reactive oxygen species (ROS), for explaining the cellular mode of action of HCN in dormancy alleviation. Beneficial HCN effect on germination of dormant embryos is associated with a marked increase in hydrogen peroxide and superoxide anion generation in the embryonic axes. It is stimulated by the ROS-generating compounds methylviologen and menadione but suppressed by ROS scavengers. This increase is induced from an inhibition of catalase and superoxide dismutase activities and also involves activation of NADPH oxidase. However, it is not related to lipid reserve degradation or gluconeogenesis and is not associated with marked changes in the cellular redox status controlled by the glutathione/glutathione disulfide couple. The expression of genes related to ROS production (*NADPHox, POX, AO1*, and *AO2*) and signalling (*MAPK6*, nSer/ThrPK, *CaM*, and *PTP*) is differentially affected by dormancy alleviation either during after-ripening or by HCN treatment and the effect of cyanide on gene expression appears to be mediated by ROS. It was also concluded that HCN and ROS both activate similarly *ERF1*, a component of the ethylene signaling pathway. They proposed that ROS plays a key role in the control of sunflower seed germination and are second messengers of cyanide in seed dormancy release.

In conclusion, it may be stated that very little research has been undertaken to develop an efficient technique for breaking the dormancy of tree species. Most foresters use traditional techniques for the seed germination of trees which are inefficient. Therefore, concerted research inputs need to be addressed in this aspect for the efficient propagation and reforestation of tree species. In most cases only traditional techniques are used by forest managers and these are not efficient.

We adopted different treatment techniques for breaking the seed dormancy of few native plants of north-east Mexico, as shown in Table 23.1.

Table 23.1 Techniques for breaking the seed dormancy of native plants, north-east Mexico.

Number	Treatment technique	Scientific name	Family	Type
1	H_2SO_4 (75%, v/v)	*Diospyros palmeri* Eastw.	Ebenaceae	Tree
2	H_2SO_4 (60%, v/v)	*Karwinskia humboldtiana* R. and S. Zucc.	Rhamnaceae	Shrub
3	Hot water	*Parkinsonia texana* A. Gray, S. Watson	Caesalpiniceae	Tree
4	H_2SO_4 (40%, v/v)	*Acacia shaffneri* (S.) Wats.	Fabaceae	Tree
5	H_2SO_4 (40%, v/v)	*Celtis pallida* Torr.	Ulmaceae	Shrub
6	H_2SO_4 (20%, v/v)	*Manfreda longiflora*	Asparagaceae	Shrub
7	Hot water	*Capsicum annuum* L. var. *Glabriusculum*	Solanaceae	Shrub
8	Hot water	*Havardia palllens* Benth.	Fabaceae	Shrub
9	H_2SO_4 (20%, v/v)	*Leucaena leucocephala* Lam. De Wit	Fabaceae	Shrub

Bibliography

Baskin J.M. **2004**. Tropical tree seed manual (review). *Native Plants Journal* **5**(1):99–100.

Baskin J.M., Baskin C.C. **2004**. A classification system for seed dormancy. *Seed Science Research* **14**(1):1–16.

Debeaujon I., Léon-Kloosterziel K.M., Koornneef M. **2000**. Influence of the testa on seed dormancy, germination, and longevity in *Arabidopsis*. *Plant Physiology* **122**(2):403–414.

Eira M.T.S., Caldas L.S. **2000**. Seed dormancy and germination as concurrent process. *Revista Brasileira de Fisiologia Vegetal* **12**: 85–104.

Evans E., Frank A. **1999**. *Overcoming Seed Dormancy: Trees and Shrubs. Horticulture Information Leaflet HIL-8704. North Carolina Cooperative Extension Service*. North Carolina University.

Finch-Savage W.E., Leubner-Metzger G. **2006**. Seed dormancy and the control of germination. *New Phytologist* **71**(3):501–523.

González-García M.P., Rodríguez D., Nicolás C., Rodríguez P.L., Nicolás G., Lorenzo O. **2003**. Negative regulation of abscisic acid signaling by the *Fagus sylvatica* FsPP2C1 plays a role in seed dormancy regulation and promotion of seed germination. *Plant Physiology* **133**(1):135–44.

Jacobsen J.V., Pressman E. **1979**. A structural study of germination in celery (*Apium graveolens* L.) seed with emphasis on endosperm breakdown. *Planta* **144**(3):241–248.

Kyereh B., D . Swaine M.D., Thompson J. **2001** Effect of light on the germination of forest trees in Ghana. *Journal of Ecology* **87**(5): 772–783.

Liu Y., Ye N., Liu R., Chen M., Zhang J. **2010**. H_2O_2 mediates the regulation of ABA catabolism and GA biosynthesis in *Arabidopsis* seed dormancy and germination. *Journal of Experimental Botany* **61**(11):2979–2990.

Oracz K., El-Maarouf-Bouteau H., Kranner I., Bogatek R., Corbineau F., Bailly C. **2009**. The mechanisms involved in seed dormancy alleviation by hydrogen cyanide unravel the role of reactive oxygen species as key factors of cellular signaling during germination. *Plant Physiology* **150**(1):494–505.

Raich J.W., Khoon G.W. **1990**. Effects of canopy openings on tree seed germination in a Malaysian diptero carp forest. *Journal of Tropical Ecology* **6**:203–217.

Vázquez-Yanes C., Orozco-Segovia A. **1993**. Patterns of seed longevity and germination in the tropical rainforest. *Annual Review of Ecology and Systematics* **24**:69–87.

Vázquez-Yanes C., Smith H. **1982**. Phytochrome control of speed germination in the tropical rain forest pioneer trees *Cecropia obtusifolia* and *Piper auritum* and its ecological significance. *New Phytologist* **92**:477–485.

Zoghi Z., Azadfar D., Kooch Y. **2011**. The effect of different treatments on seeds dormancy breaking and germination of tree species. *Annals of Biological Research* **2**(5):400–406.

CHAPTER 24

Root growth

24.1 Root system of trees

The root system, its distribution and root depth contribute greatly to the growth of trees in a forest ecosystem, depending on the soil type's nutrients and moisture. Roots serve many purposes for a tree. They absorb nutrients and supplies so the tree can grow. They provide support for the tree to stand tall and to be strong. Tree roots absorb water and nutrients from the soil, store carbohydrates and support the trunk and crown.

A great variability exists among species in rooting depth, length of tap root, mode of distribution and extension of lateral roots (primary, secondary, tertiary etc). The lateral roots of some species are extended in the upper horizons of the soil and some are inclined, reaching to different depths of the soil profile. This has been described and illustrated nicely by a few authors.

Very little research progress has been attained on the root system of trees in a forest, owing to the difficulty in observing root growth and root structure within a soil horizon. Similar to the great variability observed in canopy and crown structure in the aerial environment, the variability in rooting pattern, distribution of lateral roots and root depths of various plant species is not fully unveiled. It is assumed that different species spread their roots in different profiles and depths of soil horizon and absorb nutrients without any competition from other species. The galaxy of this root world in different profiles of soil horizon is not yet unveiled.

In a specialised book, the root systems of trees are classified such as: (a) deep tap root with lateral roots arising horizonantally or inclined manner, (b) lateral roots arising from tap roots spread superficially in the upper soil horizon, (c) lateral roots grow downwards to much deeper layers, (d) lateral roots are highly branched distributed in different soil horizons and so on.

Most studies on the root systems of trees are undertaken after excavation; and a few are inside the root growth chamber. Recently, the techniques of laser imaging and radioisotopes were applied in tree root research. In the following we mention a few advances in tree root research.

There exists a genotypic variability in the growth of roots. Roots extend radially in every direction to a distance equal to at least the height of the tree (assuming no physical barriers) and grow predominantly near the soil surface. Typically 90% of all roots, and virtually all the large supporting

Autoecology and Ecophysiology of Woody Shrubs and Trees: Concepts and Applications, First Edition.
Edited by Ratikanta Maiti, Humbero Gonzalez Rodriguez and Natalya Sergeevna Ivanova.
© 2016 John Wiley & Sons, Ltd. Published 2016 by John Wiley & Sons, Ltd.

roots, are in the upper 60 cm of the soil. Therefore, soil disturbance within the rooting area should be avoided, whenever and wherever possible, as this can have an adverse effect on root health and tree stability.

Initially, a germinating seed produces a single root, the radicle or taproot, which grows vertically downwards. Elongation is most rapid during the first two or three years but it decreases with tree age and increasing soil depth. Horizontally growing lateral roots (laterals) are formed at an early stage, later becoming largely responsible for structural support. Development of the taproot then decreases with the maturity of the taproot. Lateral roots near the soil surface thicken over successive years, becoming large woody roots to build the root system framework of a mature tree. There may extend to 30 cm or more in diameter close to the stem. They taper rapidly until at 2–3 m distance they are usually only 2–5 cm in diameter, thereby offering rigidity and physical strength. Beyond this zone, lateral roots extend outwards in a broad zone for many metres, without appreciable further decrease in size, typically maintaining a diameter of 1–2 cm. Roots are associated with much finer, thread-like mycorrhizae. Mycorrhizae are symbiotic fungi which grow on or in roots, an association which is mutually beneficial to both the tree and the fungus. They are extremely efficient in nutrient absorption, especially phosphorus, and many trees cannot survive without them. Fine roots and their mycorrhizae are jointly responsible for moisture and nutrient uptake, whilst the perennial woody roots primarily act as conducting nutrients through vessels to and from the trunk.

24.2   Root distribution

The variability of soil conditions and the presence of barriers reduce its growth and root distribution.

Soil bulk density increases and aeration decreases with increasing soil depth and root numbers and size decline sharply with depth. Thus below 1 m it is rare to find many roots which are larger than a few millimetres in diameter.

24.3   Root depth

Maximum root depth varies greatly, from only 10–20 cm in waterlogged soils to exceptionally long in loose, well-aerated soils or fissured rock.

All trees can develop a deep root system (2–3 m deep) if soil conditions allow. Some trees have a deeper system than others under the same conditions. Average root depths are typically in the range 1–2 m.

24.4   Root spread

Root spread is not confined to the area delineated by the spread of the branches. Excavation indicates that roots can grow for a considerable distance beyond the branch spread, extending outwards for a distance equivalent to at least the tree's height, and in some cases up to three times the tree's height.

Roots distant from the trunk are usually very close to the soil surface. Obstacles in the soil such as rocks, kerbs and building foundations provide a physical barrier to root extension.

24.5   Factors affecting root distribution

Root growth decreases rapidly with increasing soil bulk density. Oxygen is essential for root growth. Oxygen supply to roots is controlled by soil structure and texture. Poor soil aeration inhibits the growth of new roots, leading to the death and decay of a large proportion of the existing root system. The roots of trees proliferate in areas of moist soil, rich in nutrients, especially nitrogen and phosphorus. In general, soils with low fertility produce trees with poorly developed root systems and sites with higher fertility produce trees with well developed root systems.

24.6   Tree root structure and function

The different terminologies used to describe tree roots are very diverse and not standardised. In their compendium of root system terminology, Sutton and Tinus (1983) defined more than 2200 root terms, illustrating the wide variety of ways to describe tree roots. Root classification has historically been based on their anatomical or functional characteristics (Sutton and Tinus, 1983).

Akersen (1983) studied the growth quality and distribution of a fruit tree root system. They reported that the production of new white roots varied from year to year with root age. The production and root length increased further, but was affected by soil management. The growth and functioning of the root system under field conditions depended upon the production and integration of a range of root types.

24.7   Root growth

The root system of a tree species and its variability help to give support and to absorb water and nutrients from different soil profiles, depending on the characteristics of each species. Due to the difficulty of observing/extracting roots, very little progress has been achieved in tree root research.

According to Writer (2007), similar to the importance of the above-ground parts of trees the role of roots is well appreciated by forest scientists. Consequently, while researchers literally have been pondering roots for centuries, there is very little direct documentation on tree root activity in winter. Forest scientists continue to measure the roots of container-grown trees, painstakingly excavating living roots in forests and, at present, are beginning to use modern imaging technology to watch roots grow in place.

These ecologists describe root activity as periodic, with maximum growth in early summer especially observed in deciduous species and pulses of additional growth occurring occasionally in earlyautumn. They contemplate that not all roots grow at the same time. Even within a single tree, some roots may be active while others are not. Tree roots in some regions are thought to spend the winter in a condition of dormancy. This means they are not dead but rather they overwinter in a resting phase with essential life processes continuing at a minimal rate. Full-on root growth resumes in spring, shortly after the soils become free of frost, usually sometime before bud break.

The Forestry Commission reports the influence of soils and species on tree root depth. It is stated that, though there are numerous publications on the root plate dimensions of trees, few relate the root depth and spread to the soil types in which the trees grow. It is well documented that different soil types and their properties are an important factor in determining the rooting habit of a tree.

Herein we narrate a few advances on research on tree roots.

Bevington and William (1985) studied the annual pattern of root growth of "Valencia" orange [*Citrus sinensis* (I) Osh.)] trees and rootstocks of rough lemon (*C. jambhiri* Lush.) and Carrizo citrange [*Poncirus trifoliala* (L.) Rar. × *C. sinensis*] in relation to shoot growth, soil temperature and water stress, using Plexiglas-walled root observation chambers. The chambers were filled with a reconstituted profile of a fine sand soil and installed below-ground at a field site in central Florida. The chambers periodically were raised above-ground to record root growth. Under non-limiting soil water conditions, continuous root extension growth was observed from February to November. The overall seasonal trend in root growth showed significant correlation with soil temperature. The most intense root growth was observed when soil temperatures was abovc 27 °C and was limited at soil temperatures below 22 °C. During periods of shoot elongation, the number of growing roots and the rate of root elongation declined. When the soil water content decreased intentionally, root growth was checked at a soil water potential of −0.05 MPa. After rcwatering, there was a lag period of two days before root growth ceased.

A variation exists in the pattern of root systems among species. Zhao et al. (2000) investigated the vertical patterns of root systems of *Pinus tabulaeformis, Robinia pseudoacacia, Platycladus orientalis, Pinus sylvestris* var. *mongolica, Pinus armandi* and *Prunus armeniaca* var. *ansu* planted in the Weibei Loess with a soil auger. Site conditions showed a significant effect on the vertical root distribution of *R. pseudoacacia*, of which soil moisture was the key factor. Soil species and soil structure also had a great effect on the distribution. *P. tabulaeformis* attained a maximum rooting depth at its young stage (eight years old), but the root density increased with age. A large variation was observed in vertical root distribution among the tree species, with *R. pseudoaccia* attaining the deepest root system, which negated the opinion that *R. pseudoacacia* is a shallow-rooted tree species. According to the vertical root distribution of the tree species, the productivity of these species in the south-facing site of yellow loess soil is arranged in order of *R. pseudoacacia* > *Prunus armeniaca* var. *ansu* > *P. tabulaeformis* > *P. sylvestris* var. *mongolica* > *P. armandi* > *Platycladus orientalis*.

The root system, its distribution and root growth play an important role in the growth and development of trees. Studies have been undertaken on the distribution of lateral roots, root depth and effect of soil types on root distribution and growth.

Lateral roots. During winter, many of a tree's lateral roots stop growing but the growth of lateral roots can extend well beyond the canopy perimeter. There is considerable variation between species in the extent of lateral root growth, but many have been recorded with a radius of > 20 m.

Root depth. Tree roots do not occur in large quantities at much deeper depths (e.g. > 2 m) in the soil profile. Root depths depend on nutrients and moisture content. In shallow soils over hard rock, the root depths are shallow, at less than 1 m. In loose deep well-drained soils with large pore spaces, roots may grow deeper as the soil provides less resistance to root penetration.

Intermediate loamy soils retain more moisture and allow considerable root development. These soils may be seasonally waterlogged. Soils with a large particle size that are restricted by an impervious layer may be seasonally waterlogged. Podzols, with a cemented iron pan formed within 1 m of the soil surface, are the main soil type.

Soils with moisture-retaining upper horizons are seasonally waterlogged in the top 40 cm due to poor, slowly permeable surface horizons. In such soils, there is little necessity for deep root development.

24.8 Organic rich soils

The organic rich soils include peat soils of varying type and origin. A distinction has been made between those that are drained and those that are predominantly waterlogged.

The Forestry Commission reports the influence of soils and species on tree root depth. Though there are numerous publications on the root plate dimensions of wind-thrown trees, few relate the root depth and spread to the soil types in which the trees were growing. It is well known that different soil types and their properties are important factors in determining the rooting habit of a tree. Rooting depths may differ between species. There are many site, climatic, silvicultural and biological factors that will influence tree rooting habit. It is mentioned that the rooting environments (soil, geology and hydrology) are the largest constraints. Published data of root dimensions for mature trees that include observations on this rooting environment are limited.

Day et al. (2010) put forward contemporary concepts of a root system. The tropical forest pioneer trees *Cecropia obtusifolia* and *Piper auritum* germinate and become established in large light gaps of the forest canopy in the rain forest of south-eastern Mexico. Germination of the seeds of both species is under photocontrol and is stimulated when the red:far-red (R:FR) ratio of the incident light increases due to a reduction of the green canopy density. Exposure to simulated light canopies delayed and reduced germination. The light environment inside the forest inhibited germination totally. Experiments with alternate R and FR light treatments indicated the need for long periods of exposure to R light for germination and showed a strong reversibility of R light stimulation by FR light in both species. This property of the seeds may be related to the detection of light gap size and its differentiation from the normal sunflecks of the forest.

Livesstey et al. (2000) studied the competition, distribution and dynamics of fine roots, root length and biomass in tree rows in an agroforestry system with maize. It was assessed that maize root length decreased with greater proximity to the tree root rows. Differential maize root length and biomass suggest that competition for soil resistance and root length are more important for size than genetics.

Amato et al. (2008) studied in situ detection of tree root distribution and biomass by multi-electrode resistivity imaging. Soil electrical resistivity measured by geoelectrical methods has the potential to detect below-ground plant structures, but the quantitative relationships of these measurements with root traits have not been assessed. They tested the ability of two-dimensional (2-D) DC resistivity tomography to detect the spatial variability of roots and to quantify their biomass in a tree stand. A high-resolution resistivity tomogram was generated along a 11.75 m transect under an *Alnus glutinosa* (L.) Gaertn. stand based on an alpha-Wenner configuration with 48 electrodes spaced 0.25 m apart. Data were processed by a 2-D finite-element inversion algorithm and corrected for soil temperature. Data acquisition, inversion and imaging were completed in the field within 60 min. Root dry mass per unit soil volume (root mass density; RMD) was estimated destructively on soil samples collected to a depth of 1.05 m. Soil sand, silt, clay and organic matter contents, electrical conductivity, water content and pH were measured on a subset of samples. The spatial pattern of soil resistivity closely matched the spatial distribution of RMD. Multiple linear regression showed that only RMD and soil water content were related to soil resistivity along the transect. Regression analysis of RMD against soil resistivity demonstrated a highly significant logistic relationship ($n = 97$). This was confirmed on a separate dataset ($n = 67$), revealing that soil resistivity was quantitatively related to below-ground tree root biomass. This relationship offers a basis for developing quick non-destructive methods for detecting root distribution and quantifying root biomass, as well as for optimising sampling strategies for studying root-driven phenomena.

24.9 Conclusion and research needs

The root system and its distribution play a great role in the growth and productivity of forests. In the context of the little literature on root studies, it can be stated that very little progress has been achieved due to study difficulty in situ. Most studies are concentrated on a study of the root system and root depth of tree species after laborious and costly excavation process. Some studies are undertaken in root chambers, which reflects general growth but not real. However, this type of study is important for assessing comparative root growth between different species. It is possible to grow different plant species after breaking seed dormancy in PVC tubes (two halves joined and tied with a wire) for two months and study the roots system of different species. Concerted research inputs using imaging and radioisotopes could generate real pictures of the root systems of different species in different profiles in the root world. There is a necessity to study the distribution of lateral roots and root depth of different species in soil rofiles in a forest ecosystem.

Bibliography

Akinson D. **1983**. The growth activity and distributin and distribution of the fruit tree root system. *Plant and Soil* **71**:23–35.

Amato M., Basso B., Celano G., Bitella G., Morelli G., Rossi R. **2008**. *In situ* detection of tree root distribution and biomass by multi-electrode resistivity imaging. *Tree Physiology* **28**(10):1441–1448.

Bevington K.V., William S.C. **1985**. Annual root growth pattern of young citrus trees in relation to shoot growth, soil temperature, and soil water content. *Journal of the American Society of Horticulturae Science* **110**(6):840–845.

Day S.D., Wiseman P.E., Dickinson S.B., Harris J.R. **2010**. Contemporary concepts of root system architecture of urban trees. *Arboriculture and Urban Forestry* **36**(4):149–159.

Garwood N.C. **1983**. Seed germination in a seasonal tropical forest in Panama: A community study. *Ecological Monographs* **53**(2):159–181.

Livesstey S.L., Gregory P.J., Burresh R.J. **2000**. Competition in tree row agroforestry systems. 1. Distribution and dynamics of fine root length and biomass. *Plant and Soil* **227**(1/2):149–161.

Sutton R.F., Tinus R.W. **1983**. Root and root system terminology. *Forest Science* **24**(4):137.

Zhao Z., Li P., Wang N. **2000**. Distribution patterns of root systems of main planting tree species in Weibei Loess Plateau. *Ying Yong Sheng Tai Xue Bao* **11**(1):37–39.

CHAPTER 25

Features of boreal forest in Russia: A special study

Natalya Sergeevna Ivanova, Ekaterina Sergeevna Zolotova and Maria Victorovna Yermakova

Botanical Garden of Ural Branch of the Russian Academy of Sciences, Yekaterinburg, Russia

25.1 Distribution and general characteristics of the boreal forests

The term "boreal forests" refers to coniferous forests of the northern hemisphere, which grow in conditions of cold temperate and temperate climate. The boreal forest zone is spread out between 50 and 70°N (July isotherms 13 and 18°C; Federal Forestry Agency, 2015). The total forestland area of the boreal zone is 1214 million hectares, out of which 920 million hectares are closed forests (75.8% of the total area of boreal forests). The land area of the boreal forests in North America and Eurasia is 30% of the total forest area of the planet. In Canada, boreal forests comprise up to 75% of all forests in the country, in the United States (Alaska) 88%, in Norway 80%, in Sweden 77%, in Finland 98% and in Russia about 67% (Spurr and Barnes, 1980; Barnes, 1991).

Twelve species of conifers grow in the boreal forests of North America, including five species of pine, three species of spruce and one species each of fir, hemlock and arborvitae. Fourteen species grow in Eurasia, of which three are pine, four are fir, three are spruce and four are larch. Species diversity has existed for the last 5000 years (Kellomaki, 2000).

Boreal forests have a regulating effect on the climate of the planet. They provide a balance of thermal radiation and play an important role in the global carbon cycle (Olsson, 2009). About 27% of the carbon contained in all of the planet's vegetation accumulates in the boreal forest, with 25–30% of the planet's soil carbon (Turetsky et al., 2005; Luyssaert et al., 2008). Together with tundra, boreal forests represent the largest reservoir of carbon on earth (Soja et al., 2007).

In Russia, boreal forests are forests that grow in the tundra and forest tundra zones, subzones of the northern and middle taiga and partially in the southern taiga. The Forest Fund of Russia is distributed between these areas in the following way (Federal Forestry Agency, 2015):

• Subzone tundra woodland has 14% of the total forest area, also 17% of the forest area and 13% of the forest, meaning the actual forests.

Autoecology and Ecophysiology of Woody Shrubs and Trees: Concepts and Applications, First Edition.
Edited by Ratikanta Maiti, Humbero Gonzalez Rodriguez and Natalya Sergeevna Ivanova.
© 2016 John Wiley & Sons, Ltd. Published 2016 by John Wiley & Sons, Ltd.

- Northern taiga subzone has 10% of the total forest area, 9% of the forest area and 8% of the forest.
- Middle taiga subzone has 33%, 38% and 41%, accordingly.
- Southern taiga subzone has 18, 20 and 20%, accordingly.

The most important conifer trees of the boreal forests of Russia are pine (*Pinus sylvestris* L.), spruce (*Picea obovata* Ledeb.) and larch (*Larix sibirica* Ledeb.). All of these tree species belong to the family Pinaceae, but they significantly differ in their biological and ecological characteristics (Bulygin, 1985).

25.2 The main forest-forming tree species of the boreal forest

Pinus sylvestris **L.** This tree reaches, depending on growing conditions, a height of 45 m with a 1 m stem diameter over a period of 200 years. The crown of fully-grown trees spreads with a rounded or flat top. The branching of both the stem and branches is monopodial. One shoot is formed annually. The stem is cylindrical. The bark is a brownish-red colour. In the upper part of the stem and in the boughs of the crown the bark is reddish-orange. The pine reaches the juvenile age between 6 and 10 years old in natural forests and between 15 and 40 years old under planting conditions. *Pinus sylvestris* reproduces solely by seed. Pollen and seeds are spread by wind. *Pinus sylvestris* is able to grow in a wide range of climatic conditions, requiring little heat; it is hardy and frost-resistant, requires light, and demands very little of soil. *Pinus sylvestris* has a very ductile

root system that can change direction depending on growth conditions. In deep, fresh, sandy soil it develops a taproot and a diversity of vertical roots, whose length reaches up to 1.5 m. High ecological plasticity allows it to grow in a wide area from tundra to steppes.

Picea obovata **Ledeb.** This tree reaches, depending on growth conditions, a height of 30 m, with a 1 m stem diameter over a period of 200 years. The bark is brown, smooth with small scales in the tree's youth. In its final years, the bark becomes scaly and rough. Shoots from the tree show various shades of colour, from cinnamon-brown to pale yellow. Its branching stems and branch patterns are monopodial. One sapling is formed annually. Its crown is thick and tapered with a pointed top. *Picea obovata* enters the juvenile age between 15 and 30 years old. Spruce reproduces solely by seed. Pollen and seeds are spread by the wind. *Picea obovata* is hardy, frost-resistant and resistant to shade. It demands very little of the soil. It is a mesophyte, but it can grow in conditions of excess moisture and in the mountains. *Picea obovata* can be found from the Urals to the Far East.

Larix sibirica **Ledeb.** This is a powerful tree which, depending on the growing conditions, may achieve a height of 45 m, with a 1.5–1.8 m stem diameter over a period of up to 450 years. The bark of young trees is thin and brownish-grey. Older trees are dark with deep, longitudinal cracks and are very thick. The crown is sprawling with thick, horizontal branches. Annual shoots are light, straw-yellow. *Larix sibirica* is characterised by exceptionally strong growth (1.0–1.5 m growth per year). It enters the juvenile stage between 8 and 25 years old. *Larix sibirica* inhabits a vast Eurasian area. *Larix* achieves highly productive performance only in fertile, well-drained and fresh soil.

25.3 Reforestation of the boreal forest

The enormous scale of the commercial use of forests in the Northern Hemisphere, fires, and windfalls has led to serious changes in their structure, function and the dynamics of their trends (Achard et al., 2006; Young et al., 2006; Wang and Cumming, 2009; Bouchard and Pothier, 2011; De Groot et al., 2013). The land area of coniferous forests on all continents has been catastrophically reduced, as they are replaced by deciduous trees; and both a decrease in species composition and a fall in the stability of nature protection functions have been observed (Forster, 1988; Achard et al., 2006; Young et al., 2006; Ivanova, 2014). Reforestation and sustainable forest management is becoming a severe problem. The processes of the emergence and development of new generations of trees determine the stability of forest ecosystems and "restoration program" and age-related changes of vegetation after significant impacts (Sannikov et al., 2004). In this regard, a detailed study on the quantitative level of natural regeneration processes is of the greatest importance. Now a sufficient number of studies on the number and viability of the undergrowth of coniferous plants in open habitats (felling and slash areas) have been accumulated under the canopy of natural forests and their derivatives communities. Features of renewal in Western and Central Europe are considered in numerous publications (Wagenknetcht and Belitz, 1959; Schmidt-Vogt, 1967, 1977; Blankmeister and Hengst, 1971; Thomasius, 1990; Ericsson et al., 2000, Sannikov et al., 2011). Renewal processes researched in Scandinavia have been documented by a number of researchers (Kuusela, 1990; Leikola, 1990) and in Canada by Fitzsimmons (2003). The widest study of the characteristics of natural regeneration of coniferous woody plants is in Russia. This issue has been studied by Morozov (1920, 1928), Karpov (1965, 1969), Rysin and Savelyeva (2002), Sannikov et al. (2004) and many others. For the Urals, the geographical differentiation zone of natural reforestation is most fully represented in the works carried out by Isaeva and Luganskiy (1974, 1981). It is noted that in spruce forests natural regeneration is considerably worse than in pine ones. The optimal conditions for natural regeneration of spruce can be found in the subzones of middle and southern taiga (Isaeva and Luganskiy, 1974). In the subzone of dark coniferous-deciduous forests spruce regenerates significantly worse and not always satisfactorily (Zubareva, 1970). The problem of natural regeneration in clear-cuts in the Urals is illustrated in the works of Sannikov (1960), Shevelev (1965) Isaeva (1968a,1968b, 1975); Isaeva et al. (2009), Maslakov and Kolesnikov (1968), Muratov and Amirkhanova (1997), Davydychev (2006) and Gorichev et al. (2009). This work reveals that reforestation in clear-cuts is difficult: they do not renew a large part of the original trees, not only conifers, but also deciduous tree species. Numerous studies have shown that the undergrowth of prior generations (Danilik, 1968; Isaeva, 1968a,1968b; Kolesnikov et al., 1970; Ivanova and Andreev, 2008a-d) plays a huge role in the reforestation of clear-cuts in dark woods and the regeneration of pine forests due to fires (Sannikov, 1992). Most of these studies were carried out in connection with the problem of reforestation in clear-cuts, burned areas and for the limited purpose of evaluating the success of regeneration by taking into account the number of seeds and amount of new growth in different regions and forest types. This study served as the basis for planning and carrying out of clear-cutting, which are measures to promote natural regeneration, but say little for the identification of habitat (Sannikov, 1992). Information about the peculiarities of reforestation activities in different natural zones is still controversial, and the influence of some factors is not fully established. This is due to the variety of natural conditions and active factors, and the complexity and non-linearity of interdependencies in the ecosystem (Petrova et al., 2013). Therefore, in order to preserve the planet's biological resources, it is necessary to further develop and improve methodological approaches to the study and analysis of the reforestation of the boreal forest, as the main component of stabilising the biosphere.

25.4 Field study

We conducted the study in two landfills: Western low mountains (South Urals, Russia) and the Zauralskaya hilly piedmont province (Middle Urals, Russia). In these districts, mountain forests are heterogeneous with regard to forest typological and are highly fragmented by cutting. As a result, a complex vegetation mosaic has been formed, which is ideal for the purposes of our research. We conducted our research in the territory of Katav-Ivanovsk forest, in the Chelyabinsk region (Southern Urals, Russia) at 54°33′–54°40′N; 57°48′–57°55′E. These forests are typical for the Yuryuzan–Verkhneaysk province of the mountain taiga and mixed forests of the Ural forest region (Kolesnikov, 1969) with altitudes of 400 to 800 m asl. The local climate is cold-temperate and quite humid. The average annual temperature is 2.1 °C, while the mean temperature in January is around −12.5 °C. The average temperatures from May to July is 14.4 °C. The average duration of the frostless season is 120 days. Annual precipitation is 580–680 mm on average, with an average amount of snow between 56 and 64 cm. The Zauralskayahilly piedmont province of Sverdlovsk region (at 57°00′–57°10′N; 60°10′–60°30′E) is divided into foothills formed by alternating meridional hills and ridges with wide, intermountain elongated lowlands, in which there are large lakes surrounded by moors (Kolesnikov et al., 1974), with altitudes of 200–500 m asl. The hills have soft contours, with a blunt, broad peak, while the slopes are long and flat. The climate is moderately cold and moderately humid. The frostless period lasts 90–115 days, the average annual temperature is 1 °C, and the average snowfall is 40–50 cm (Kolesnikov et al., 1974). The main factors of soil formation in the Urals are mountainous terrain, continental climate, widespread soils both ancient and young forming, which combined with a great diversity of vegetation alters the soil characteristics. Mountain soils are characterised by a relatively small depth, light mechanical composition (dominated by light and medium loam), different degrees of skeleton (skeletal soil) increasing with decreasing depth of the soil. In the soil distribution of the Zauralskaya hilly piedmont province of the Middle Urals, dependence on the terrain is clearly visible. The tops and upper thirds of the steep slopes, where the thickness of eluvial deposits is least, are occupied by shallow, under-developed soils with a high degree of skeleton and relatively light texture. Brown non-podzolised and podzolised soils are confined to the middle and lower thirds of the gentle slopes. Flat-topped low ridges, gentle, well-drained downward slopes are occupied by sod pale-yellow podzolic soil, which differs in thickness soil profile and degree of podzolisation (Firsova and Rzhannikova, 1972). The soil of the gentle slopes, intermountain depressions and lows often has signs of glycification and wetland and is characterised by greater depth and moderate skeleton (Kolesnikov, 1969).

25.5 Methodological approaches and methods

Our work is based on the methodological approach of geo-genetic (geodynamical) forest typology (Ivanova and Zolotova, 2014). A geo-genetic classification is a classification based on forest origin and evolution patterns which takes into account all of the developmental stages of the forest ecosystem and can be used to predict their future changes. Units of

forest cover are always given with a site type index, a three-digit number defining the position of an area in space; that is its ecological address. Unlike floristic classification, genetic typology provides an exact "address" (geographical) for each forest type. Forest typology reflects two types of changes: the differentiation of habitat and the corresponding differentiation of forests (Ivanova and Zolotova, 2014).

This work is based on the principles of geodynamical forest typology (Kolesnikov et al., 1974; Ivanova and Zolotova, 2014), the method of sample plots and the generally accepted methods of botanic and soil studies. A survey and the appraisal of tree stands were carried out on sample plots (0.5 ha), changes in young growth were recorded and subordinate layers were comprehensively studied (Forest Communities, 2002). soil profiles were laid and soil morphology was described by Rozanov (2004). The physical and chemical properties of the soil were determined in the laboratory.

Sample plots included no less than 200 woody plants. Diameters were measured and height density was determined. In the Western low mountains of the Southern Urals, tree layer characteristics were provided by Andreev (1998).

In order to reveal the history of the forest stands, we determined the age of trees on plots. Age was determined from core samples. Undergrowth was measure for 40 subplots, each 5×5 m. Grounds were arranged in parallel rows. We identified the categories of undergrowth according to the following living conditions: viable, non-viable and deceased. For each instance of undergrowth we determined the age as follows: small undergrowth (height 50 cm) by verticils (in each instance);

larger undergrowth by verticils accounting for the correct definition of spilite at the root collar for a sample of plants (three or four units for each category of vitality and 25 cm in height).We morphologically described the soil profile, using the method of Rozanov (2004). Soil was named in accordance with traditional classifications (Classification and Diagnosis, 1977) to ensure the continuity of research in the study region (Firsova and Rzhannikova, 1972). Soil samples were taken over the horizon to determine physical and chemical properties. We studied skeletal soils and the distribution of particle fractions, density, the density of the solid phase, total porosity, hydroscopic and maximum hydroscopic moisture, wilting point, actual and potential soil acidity (pH_{H2O}, pH_{KCl}) and content-flowing potassium in the laboratory. To evaluate different parameters of the studied habitats we used both ecological and floristic approaches (Westhoff and van der. Maarel, 1978) and a range of ecological scales (Tsyganov, 1983). The experimental study of the activity of seed germination the main coniferous tree species of the Urals: *Pinus sylvestris* L., *Picea obovata* Ledeb. and *Larix sibirica* Ledeb. was carried out in laboratory conditions in accordance with the requirements of GOST (1997). After soaking, seeds were dried to floatability and then evenly (10×10 pieces) set out in a three-fold repetition on the soil surface, placed in a cell layered 2.5 cm. Humus horizons of brown mountain-forest and soddy pale ash-containing soils of three forest types were taken: *Dicrano-Pinetum sylvestris* Preising et Knapp ex Oberdorfer (1957), *Tilio-Pinetum sylvestris* Kolesnikov et al. (1973), *Bupleuro longifolii-Pinetum sylvestris*

Fedorov ex (Ermakov et al. 2000). During the period of germination we maintained constant, optimal soil moisture at 65% and a temperature of 22 °C. The energy of germination was determined on day 7, and seed germination on day 14. Data analysis was carried out on the basis of modern statistical methods, using a standard package of Excel and Statistica Khalafyan (2010).

25.6   Results and discussion

25.6.1   Natural regeneration of *Picea obovata* Ledeb. on the southern border of the range (the mountain forests of the Southern Urals)

The aim of our study was to determine the characteristics of the natural regeneration of *Picea obovata* in indigenous forests on cuttings and under the canopy of birch and aspen forests in the most common conditions in the Southern Urals. Figure 25.1 shows a Pine fir forest in the mountains of the Middle Urals, Russia.

25.6.2   *Picea obovata* Ledeb. regrowth under the canopy of dark-coniferous forests

Out of all the varieties of coniferous forest in our target region, natural spruce is a component of two eco-dynamic series. They differ greatly in the dominant types of microhabitats and ground cover. By "natural dark coniferous forests" we refer to 140–160 year old coniferous forests which have been only slightly affected by logging and fires and where the original structure and subordinate layers have been preserved (Ivanova, 2014). Fires led to a change in the original type of substrate and

Figure 25.1 Pine fir forest in the mountains of the Middle Urals (Russia). (See insert for color representation.)

grass and moss layer and, at the same time, they led to the death of younger generations of woody species. Moderate grazing influenced the formation of specific microhabitats: sparse grasses and a multiple bush and grass layer. In general, post-fire and post-clearcut coniferous forests are characterised as lower (compared to sub-root spruce) in absolute fullness stand, a large part birch and pine in the tree layer (Ivanova and Andreev, 2008a).The number of spruce in post-clearcut and post-fire coniferous forests can either decrease or increase in comparison with natural dark coniferous forests (Figure 25.2). The main factor that reduces the number of spruce, under the canopy of dark coniferous forests, are atypical fires that lead to the death of younger generations.

The age structure of a tree population is a good indicator of the stability of communities. The main condition of dynamic stability is considered the uniformly damped distribution of the number of individual trees by age. All studied dark coniferous forests (natural, post-cutdown and post-fire) have shown a continuous emergence of new generations of spruce (Figure 25.2). That is, despite the differences in stand structure and subordinate layers, the history of the formation of communities, under a canopy of dark coniferous forests, preserves the conditions for the emergence and survival of *Picea obovata*, which can ensure the formation of a stable, uneven-aged stand. In this way, the ability to heal the size and population structure of *P. obovata* in all studied conifer forests is retained and, therefore, the ability to auto-regulate the dynamics of the forest ecosystems is retained, supplying it with sustainable development. However,

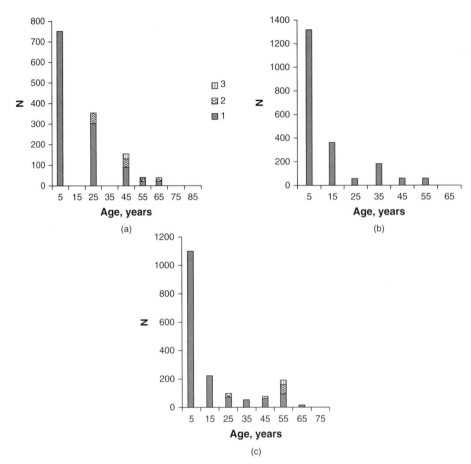

Figure 25.2 The age structure during the regrowth of *Picea obovata* Ledeb. in mountainous coniferous forests of the Southern Urals: N – number of regrowths per hectare; 1 – viable regrowth, 2 – inhibition regrowth, 3 – dead regrowth. (a) A native forest (140–160 year old stand), (b) dark coniferous forest under the influence of moderate grazing (50 year old stand), (c) dark coniferous forest after fire (55 year old stand).

the distribution of dark coniferous forests in the mountains of the Southern Urals has greatly been reduced by clear-cutting. Today coniferous forests cover only 16% of this area (Ivanova, 2014).

25.6.3 Picea obovata Ledeb. regrowth under the canopy of birch and aspen.

For the western low mountains of the Southern Urals we analysed the dynamics of the population structure of *Picea obovata* during the formation of short-term secondary birch, long-term secondary birch and stable-term secondary aspen. We revealed a serious imbalance in determining the age structure of populations of conifer species. In short-term secondary birch the initial recovery age shifts and the emergence of new generations of *P. obovata* are completely suppressed. In the formation of tree communities,

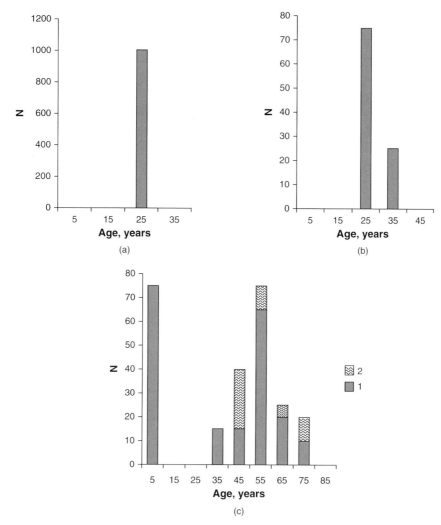

Figure 25.3 The age structure of *Picea obovata* Ledeb. regrowth in short-term secondary birch forests of the Southern Urals: (a) young (stand is 5 years old), (b) stand is 20 years old, (c) stand is 80 years old. N – number of regrowths per hectare; 1 – viable regrowth, 2 – inhibition regrowth, 3 – dead regrowth.

only the preliminary generation of *P. obovata* is involved. At later stages of recovery, age shifts restore the ability to form new generations of coniferous species, multilayers of tree stands and age differences in undergrowth, but failures in regeneration remained (Figure 25.3).

In long-term secondary birch (throughout their formation) the age structure of populations of species of conifer is severely impaired: new generations of *P. obovata* appear unstable, in most cases only a few are marked as numerically small generations. *Picea obovata*, located in the main layer of the tree stand acts as a source

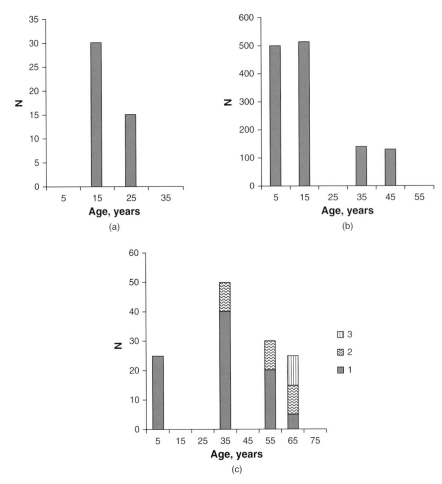

Figure 25.4 The age structure of the regrowth of *Picea obovata* Ledeb. in long-term secondary birch forests of the Southern Urals: (a) stand age is 20 years old, (b) stand is 35 years old, (c) stand is 100 years old). N – number of regrowths per hectare; 1 – viable regrowth, 2 – inhibition regrowth, 3 – dead regrowth.

of semination .The restoration of conifer species tree stand dominance is greatly retarded and is possible only after the natural decay of birch trees over 120 years of age. Long-term secondary birch occupies more than 60% of the area (Ivanova, 2014). Accordingly, after clear-cutting, low rates of indigenous conifer forest regeneration are observed in vast areas (Figure 25.4). In the case of the formation of stable-term secondary aspen, we did not observe trends of the forest ecosystem returning to its original state (moss spruce forests): there is no regeneration of dark coniferous forest stands or the original small grass and moss cover. The dynamics of the grass–shrub tier, in the formation of stable derivative aspen, move in the direction of the change in the cover of cereal grasses and motley grass and on to tall ones that completely suppress the natural regeneration processes of *P. obovata*. The regeneration of indigenous conifer forests is stretched indefinitely; it is combined with widespread resistance derivative aspen (which occupies more than 25% of the area) and this seriously undermines the position of coniferous forests in this region.

25.7 The main forest types of the Middle Urals

On the tops and upper halves of the hill slopes, humidity is most dependent on weather conditions and is highly variable. *Dicrano-Pinetum sylvestris* Preising et Knapp ex Oberdorfer (1957) grows in these conditions (Table 25.1). *Dicrano-Pinetum* *sylvestris* is characterised by a poor species composition of higher plants, a sparse, stunted grass–shrub level and a good natural regeneration of *Pinus sylvestris*. According to the ecological and floristic classification of these forests, they belong to the class *Vaccinio-Piceetea* (boreal dark coniferous and light coniferous forests), in the Union *Dicrano-Pinion* Libbert (1933), Matuszkiewicz (1962).

On gentle slopes (in the context of sustainable humidification) forest types are formed at the junction of two vegetation classes: *Vaccinio-Piceetea* and *Brachypodio Pinnati-Betuletea*. They display serious differences in floristic composition from typical European oligotrophic *Dicrano-Pinion* Union pine forests and are highlighted with the new *Brachypodio pinnatae-Pinenion sylvestris* suball. nov. (Shirokih and Martynenko, 2012; Ivanova and Zolotova, 2013).

Table 25.1 Characteristics of indigenous vegetation in the southern taiga forests of the Middle Urals.

Forest type	Woody plant species	Stand composition (%)	Age (years)	Average height (m)	Average diameter at 1.3m (cm)	Stand density (m² ha⁻¹)
Dicrano-Pinetum sylvestris	*Pinus sylvestris* L.	90	160	24.3	40.9	35.7
	Larix sibirica Ledeb.	8–9	160	27.5	48.4	2
	Betula pubescens Ehrh., *B. pendula* Roth.	1–2	100	20	21.4	0.3
Tilio-Pinetum sylvestris	*Pinus sylvestris* L.	96–98	140	25.5	35.2	30
	Betula pubescens Ehrh., *B. pendula* Roth.	1–2	120	20	22.2	1
	Tilia cordata Mill.	1–2	–	10	9.5	0.5
Bupleuro longifolii-Pinetum sylvestris	*Pinus sylvestris* L.	90	150	28.9	42.4	35.5
	Picea obovata Ledeb.	1	80	10	9.5	0.25
	Betula pubescens Ehrh., *B. pendula* Roth.	9	120	26.9	32.5	6.5

Table 25.2 Distinct and dominant species in the subordinate layer of indigenous southern taiga forests of the Middle Urals.

Forest type	Distinct species	Dominant species
Dicrano-Pinetum	Vaccinium vitis-idaea, Antennaria dioica, Vaccinium myrtillus, Rubus saxatilis, Fragaria vesca	Calamagrostis arundinacea, Vaccinium myrtillus, Rubus saxatilis
Tilio-Pinetum	Tilia cordata, Lathyrus vernus, Vaccinium myrtillus, Vaccinium vitis-idaea, Carex digitata, Brachypodium pinnatum, Rubus saxatilis	Vaccinium myrtillus, Vaccinium vitis-idaea, Calamagrostis arundinacea
Bupleuro Pinetum	Aegopodium podagraria, Lathyrus vernus, Heracleum sibiricum, Viola mirabilis	Calamagrostis arundinacea, Brachypodium pinnatum, Rubus saxatilis

Table 25.3 Characteristics subordinate layers of indigenous southern taiga forests of the Middle Urals. CV is coefficient of variation (%).

Forest type	Projected cover (%)			Absolutely dry biomass $(g\,m^{-2})$			Number of species per $1\,m^2$		
	Mean	Maximum	CV	Mean	Maximum	CV	Mean	Maximum	CV
Dicrano-Pinetum	51.4	94.5	41.9	116.4	243.6	52.6	8	11	18.0
Tilio-Pinetum	57.6	78.0	22.2	69,7	81.6	78	17	21	20.0
Bupleuro Pinetum	86.3	100	19.5	89.8	113.1	12.6	28	31	9.3

The most productive and common types of forest in these conditions are *Tilio-Pinetum sylvestris* forests (Kolesnikov et al., 1974). Table 25.2 shows the characteristics of the species composition.

Humid, occasionally wet habitats feature pine forest *Bupleuro longifolii-Pinetum sylvestris*. In *Bupleuro longifolii-Pinetum sylvestris*, in comparison to other forest types, the signs of class *Brachypodio Pinnati-Betuletea* (boreal light-pine, small-leaved grass forests of Western, Central Siberia and the Urals) are featured in Union *Trollio europaea-Pinion sylvestris*

(pine-birch forests on fertile grassy and well-enriched moisture; Ermakov, 2003). Characteristics of subordinate layers are given in Table 25.3.

25.7.1 Features of natural and artificial regeneration of *Pinus sylvestris* L. and *Picea obovata* Ledeb. on clear-cuts in the Zauralsky (Transuralian) hilly piedmont province of the Middle Urals

Natural pine saplings can be found on the tops and upper slopes of hills, flat

Table 25.4 Density and dendrometric parameters of natural and artificial pine young stands.

Woody plant species	Density (thousand units ha⁻¹)	Diameter at mid-height of the tree (cm)	Height (cm)
Natural young stands in forest type *Dicrano-Pinetum sylvestris*			
Pinus sylvestris	71.3	2.1 ± 0.12	316.2 ± 9.54
Artificial young stands in forest type *Dicrano-Pinetum sylvestris*			
Pinus sylvestris (artificial)	2.7	4.4 ± 0.09	403.9 ± 5.28
Pinus sylvestris (natural)	0.9	2.1 ± 0.31	185.0 ± 25.75
Larix sibirica	0.2	2.0 ± 0.35	300.0 ± 0.05
Betula pendula, B. pubescens	2.8	1.2 ± 0.11	203.0 ± 9.23
Populus tremula	1.2	1.1 ± 0.22	162.0 ± 25.77
Tilia cordata	0.2	0.5 ± 0.1	140.0 ± 20.0
Forest type *Bupleuro longifolii-Pinetum sylvestris*			
Pinus sylvestris (artificial)	1.2	3.3 ± 0.08	329.3 ± 5.7
Pinus sylvestris (natural)	4.3	2.6 ± 0.25	300.0 ± 16.48
Betula pendula, B. pubescens	5.2	1.8 ± 0.14	345.1 ± 21.22
Populus tremula	0.3	0.9 ± 0.21	216.7 ± 101.38
Salix spp.	0.3	1.2 ± 0.13	255.0 ± 34.28
Tilia cordata	0.3	2.3 ± 0.25	282.5 ± 1.81

watersheds, and gentle southern slopes in the forest type *Dicrano-Pinetum sylvestris* (Table 25.4). In the lower part of the slope, in the forest type *Bupleuro longifolii-Pinetum sylvestris*, regeneration in clear-cuts is dominated by deciduous woody plants and there are replacement pines on birch (Ivanova and Bystrai, 2010; Yermakova and Ivanova, 2011). Artificial young stands of *Pinus sylvestris* (Table 25.4) are characterised by a lower density and a larger than average height and diameter, compared with natural young *Pinus sylvestris*. In the forest type *Bupleuro longifolii-Pinetum sylvestris*, where the natural regeneration of *Pinus sylvestris* is suppressed, forest plantation is the main way to restore the original forests.

The characteristics of artificial young stands of *Picea obovata* are shown in Table 25.5. In the forest type *Dicrano-Pinetum sylvestris*, polydominant saplings of natural origin are formed. In the forest type *Bupleuro longifolii-Pinetum sylvestris*, the underwood species composition is poor. This are dominated by deciduous woody plants. Thereby, reforestation in different forest types of the Middle Urals is different. The density and species composition of the regrowth of woody plants in clear-cuts and silvicultured areas varies widely. The regeneration of coniferous forests is successful only in the *Dicrano-Pinetum sylvestris* type of forest. For forest type *Bupleuro longifolii-Pinetum sylvestris*, the forecasting paths and timing regeneration of coniferous tree species differ, as in natural regeneration and forest cultures. Further studies of the factors of reforestation in these conditions are required.

Table 25.5 Density and dendrometric parameters of artificial young stands of *Picea obovata*.

Woody plant species	Density (thousand units ha⁻¹)	Diameter at mid-height of the tree (cm)	Height (cm)
Forest type *Dicrano-Pinetum sylvestris*			
Picea obovata (artificial)	1.9	1.6 ± 0.18	156.6 ± 12.76
Picea obovata (natural)	0.3	0.7 ± 0.25	63.3 ± 20.28
Pinus sylvestris	1.3	0.7 ± 0.05	65.6 ± 5.91
Betula pendula, B. pubescens	6.2	0.4 ± 0.04	97.2 ± 8.59
Populus tremula	0.1	0.9 ± 0.01	220.0 ± 5.1
Salix spp.	4.0	0.9 ± 0.091	159.2 ± 8.31
Forest type *Bupleuro longifolii-Pinetum sylvestris*			
Picea obovata (artificial)	2.9	1.3 ± 0.1	131.4 ± 9.35
Betula pendula, B. pubescens	0.3	0.5 ± 0.1	125.0 ± 0.45
Salix spp.	4.7	0.9 ± 0.07	175.8 ± 8.6

25.7.2 Germination and seedling growth of *Pinus sylvestris* L., *Picea obovata* Ledeb. and *Larix sibirica* Ledeb. on soil of different types of boreal forests

The variety of natural conditions and operating factors and the complexity and non-linearity of interdependencies in ecosystems make it difficult to study the processes of natural regeneration. To separate the influence of factors, under natural conditions, is extremely difficult. The problem is complicated by the low seed germination of woody plants in natural ecosystems: the germination of pine and spruce is only 1–20% and 1–10% of larch, whereas germination under laboratory conditions is 80–95%. Therefore, the best results for the study of reforestation factors are provided by controlled experiments. However, to date, there have been no laboratory experiments that allow us to distinguish the effect of different types of forest soil and to evaluate the role of this factor on the germination, survival and growth of forest formers.

For this experiment, we took brown mountain-forest and soddy pale-yellow podzolic soil, under the canopy of old-growth trees (140–160 years) in forest and relevant clear-cutting (which held the burning of forest residues), in the southern taiga district of the Zauralskaya (Transuralian) hilly piedmont province (the Middle Urals). Samples of brown mountain-forest and soddy pale-yellow podzolic soil were selected from the humus horizon in three forest types. These forest types differ in their position in the landscape and their pattern of moisturising (periodically dry, stable fresh, fresh, periodically moist habitats).

The common soils on the tops and the upper half of slopes of the hills of the Middle Urals, under the canopy of pine forests, are under-developed and typical brown mountain-forest soils with a poor differentiation profile on the horizon, a highly skeletal soil with low thickness (30–55 cm). The forest floor of typical brown mountain-forest soil, in periodically dry habitats, is about 2 cm, its dark grey accumulative horizon is 9 cm in depth and has a light-brown alluvial horizon. The soil is sandy loam. The density of the upper to the lower horizon ranges from 0.9 to 1.6 g cm^{-3}, the total porosity 65.3 to 40.1% and the solid phase density is 2.6 g cm^{-3}. The hygroscopic moisture depth varies from 2.0 to 1.04% and the wilting point varies from 5.13 to 2.89%. Brown mountain-forest soil dry habitat locations have a slight water extraction reaction. From the upper to the lower elements of the relief, in connection with the increase in soil depth and the increase in its buffer properties, the stability of the water regime in drained areas increases. The pine *Dicrano-Pinetum sylvestris*, growing on the upper slopes and tops of low hills, is characterised by typical brown mountain-forest soil, that has a sandy loam composition, a thickness profile of 50 cm and a low skeleton. The depth of the forest floor is

3 cm and the humus horizon is 12 cm. It is dominated by brown tones, the intensity of which attenuates with depth. The density of the upper to the lower horizon ranges from 1.1 to 1.6 g cm^{-3}, the total porosity ranges from 58.1 to 40.2% and the density of solids is 2.5–2.6 g cm^{-3}. Hygroscopic moisture content has a depth varying from 1.5% to 0.4% and a wilting point from 5.55% to 1.43%; the actual acidity (pH$_{H2O}$) ranges from 5.24 to 5.57.

Fresh, occasionally wet habitats are characterised by soddy pale-yellow podzolic soil with a high soil depth and are low skeleton and fairly high density. In the forest type *Bupleuro longifolii-Pinetum sylvestris* the soil profile can be up to 90 cm, the humus horizon 8 cm and the humus-illuvial (podzolic) 30 cm. The texture of the upper horizon is sandy loam, the alluvial horizon is medium loamy and the lower horizon is sand. The density of the upper to the lower horizon ranges from 1.1 to 1.6 g cm^{-3}, the total porosity ranges from 57.8 to 38.7% and the density of solids is 2.5–2.6 g cm^{-3}. The hygroscopic

Table 25.6 The ecological characteristic of richness of soil conditions of the forest and clear-cuttings of the Middle Urals (Tsyganov, 1983).

Successional status	Soil moisture (HD)	Trophicity soil (TR)	Richness of soil nitrogen (NT)	Soil acidity (pH; RC)	Soil moisture variability (fH)
Forest/clear-cutting	Forest type – *Dicrano-Pinetum sylvestris*				
	12.5/12.1	5.1/5.5	4.5/5.0	5.5/6.3	4.7/5.2
	Forest type – *Tilio-Pinetum sylvestris*				
	12.7/11.9	5.2/5.9	4.6/5.6	5.9/6.4	4.9/5.4
	Forest type – *Bupleuro longifolii-Pinetum sylvestris*				
	12.6/12.5	5.8/6.3	5.1/5.3	6.8/7.4	5.4/5.5

Note: We used a GPA of botanical descriptions on edaphic environmental factors. The list of D.N. Tsyganov's scales and ranges (Tsyganov, 1983) is: the scale of soil moisture (HD) from 1 to 23 points, the scale of trophicity soil (TR) from 1 to 19 points, the scale of soil acidity (RC) from 1 to 13 points, the scale of the richness of soil nitrogen (NT) from 1 to 11 points, the scale of soil moisture was variable (fH).

moisture of soddy pale-yellow podzolic soil has a varying depth from 0.94 to 0.26%, wilting points from 3.97 to 1.63% and an actual acidity (pH_{H2O}) from 5.40 to 5.27. Utilising the range of ecological tools (Tsyganov, 1983), we evaluated edaphic factors of the studied habitats of the Zauralskayas hilly piedmont province. According to the average score of botanical descriptions, the soils of the studied forests are close to each other. On the moisture scale (Table 25.6) humid forests belong to the forest–meadow ecological group (around 13 points), with poorly varied moisturising soil (about 5 points). By the salt regime of the soil (trophicity soil) and acidity, the soil in pine forests *Dicrano-Pinetum sylvestris* and *Tilio-Pinetum sylvestris* is poor, acidic and weakly acidic, and the soil in pine forests *Bupleuro longifolii-Pinetum sylvestris* is not rich, slightly acidic soil. According to the scale of the richness of soil nitrogen, the studied soil is poor (about 5 points; Table 25.6).

25.7.2.1 Forest soil after clear-cuttings

In brown mountain-forest and soddy pale-yellow podzolic soil of the Middle Urals, after clear-cutting, the forest floor depth and the degree of decomposition is reduced, while the humus horizon remains unchanged or slightly increased. The greatest changes in physical and chemical properties have been identified for the soil humus horizon of pine forests. The other forest type differences, concerning soil properties, between the forest and clear-cutting are only defined for content-flowing (exchange) and potassium (Table 25.7).

According the range of environmental scales by Tsyganov (1983) forest soil after clear-cuttings are close to each other but different from forest soil (Table 25.6). Under the influence of clear-cutting the most significant changes are noted for brown mountain-forest soil, especially for forest type *Tilio-Pinetum sylvestris*, with changes in soil moisture, salt regime and nitrogen richness. The soil of periodic

Table 25.7 Some physical and chemical properties of soil humic layer, taken from the experiment.

Successional status	Skeleton (%)	Density (g cm^{-3})	Total porosity (%)	Wilting point (%)	pH_{H2O}	pH_{KCl}	K_2O (mg kg^{-1})
Forest	Forest type – *Dicrano-Pinetum sylvestris*						
	50.3	0.9	65.3	5.13	5.10	4.06	292.5
	Forest type – *Tilio-Pinetum sylvestris*						
	4.7	1.1	58.1	5.55	5.24	3.95	245.0
	Forest type – *Bupleuro longifolii-Pinetum sylvestris*						
	9.1	1.1	57.8	3.97	5.40	4.34	276.5
Clear-cutting	*Dicrano-Pinetum sylvestris*						
	36.1	1.0	59.8	5.81	5.22	4.06	459.0
	Tilio-Pinetum sylvestris						
	10.3	1.0	57.4	4.09	5.16	3.89	164.5
	Bupleuro longifolii-Pinetum sylvestris						
	2.5	0.7	71.1	13.12	5.32	5.07	1375.0

dry and stably fresh habitats with brown mountain-forest soil is weak, with slightly variable moisture, a salt content from poor to quite rich, acidic or weakly acidic and providing a poor to fairly rich amount of nitrogen (Table 25.6). The clear-cutting of fresh, occasionally wet habitats with soddy pale podzolic soil differs slightly from the original forest (Table 25.6). The soil is weak, with slightly variable moisture, a salt content from poor to quite rich, slightly acidic and poor in nitrogen.

25.7.2.2 Dependence of germination energy and woody plant germination on soil conditions

An experimental study of the growth characteristics in seedlings of *Pinus sylvestris* L., *Picea obovata* Ledeb and *Larix sibirica*

Ledeb was conducted in the humus horizon of brown mountain-forest and soddy pale ash-containing soil of different types of forests: *Dicrano-Pinetum sylvestris*, *Tilio-Pinetum sylvestris* and *Bupleuro longifolii-Pinetum sylvestris* (in two versions: indigenous forest; clear-cutting). The experimental conditions allowed us to avoid the effects of vegetation, moisture, temperature and other factors on the initial stages of growth and development of the studied tree species and to investigate the effects of clear-cutting, not only on soil properties, but also on the germination and vigour of major conifer species.

The three most common forests types of the Middle Urals in the stage of "seed germination" for *Pinus sylvestris* L. (Figure 25.5, Tables 25.8 and 25.9)

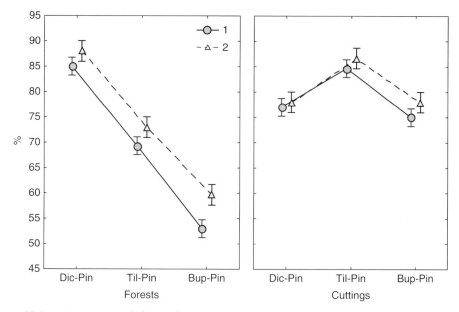

Figure 25.5 Activity rates of the seed germination of *Pinus sylvestris*. Dashed lines represent the average value, solid lines represent the 95% interval; 1 – Germination energy, 2 – seed germination; Dic-Pin – *Dicrano-Pinetum sylvestris*, Til-Pin – *Tilio-Pinetum sylvestris*, Bup-Pin – *Bupleuro longifolii-Pinetum sylvestris*. (See insert for color representation.)

Table 25.8 LSD test of comparison of *Pinus sylvestris* germination energy (significance level $p \leq 0.05$).

Experiment number	Forest type	Successional status	Experiment number (mean, %)					
			1 (85.0)	2 (77.0)	3 (69.3)	4 (84.7)	5 (53.0)	6 (75.0)
1	Dicrano-Pinetum sylvestris	Forest		+	+	−	+	+
2		Cutting	+		+	+	+	−
3	Tilio-Pinetum sylvestris	Forest	+	+		+	+	+
4		Cutting	−	+	+		+	+
5	Bupleuro longifolii-Pinetum sylvestris	Forest	+	+	+	+		+
6		Cutting	+	−	+	+	+	

Note: + Statistically significant difference.

Table 25.9 LSD test of comparison of *Pinus sylvestris* seed germination (significance level $p \leq 0.05$).

Experiment number	Forest type	Successional status	Experiment number (mean, %)					
			1 (88.0)	2 (78.0)	3 (73.0)	4 (86.7)	6 (59.7)	6 (78.0)
1	Dicrano-Pinetum sylvestris	Forest		+	+	−	+	+
2		Cutting	+		+	+	+	−
3	Tilio-Pinetum sylvestris	Forest	+	+		+	+	+
4		Cutting	−	+	+		+	+
5	Bupleuro longifolii-Pinetum sylvestris	Forest	+	+	+	+		+
6		Cutting	+	−	+	+	+	

Note: + Statistically significant difference.

revealed the following differences in forests and cuttings:

1 Differences between the forest and cuttings are statistically significant for all the studied forest types, for germination energy and for seed germination.

2 Changes in germination energy and seed germination in forest soil and forest soil after clear-cuttings are multidirectional. A statistically significant reduction in seed germination and germination energy in forest soil after clear-cuttings (compared with forest soil) was observed for the forest type *Dicrano-Pinetum sylvestris*. For the forest types *Tilio-Pinetum sylvestris* and *Bupleuro longifolii-Pinetum sylvestris* we found an inverse trend: an increase in

the studied parameters on forest soil after clear-cuttings.

3 Forest soil revealed a statistically significant decline in seed germination in a generalised topological profile *Dicrano-Pinetum sylvestris* to *Bupleuro longifolii-Pinetum sylvestris*.

4 Forest soil after clear-cuttings were found to be dependent on seed germination in a topological profile.

In the stage of "seed germination" for *Picea obovata*, we identified the following features (Figure 25.6, Tables 25.10 and 25.11):

1 Differences between forest and clear-cutting on the germination energy and seed germination capacity were statistically significant for all the studied forest types. Germination energy and seed germination in forest soil after

clear-cuttings were significantly lower compared to the forests.

2 With *Picea obovata*, among all variations of the experiment, we observed the highest rates of germination energy and germinating capacity, on the basis of pine *Bupleuro longifolii-Pinetum sylvestris*.

3 In forest soil after clear-cuttings, we observed the highest rates of germination energy in spruce (like pine) for pine forest *Tilio-Pinetum sylvestris*.

4 Seed germination in forest soil after clear-cutting increases in generalised topological profile.

For *Larix sibirica* we identified the following characteristics of seed germination in forest soil and in forest soil after clear-cutting for the three most common forests types of the Middle Urals (Figure 25.7; Tables 25.12 and 25.13):

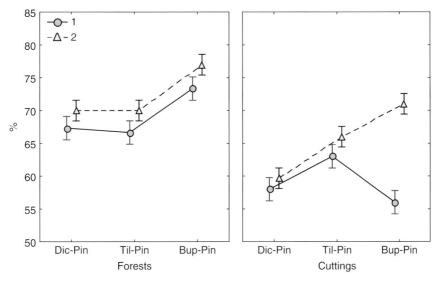

Figure 25.6 Activity rates of seed germination of *Picea obovata*. Dashed lines represent the average value, solid lines represent the 95% interval; 1 – germination energy, 2 – seed germination; Dic-Pin – *Dicrano-Pinetum sylvestris*, Til-Pin – *Tilio-Pinetum sylvestris*, Bup-Pin – *Bupleuro longifolii-Pinetum sylvestris*. (See insert for color representation.)

Table 25.10 LSD test of comparison of *Picea obovata* germination energy (significance level $p \leq 0.05$).

Experiment number	Forest type	Successional status	Experiment number (mean, %)					
			1 (67.3)	2 (58.0)	3 (66.7)	4 (63.0)	5 (73.0)	6 (56.0)
1	Dicrano-Pinetum sylvestris	Forest		+	−	+	+	+
2		Cutting	+		+	+	+	−
3	Tilio-Pinetum sylvestris	Forest	−	+		+	+	+
4		Cutting	+	+	+		+	+
5	Bupleuro longifolii-Pinetum sylvestris	Forest	+	+	+	+		+
6		Cutting	+	−	+	+	+	

Note: + Statistically significant difference.

Table 25.11 LSD test of comparisons of *Picea obovata* seed germination (significance level $p \leq 0.05$).

Experiment number	Forest type	Successional status	Experiment number (mean, %)					
			1 (70.0)	2 (59.7)	3 (70.0)	4 (66.0)	5 (77.0)	6 (71.0)
1	Dicrano-Pinetum sylvestris	Forest		+	−	+	+	−
2		Cutting	+		+	+	+	+
3	Tilio-Pinetum sylvestris	Forest	−	+		+	+	−
4		Cutting	+	+	+		+	+
5	Bupleuro longifolii-Pinetum sylvestris	Forest	+	+	+	+		+
6		Cutting	−	+	−	+	+	

Note: + Statistically significant difference.

1 In all types of forests we found a significant decline in germination energy and germination capacity in forest soil after clear-cuttings, compared with the soil of forests. This feature is particular to *Larix sibirica* and is clearly expressed in comparison with *Pinus sylvestris* and *Picea obovata*.

2 In summary, the topological profile of *Dicrano-Pinetum sylvestris* to *Bupleuro longifolii-Pinetum sylvestris* did not reveal a clear trend to increase or decrease the germination of seeds with forest soil or with forest soil after clear-cutting.

To evaluate the interference of specific features, successive status and forest type, we conducted a factor analysis. According to the results of the factorial analysis, for each tree species two factors were allocated; F1 and F2 (Table 25.14).

As seen in Table 25.14, *Pinus sylvestris* displays a strong correlation between

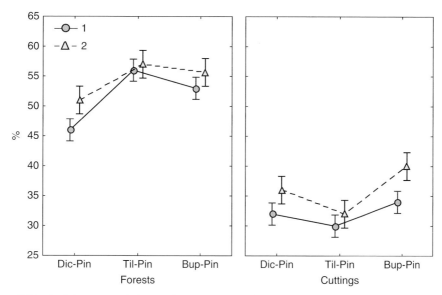

Figure 25.7 Activity rates of the seed germination of *Larix sibirica*. Dashed lines represent the average value, solid lines represent the 95% interval; 1 – germination energy, 2 – seed germination; Dic-Pin – *Dicrano-Pinetum sylvestris*, Til-Pin – *Tilio-Pinetum sylvestris*, Bup-Pin – *Bupleuro longifolii-Pinetum sylvestris*. (See insert for color representation.)

Table 25.12 LSD test of comparison of *Larix sibirica* germination energy (significance level $p \leq 0.05$).

Experiment number	Forest type	Successional status	Experiment number (mean, %)					
			1 (46.0)	2 (32.0)	3 (56.0)	4 (30.0)	5 (53.0)	6 (34.0)
1	*Dicrano-Pinetum sylvestris*	Forest		+	+	+	+	+
2		Cutting	+		+	−	+	−
3	*Tilio-Pinetum sylvestris*	Forest	+	+		+	+	+
4		Cutting	+	−	+		+	+
5	*Bupleuro longifolii-Pinetum sylvestris*	Forest	+	+	+	+		+
6		Cutting	+	−	+	+	+	

Note: + Statistically significant difference.

germination energy and seed germination and the type of wood that appears in the first factor. Data for the second factor indicates a lesser role in successive status. *Picea obovata*, unlike *Pinus sylvestris*, has a strong influence on the energy status of the succession of germination and is less significant for germination that appears in factor one. The second factor revealed the dependence of seed germination on forest type. *Larix sibirica* has a strong correlation of germination energy and seed germination,

Table 25.13 LSD test of comparisons of *Larix sibirica* seed germination (significance level $p \leq 0.05$).

Experiment number	Forest type	Successional status	Experiment number (mean, %)					
			1 **(51.0)**	**2** **(36.0)**	**3** **(57.0)**	**4** **(32.0)**	**5** **(55.7)**	**6** **(40.0)**
1	*Dicrano-Pinetum sylvestris*	Forest		+	+	+	+	+
2		Cutting	+		+	+	+	+
3	*Tilio-Pinetum sylvestris*	Forest	+	+		+	−	+
4		Cutting	+	+	+		+	+
5	*Bupleuro longifolii-Pinetum sylvestris*	Forest	+	+	−	+		+
6		Cutting	+	+	+	+	+	

Note: + Statistically significant difference

Table 25.14 The results of the factorial analysis of data on germination energy and seed germination main forest-forming species of the Middle Urals.

Variable	Tree species					
	Pinus sylvestris		*Picea obovata*		*Larix sibirica*	
	F1	**F2**	**F1**	**F2**	**F1**	**F2**
Successional status (forest-cutting)	0.143	0.953*	−0.968*	−0.026	−0.982*	0.064
Forest type	0.882*	0.244	−0.071	0.990*	0.059	0.998*
Germination energy (%)	0.906*	0.383	0.940*	0.155	0.985*	0.121
Seed germination (%)	0.902*	0.342	0.628	0.751*	0.984*	0.125
Total variance (%)	66.6	25.0	64.6	30.0	73.4	25.0

Note: *– Statistically significant factors

with successional status appearing in factor one. Forest type, associated with the second factor, is slightly correlated with germination energy and the germination of seeds.

Thus, in the conditions of southern taiga forests of the Middle Urals, as a result of our controlled study, we found a significant effect of clear-cutting, not only on soil properties, but also on the germination and germination energy of tree species, primarily for *Picea obovata* and *Larix sibirica*.

The differences in germination energy and germination of major conifer species, both in forest soil and in forest soil after clear-cuttings, were statistically significant for all studied forest types and forest forming. Determined responses of the germination and germination energy of *Pinus sylvestris*, *Picea obovata* and *Larix sibirica* to changes in the soil in the topological profile and in particular in the anthropogenic factor (clear-cutting) explain the actuality of new integrated interdisciplinary research with a goal to predict reforestation on clear-cuttings in different forests types of the Middle Urals and the development of sustainable forest management.

25.8 Conclusions

The boreal forests of North America and Eurasia account for 30% of the total forest area of the planet. They play a significant role in maintaining the stability of the biosphere and human life support. However, an enormous amount of commercial use, fires and windfalls have led to a dramatic reduction in the land area of coniferous forests on all continents, the depletion of species composition and a drop in stability and nature protection functions.

In this work, we provide an overview of the study of the patterns of renewal of the main forest tree species in the coniferous forests of Russia. It has been revealed that, despite numerous studies in the boreal zone, the problem of a lack of information and conflicting information concerning the features of reforestation in different natural zones and ecotypes still persists, and the influence of individual factors has not been fully established.

Extensive studies, carried out by the authors for more than 20 years (1991–2014), are based on a combination of approaches and methods of geo-dynamic forest typology, forest science, soil science and botany, showing that the mountain forests of the Urals, which are located on the border of Europe and Asia, are unique, natural beauties and, due to their sharply expressed ecotone effect, the effects, above all else, of both anthropogenic impacts and climate changes can be traced.

Using the research concept of splitting indigenous forest types into a series of dynamic rows by restoration and development of ecosystems allowed us to get valuable information on the age structure of the undergrowth of *Picea obovata* for different types of regeneration (recovery of conifer forests, the formation of secondary short, long and stable derivatives of birch and aspen) with a minimum number of plots placed in the Ural mountains. This information is required to assess the pace and stability of the regeneration of natural coniferous forests, in terms of human impact and climate change. By the method of topological profiles we revealed the particularities of both the natural and artificial regeneration of *Pinus sylvestris* and *Picea obovata*, which provided us with an explanation of their dominance in different habitats of the Middle Urals.

An experimental study of the growth characteristics of seedlings of *Pinus sylvestris, Picea obovata* and *Larix sibirica* was carried out on the humus horizon of brown mountain-forest and soddy pale-yellow podzolic soil of different forest types of the Middle Urals: *Dicrano-Pinetum sylvestris, Tilio-Pinetum sylvestris* and *Bupleuro longifolii-Pinetum sylvestris* (in two versions: an indigenous forest and a clear-cutting). A laboratory experiment allowed us to exclude the influence of vegetation, moisture, temperature and other factors in the initial stages of the growth and development of the studied tree species and to investigate the influence of clear-cutting, not only on soil properties, but also on the germination, germination energy of major conifer species. The study revealed that differences in germination energy and seed germination in *Pinus sylvestris, Picea obovata* and *Larix sibirica* are statistically significant in forest soils and in forest soils after clear-cutting for all the studied forest types and tree species. The results of our research contribute to the development of the theory of the forest-forming process in the boreal zone and can be used in the planning of forestry and environmental projects.

Bibliography

Achard F., Frédéric A., Danilo M., Stibig H.J., Aksenov D, Laestadius L., Zengyuan Li, Popatov P., Yaroshenko F. **2006**. Areas of rapid forest-cover change in boreal Eurasia. *Forest Ecology and Management* **237**(1/3):322–334.

Andreev G.V. **1998**. Analysis of the typological structure of the forest lands within the southern Ural province of south taiga and mixed forests. In: *Conference on Contemporary Issues of the Population, Historical and Applied Ecology, Ekaterinburg.* Pp. 231–232.

Barnes B.V. **1991**. Deciduous Forests of North America. In: E. Röhrig and B. Ulrich, (eds.). *Ecosystems of the World 7. Temperate Deciduous Forests.* Elsevier, New York. Pp. 219–344.

Blankmeister J., Hengst E. **1971**. Die Fichte im Mittelgebirge. *Radebeul* **1971**:287.

Bouchard M., Pothier D. **2011**. Long-term influence of fire and harvesting on boreal forest age structure and forest composition in eastern Quebec. *Forest Ecology and Management* **261**(4):811–820.

Bulygin N.E. **1985**. *Dendrology.* Agropromizdat, Moscow. 280 pp.

Classification and Diagnosis. **1977**. Classification and Diagnostic System of Soils of the USSR *(in Russian)*. Kolos Press, Moscow. 221 pp.

Danilik V.N. **1968**. The experiment on carrying out selection and gradual fellings in mountain woods of Southern and Average Urals Mountains. *Woods of the Ural Mountains and Their Economy* **1**:179–204.

Davydychev A.N. **2006**. Natural renewal dark coniferous forests of Southern Urals Mountains (on an example of the South Ural state

natural reserve). In: *The MGUL-WOOD Bulletin* **3**:46–54.

De Groot W.J., Cantin S., Flannigan M.D., Soja A.J., Gowman L.M., and Newbery A. **2013**. A comparison of Canadian and Russian boreal forest fire regimes. *Forest Ecology and Management* **294**:23–34.

Ericsson S., Östlund L., Axelsson A.L. **2000**. A forest of grazing and logging: Deforestation and reforestation history of a boreal landscape in central Sweden. *New Forests* **19**(3):227–240.

Ermakov N.B. **2003** A variety boreal vegetation of Northern Asia. In: *Hemiboreal Woods. Classification and Ordinance,* Ermakov N.B. (Ed.). The Siberian Branch of the Russian Academy of Science, Novosibirsk. Pp. 9–32.

Federal Forestry Agency of the Russian Federation. **2015**. Available from http://www.rosleshoz.gov.ru/terminology/b/61.

Firsova V.P., Rzhannikova G.K. **1972**. Soils of a southern taiga and coniferous-broad-leaved woods of Urals Mountains and Zauralye. In: Wood Soils of a Southern Taiga of Urals Mountains and Zauralye. *Works of the Institute of Ecology of Plants and Animals* **85**:3–87.

Fitzsimmons M. **2003**. Effects of deforestation and reforestation on landscape spatial structure in boreal Saskatchewan, Canada. *Forest Ecology and Management* **174**(1/3):577–592.

Forest Communities. **2002**. Forest Communities Study Methods *(in Russian)*. St. Petersburg State University, Saint Petersburg. 240 pp.

Forster D.R. **1988**. Disturbance history, community organization and vegetation dynamics of the old-growth Pisgah Forest, south-western New Hampshire, USA. *Journal of Ecology* **76**(1):105–134.

Gorichev J.P., Davydychev A.N, Kulagin A.U., Alibaev F.H. **2009**. Charge potential of derivative forests of the South Ural state natural reserve. *News of the Samara Centre of Science of the Russian Academy of Sciences* **11**(1):372–376.

GOST **1997**. *Seeds of trees and shrubs. Methods to determine germination.* GOST 13056.6-97, GOST, Moscow.

Isaeva R.P. **1968a**. The survival abilit and growth fur-tree of undergrowth a on concentrated cuttings down at in front of Ural. *Woods of the Urals Mountains and Their Economy* **1**:205–232.

Isaeva R.P. **1968b**. The stability and efficiency fir seed trees in taiga zone in front of Ural. *Woods of the Urals Mountains and Their Economy* **1**:233–245.

Isaeva R.P. **1975**. The special features of formation of young growths on the continuous concentrated cuttings down in dark coniferous forests. *Woods of the Urals Mountains and Their Economy* **8**:59–69.

Isaeva R.P., Lougansky N.A. **1981**. Forest regeneration in Urals Mountains. *Forest Husbandry* **10**:38–40.

Isaeva R.P., Lougansky N.A., Lougansky V.N. **2009**. Nature and basic regularities of natural reforestation in Sverdlovsk region forests. *Woods of Russian Mountains and Their Economy* **1**(31):4–14.

Isaeva R.P., Lugansky N.A. **1974**. Natural forest regeneration in subareas of a southern taiga and dark coniferous and broad-leaved forests of Urals Mountains. In: *Process of Forest Regeneration in Urals Mountains and Zauralye*, Isaeva R.P., Lugansky N.A. (Eds.), Sverdlovsk, Pp. 94–128.

Ivanova N. **2014**. Differentiation of forest vegetation after clear-cuttings in the Ural Mountains. *Modern Applied Science* **8**(6):195–203.

Ivanova N.S., Andreev G.A. **2008a**. The natural regeneration of *Picea obovata* Ledeb. and *Abies sibirica* Ledeb. Cenopopulations in dark-coniferous forests of southern Urals. *Agrarian bulletin of Ural* **6**:82–86.

Ivanova N.S., Andreev G.A. **2008b**. The natural regeneration of *Picea obovata* Ledeb. and *Abies sibirica* Ledeb. Cenopopulations in short-term secondary birch forests of southern Urals Mountains. *Agrarian Bulletin of Ural* **7**:75–77.

Ivanova N.S., Andreev G.A. **2008c**. The natural regeneration of *Picea obovata* Ledeb. and *Abies sibirica* Ledeb. Cenopopulations in long-term secondary birch forests of southern Urals Mountains. *Agrarian bulletin of Ural* **8**:74–76.

Ivanova N.S, Andreev G.A. **2008d**. The permanent secondary aspen forests of the western

low mountains of the southern Urals. *Agrarian Bulletin of Ural* **10**:91–93.

Ivanova N.S., Bystrai G.P. **2010**. Model of the formation of the tree layer structure on cuttings. Part 1.The control parameters. *Agrarian Bulletin of Ural* **5**:85–89.

Ivanova N.S., Zolotova E.S. **2013**. Biodiversity of the natural forests in the Zauralsky hilly piedmont province. *Modern Problems of Education and Science* **9**:65–67.

Ivanova N.S., Zolotova E.S. **2014**. Development of forest typology in Russia. *International Journal of Bio-resource and Stress Management* **5**(2):298–303.

Karpov V.G. **1965**. *Structure and Dynamics of Dark Coniferous Forests*. Leningrad 154 pp.

Karpov V.G. **1969**. *Experimental Phytocoenology of Dark Coniferous Taiga*. Leningrad, 236 pp.

Kellomaki S. **2000**. Forests of the boreal region: gaps in knowledge and research needs. *Forest Ecology and Management* **132**:63–71.

Kolesnikov B.P. **1969**. Wood of Sverdlovsk area. *In: Woods of the USSR. Science* **4**:64–124.

Kolesnikov B.P., Zubarev R., Smolonogov E.P., Filroze E.M. **1970**. The questions of mountain forestry in Urals Mountains. *Woods of the Urals Mountains and Their Economy* **5**:90–94.

Kolesnikov B.P., Zubareva R.S., Smolonogov E.P. **1974**. *Forest-Vegetable Conditions and Forest Types of the Sverdlovsk Area. Practical Guidance*. USC of AS the USSR, Sverdlovsk, 176 pp.

Kuusela K. **1990**. Zur Bedeutung der Fichte in europaischen borealen Nadelwaldzone. *Forstwissenschaftlichen Centralblatt* **109**: 155–161.

Leikola M. **1990**. Zur Bedeutung der Fichte im Finnland und zu ihrer Verjüngung. *Forstwissenschaftlichen Centralblatt* **109**:162–167.

Luyssaert S., Schulze E.-D., Borner A., Knohl A., Hessenmoller D., Law B.E., Ciais P. Grace J. **2008**. Old-growth forests as global carbon sinks. *Nature* **455**:213–215.

Maslakov E.L., Kolesnikov B.P. **1968**. The classification of cuttings down and natural renewal of pine woods middle-taiga subbands of flat Zauralye. *Woods of the Urals Mountains and Their Economy* **1**:246–279.

Morozov G.F. **1920**. *Foundations of the Science of Wood*. Simferopol, 319 pp.

Morozov G.F. **1928**. *The Science about Forest*. Leningrad 368 pp.

Muratov M.E., Amirkhanova A.H. **1997**. The feature of natural renewal of dark coniferous forest in Southern Urals Mountains. In: *Woods of Bashkortostan. A Modern Condition and Prospects*, Ufa. Pp. 161–162.

Olsson R. **2009**. *Boreal Forest and Climate Change*. Air Pollution and Climate Secretariat and Taiga Rescue Network, 32 pp.

Petrova I.V., Sannikov S.N., Cherepanova O.E., Sannikova N.S. **2013**. Reproductive isolation and disruptive selection as factors of genetic divergence between *Pinus sylvestris* L. populations. *Russian Journal of Ecology* **44**(4):296–302.

Rozanov B.G. **2004**. *Morphology of Soils*. The Academic Project, Moscow. 432 pp.

Rysin L.P., Savelyeva L.I. **2002**. *The fir forest of Russia*. The Science, Moscow. 335 pp.

Sannikov S.N. **1960**. The natural pine renewal on continuous cuttings down in Pripyshminsky pine forests. In: *Questions of development of a forestry in Urals Mountains. Works of the Institute of Biologies UFAS the USSR* **16**:81–106.

Sannikov S.N. **1992**. *Ecology and geography of natural renewal of a pine ordinary*. The Science, Moscow. 264 pp.

Sannikov S.N., Petrova I.V., Egorov E.V., Schweingruber F., Parpan T.V. **2011**. Genetic differentiation of *Pinus mugo* Turra and *P. sylvestris* L. Populations in the Ukrainian Carpathians and the Swiss Alps. *Russian Journal of Ecology* **42**(4):270–276.

Sannikov S.N., Sannikova N.S., Petrova I.V. **2004**. *Natural Forest Regeneration in Western Siberia (An Ekologo-Geographical Sketch)*. Publishing House of The Ural Branch of the Russian Academy, Ekaterinburg. 198 pp.

Schmidt-Vogt H. **1967**. Brotbaum Fichte. *Allgemeine Forstzeitschrift* **19**:303.

Schmidt-Vogt H. **1977**. *Die Fichte. and Die Fichte. Band 1: Taxonomie. Verbreitung. Morphologie. Okologie*. Waldge-sellschaften, Hamburg, 647 pp.

Shevelev A.A. **1965**. The natural renewal on the concentrated cuttings down of spruce-fir plantings low-mountain wood area of a subband of a southern taiga of the western slope of Average Urals Mountains (within Sverdlovsk region): The author's abstract. *Agricultural Sciences, Urals Forestry Engineering Institute* **1965**:27.

Shirokih P.S., Martynenko V.B. **2012**. Preliminary results of sin-taxonomic union Dicrano–Pinion corrections. In: *4th Russian Konference, Actual Problems of Geobotany*. Ufa. Pp. 331–335.

Soja A.J., Tchebakova N.M., French N.H.F., Flannigan M.D., Shugart H.H., Stocks B.J., Sukhinin A.I., Parfenova E.I., Chapin III, F.S., Stackhouse Jr., P.W. **2007** Climate-induced boreal forest change: Predictions versus current observations. *Global and Planetary Change* **56**(3/4):274–296.

Spurr S.H., Barnes B.V. **1980**. *Forest Ecology, 3nd edn*. John Wiley & Sons, Inc., New York, 687 pp.

Statistica Khalafyan A.A. **2010**. *The Statistical Analysis of Data: The Textbook, 2nd edn*. Binomial-Press, Moscow. 528 pp.

Thomasius H. **1990**. Vorkommen, Bedeutung und Bewirtschaftung der Fichte in der DDR. *Forstwissenschaftlichen Centralblatt* **109**: 138–151.

Tsyganov D.N. **1983**. *Phytoindication of Ecological Factors in a Subband Coniferous-Broad-Leaved Wood*. The Science, Moscow. 198 pp.

Turetsky M.R., Mack M.C., Harden J.W., Manies K.L. **2005**. Spatial patterning of soil carbon storage across boreal landscapes. *Ecosystem function in heterogeneous landscapes*. **2005**:229–255.

Wagenknetcht E., Belitz G. **1959**. *Die Fichte im norddeutsche Flachland*. Radebeul, Berlin. 121 pp.

Wang X., Cumming S.G. **2009**. Modeling configuration dynamics of harvested forest landscapes in the Canadian boreal plains. *Landscape Ecology* **24**(2): 229–241.

Westhoff V., van der . Maarel E. **1978**. The Braun–Blanquet approach. In: *Classification of Plant Communities*. R.H. Whittaker (ed.). Springer, The Hague, Pp. 287–399.

Yermakova M.V., Ivanova N.S. **2011**. The peculiar properties of structure of young growths Pinus sylvestris L. An artificial and natural origin on cuttings of average Urals Mountains. *Wood, Ecology, Wildlife Management Bulletin Mari GTU* **2**:13–23.

Young J.E, Sánchez-Azofeifa A., Hannon S.J., Chapman R. **2006**. Trends in land cover change and isolation of protected areas at the interface of the southern boreal mixed wood and aspen parkland in Alberta, Canada. *Forest Ecology and Management* **230**(1/3):151–161.

Zubareva R.S. **1970**. *Features of growth of young generations of a fir-tree and a fir in dark coniferous and broad-leaved forests of average Urals Mountains*. In: *Dynamics and a Structure of Woods in Urals Mountains*. Sverdlovsk, Pp. 135–149.

CHAPTER 26

Case study: Autoecology, biodiversity and adaptive characteristics of *Prosopis* in the Arizona region

26.1 Introduction

Mesquite is the most common shrub/tree in the desert/arid regions of the world. It belongs to the family Fabaceae. It is the dominant species in semiarid and arid lands of the world, such as North Africa, Mexico, USA, Argentina; and it is well adapted to cold, subhumid regions of the world. A good many studies have been undertaken on this plant, its adaptive characters and species biodiversity.

Three common species are known: honey mesquite (*Prosopis glandulosa*), screwbean mesquite (*P. pubescens*) and velvet mesquite (*P. velutina*). The desert region of Arizona contains *P. glandulosa* and *P. velutina*. Under severe drought and water stress, the mesquite serves as the primary food source for desert settlers. Mesquite beans are good source of food for both desert persons and animals. They can be eaten toasted or boiled. The mesquite bean can serve as coffee. Mesquite blooms, pollinated by bees, give a good quantity of honey. Mesquite beans can be stored for years.

The wood is very hard and durable and is used for construction and fencing and has served artisans in the crafting of furniture, flooring, panelling and sculptures. Two types of wood are available: yellowish wood and deep reddish which can be polished like mahogany wood.

Mesquites are drought-resistant and adapted to arid regions. They produce beans with a high nutritive value (protein in seeds 35% and flour 65% reported in the literature), and their beautiful blooms attract wildlife. Mesquites also provide shelters and nesting sites for birds in desert habitats. In Matehuala, San Luis Potosí, Mexico, people prepare cheese and sweets from mesquite pods.

A study was made by Berhanu and Tesfaye (2006) on the impact of *P. juliflora*, a multipurpose leguminous species introduced into north-east Ethiopia. The study focused on an assessment of the uses and negative impacts of *Prosopis* and an evaluation of mechanical control and prescribed burning. Various participatory rural appraisal (PRA) techniques were employed to collect ethno-botanical information on the uses of this species. *Prosopis* is used for firewood, charcoal, forage, fencing, windbreaks and other purposes.

Autoecology and Ecophysiology of Woody Shrubs and Trees: Concepts and Applications, First Edition.
Edited by Ratikanta Maiti, Humbero Gonzalez Rodriguez and Natalya Sergeevna Ivanova.

26.2   Medicinal value

Indians and settlers believed that tea made from mesquite root or bark cured diarrhoea. Boiled mesquite roots were effective in curing colic and healed flesh wounds. Mesquite leaves are crushed with water and mixed with urine to cure headache. Mesquite gums are applied to soothe failing eyes and sore throat and are used against dysentery and headaches. Mesquite flour is very useful in controlling blood sugar levels in diabetic patients. The sweetness of this flour comes from fructosen. In addition, it contains soluble fibres, such as galactomannin gum. The seeds and pods slow the absorption of nutrients, leading to a flattened blood sugar curve, unlike other other common staples.

Mesquite meal or flour can be made by finely milling the seeds and pods. Mesquite meal offers a sweet, chocolate/coffee flavour with a hint of cinnamon. Certified organic mesquite meal is 100% natural gluten free.

The gel-forming fibre permits foods to be slowly digested and absorbed over a four- to six-hour period in place of one or two hours in the case of other food, which produces a rapid rise in blood sugar. Mesquite meal is sold in stores in US deserts (Villagra et al., 2010).

Villagra et al. (2010) reported the expansion of the *Prosopis* genus from the subhumid Chaco towards colder and drier zones. They discussed the phenological, morphological and physiological features of seven *Prosopis* species native to Argentinian arid regions which are adapted to drought, salinity and other environmental stress factors in arid lands. In addition, some morphological or physiological adaptations appear to be specific to each species, for adaptation to a particular environment. The inter- and intra-specific variability contributes to their adaptation to stressful factors. This suggests that some *Prosopis* species may be a good option to be used in the restoration of degraded areas or in afforestation projects with productive potential.

26.3   Plant characteristics

The mesquites possess woody stems and branches, and have bipinnately compound leaves with two or more secondary veins; they have two rows of leaflets. They have flowers with five petals. They produce profuse large seed pods serving as a nutritious food source for wildlife. They grow wide-spreading and possess deep root systems that host colonies of nitrogen-fixing bacteria.

Depending on the species, mesquites attain a height ranging from a few feet to 10–15 feet. In a few cases the honey and velvet mesquites may reach 30–60 feet. They may possess single or multiple-branched stems. They bear stiff thorns on the smaller branches. They shed their leaves in the winter. They flower from spring into summer, in "catkins". They possess five petals and pale green or yellowish flowers which attract numerous pollinating insects. They produce pods that contain hard, long-lasting seeds which can be scarified (cut or slit) to induce germination. Mesquites possess lateral roots that extend far beyond the canopies of the plants and taproots that penetrate deep in the soil (USDA, 2010).

26.4   Control of mesquites

The mesquites can encroach into pasture lands and cause the displacement of grasses. This is highly frustrating. They compete with grass for water and light (Arizona-Sonora Desert Museum Internet site).

Different methods are used to control the spread of mesquites, such as planned burns, herbicides or physical removal. These are highly costly and cause potential environmental hazards.

Fire has been used as a management tool to control mesquite distribution for decades in southern Arizona. The mesquites may succumb to frequently repeated burns but it leads to the death of the native grasses, thereby making way for imported invasive species such as the highly aggressive Lehmann love grass.

Herbicides are usually applied by aircraft in an attempt to control the mesquites but the multiple applications could create adverse side-effects to rangeland species diversity and biomass.

Physical methods such as bulldozing, root ploughing, chaining, roller chopping or shredding has reduced mesquite density in pasturelands for brief periods, but the plants soon re-sprout from their bases.

A study in Ethiopia revealed that mechanical control (manual clearance and using bulldozers) was found to be effective when followed by proper management systems. Prescribed burning was destructive for young stands, whereas mature stands were not killed. Generally, cutting individual plants may aggravate invasion by *Prosopis* unless proper management is employed, such as repeated clearance. Thus, proper management and control of the species is urgent, using the control methods described above in cooperation with the local people. Otherwise, more areas could be invaded and tribal conflict for the remaining few grazing and farm areas free from *Prosopis* may turn into an unexpected political crisis (Berhanu and Tesfaye, 2006).

Ali Osman mentioned in Linkedin that *Prosopis* trees are the best species used to prevent desertification in Sudan. It is also used in sand-dune fixations especially in the northern part of the country. Economically, a reasonable number of families depend on the products of this tree for their livelihood. The main problem of this tree in Sudan that it spreads into irrigated schemes in which the allocated land for agricultural programs can be reduced to less than 50%, where the cost of eradication is very expensive. The government eradicated this tree from two irrigated schemes with some success.

26.5   Adaptive traits of mesquites to the desert environment

Mesquites possess a number of adaptive traits for survival in the desert environment. Their thorns are sharply pointed and strong to protect against desert herbivores. Their leaves are small and wax-coated to minimize transpiration. During extreme drought, the mesquites shed their leaves in order to conserve moisture. Their flowers are fragrant and attract the insects, especially the bees, necessary for prolific pollination. Their seeds are abundant and protected with hard seed coats which may last for decades, serving as seed banks.

Mesquites possess a deep root system ranging 150 to perhaps 200 feet below the surface. According to the Arizona-Sonora Desert Museum Internet site, a mesquite root reached over 160 feet (50 m) below the surface in a mine.

26.6 Physiological mechanism of resistance

Various studies have been undertaken on the mechanism of adaptation of *Prosopis* to arid conditions.

Villagra et al. (2010) studied seven *Prosopis* species native to Argentinian arid regions, investigating the phenological, morphological and physiological features required for tolerance to water stress, salinity and other environmental stress factors in arid lands. Some of these adaptations appear to be spread throughout the genus and should confer the capability to deal with the most common stressful factors of arid lands. However, other morphological or physiological adaptations appear to be specific to each species and could be the cause of niche differentiation between species and the occupation of particular environments within arid lands (e.g. sand dunes, saline environments). The inter- and intra-specific variability observed in their adaptation to stressful factors suggests that some *Prosopis* species may be a good option to be used in the restoration of degraded areas or in afforestation projects with productive potentials.

Felker (2009) studied photosynthesis and water relations, roots and soil relations, depth to water table N fixation and cross-inoculation in *Prosopis*. Water stress was experienced when fixing N adaptation

to low soil P levels in the field, with N fixation in natural stands.

Devinar et al. (2013) studied the halophytic shrub *P. strombulifera* by lowering an osmotic potential from 0 to −1.0, −1.9 and −2.6 MPa generated by NaCl and Na_2SO_4, and they analysed the content of abscisic acid (ABA) and related metabolites and transpiration rates. They observed that ABA content varied depending on type of salt, salt concentration, organ analysed and age of a plant. ABA content was much higher in leaves than in roots. Significant content of ABA-glucose ester (ABA-GE) was found in both roots and leaves, with a low content of phaseic acid (PA) and dihydrophaseic acid (DPA). The roots showed high ABA-GE accumulation in all treatments. The highest content of free ABA was correlated with ABA-GE glucosidase activity. The results reveal that ABA-GE and free ABA work together to create a specific stress signal.

Chávez et al. (2013) undertook modelling the spectral response of the desert tree *P. tamargo* to water stress. In the laboratory they studied the changes in reflectance under controlled water stress during the day over the whole experimental period. The canopy reflectance changes are well explained by the leaf area index (LAI) and leaf inclination distribution. It is concluded that one should consider LAI and canopy water content (CWC) as a water stress indicator of *Prosopis*.

Diurnal leaf movement and chlorophyll index performed best for understanding leaf movement and its effects, thereby detecting the changes in CWC.

It was observed that ABA and salicylic acid (SA) act as endogenous signal molecules responsible for inducing abiotic stress tolerance in plants. Devinar et al. (2013) determined the endogenous ABA and SA levels, growth parameters and chlorophyll content in leaves and roots of the halophyte *P. strombulifera* cultivated under increasing NaCl and Na_2SO_4 concentrations, at 30 and 70% relative humidity (RH) conditions. Under low RH conditions *P. strombulifera* growth was strongly inhibited and chlorophyll a and b content were decreased. NaCl-treated plant growth was also inhibited at 30% RH although levels of both hormones were not significantly increased. Taken together, the salt toxic effects on growth parameters and photosynthetic pigments were accentuated by low RH conditions and these responses were reflected in ABA and SA content.

26.7 Biodiversity

Several studies were undertaken on the biodiversity of *Prosopis*, of which a few are mentioned below:

Maundu et al. (2009) studied the impact of *P. juliflora* on Kenya's semiarid and arid ecosystems and local livelihoods. It is a small, fast-growing, drought-resistant evergreen tree of tropical American origin. Its pods are used as livestock food and are fed upon by native herbivores. It produces good timber and shade and keeps arid environments green. This is an attractive candidate for arid land environmental rehabilitation programmes. *Prosopis* produces masses of pods containing small tough smooth seeds. When pods are eaten by livestock, seeds pass easily through the gut. Once in the soil, seeds can lie dormant for a long time, until good conditions return. *Prosopis* is deep-rooted. It has

become highly invasive and is hard to control once established. Research was carried out to determine the spread of *P. juliflora* over time in Kenya and its effect on livelihoods and biodiversity. It was more aggressive in arid lands of the north and formed thorny impenetrable thickets, especially along water courses, flood plains, roadsides and in inhabited areas. It encroached upon paths, dwellings, irrigation schemes, crop farms and pastureland, significantly affecting biological diversity and rural livelihoods. In all three sites, there was significantly more plant diversity outside a *Prosopis* thicket than within it. It was evident that, in areas where *Prosopis* was well established, it was beyond the community's ability to control its expansion. Despite the stand of the affected local communities, the environmentalists, scientists and development workers are still divided on *Prosopis* matters. It is evident that Kenya's vast arid and semiarid lands totalling 80% of land area are at risk and, the longer the wait, the more difficult it will be to control or eradicate the species. A comprehensive policy on *Prosopis* is proposed.

Gichua (2014) has reported impacts of *P. juliflora* Linnaeus emend. Burkart on aspects of biodiversity and selected habitat conditions in Baringo, Kenya. It is a highly invasive species. Four hypotheses were tested to explain the effects of the invasion of *P. juliflora* on soil nutrient status, diversity of indigenous plant species, diversity of floral insect visitors of native *Acacia* spp., and plant and soil analysis revealed that *P. juliflora* density had a significant influence on total organic carbon and pH, and had a negative effect on overall native plant species richness and density of *Acacia* species. This revealed clearly that *P. juliflora* has an effect on the species richness of local species, whereby *P. juliflora* is slowly replacing the indigenous species. The results showed that *P. juliflora* is a preferred source of floral resources by bees, *Apies mellifera*, during the dry season. Allozyme analysis revealed minimal genetic distance between populations. This study shows that the invasive species has negative consequences for the local ecosystem at high densities.

Kumar and Mathur (2014) studied the impacts of invasion of a ligneous *P. juliflora*, in plant communities of reserve forests, protected forests, unprotected forests and open grazing lands in Jamnagar district in arid coastal areas of Gujarat state, India. Sites invaded by *P. juliflora* had different ligneous plant composition. In all land uses, *Prosopis*-invaded sites showed more species richness, diversity and evenness. However, the increase in species richness was due to occurrence of more weedy species along with *P. juliflora*. Even amongst the protected and undisturbed sites, the dominance of late successional species, that is *Acacia senegal, Maytenus emarginata, Zizyphus nummularia* and *A. nilotica*, was less at sites with *P. juliflora* than at sites without it. The density of *Commiphora wightii*, an endangered species, was decreased with increasing density of *P. juliflora*. Invasion of *P. juliflora* had shown demonstrable adverse impacts on plant communities in arid grazing lands.

In the context of the literature mentioned above, during my short visit I made an attempt to make an overview of a few aspects of the distribution pattern, plant architecture, ecophysiology and biodiversity of *Prosopis* at a few places in Arizona, USA.

26.8   Materials and methods

During a short visit to Arizona, I had an opportunity to travel through arid and subhumid regions in different places of this state (i.e. Nogales, Tucson, Phoenix, Patagonia, Tomestone and a few other places). During my travels, I made an overview of the vegetation on both sides of the road and observed that mesquite, *Prosopis*, is a dominant and aggressive plant in these regions. We observed the general autoecological characteristics (qualitative; i.e. variation in plant architecture, branching paters, crown, leaf canopy, phenology, population density, habitat and biodiversity of *Prosopis*). Apart from my observations, I collected some information from my family members resident in Nogales opposite to Sonora, Mexico.

26.9   Results and discussion

The present survey and observations on the autoecology and the morphological and adaptive traits of mesquite in a few arid regions coincide with the findings of various authors.

26.9.1   Plant characteristics

Prosopis spp. are thorny shrubs attaining a height of more than 15 m, depending on sites and soil strata. The stem produces thick, rough, grey-green bark in the form of scales. The plants in general are highly branched with thinner or thicker secondary branches. The tree is deeply rooted. The stems show a bending "zigzag" way with one or two stout prickly thorns. This coincides with the findings of Villagra et al. (2010).

Leaves are bipinnately compound, stiff, velvet or thin with a shining leaf surface in some, reflecting the sun's rays. The flowers are fragrant golden-yellow, borne in dense spikes called catkins. The fruits are flat or irregularly curved green pods which turn yellow upon ripening. The number of seeds per pod varies depending on the locality, with up to 10 seeds or more in the pod. Seeds are covered with a hard seed coat, highly durable for years, as reported in the literature.

The plant reproduces through seeds. The seeds are passed through the digestive tract of grazing animals, such as goats, cattle, camels and some wild herbivores. The seeds are spread along water courses during periods of rain.

26.9.2 Plant architecture

Prosopis shows a large variability in plant architecture, such as crown shape, leaf canopy, leaf orientation and branching pattern, depending on the habitat, density of populations, soils, nutrient availability and so on. Very little information is available on this aspect.

The plant possesses velvet or thin bipinnate leaves with an open canopy exposed to the sun's rays, falling vertically and laterally thereby with a great capacity of capturing light for efficient photosynthesis and translocation of photosynthates, possibly with a high production of biomass and wood density. Virtually the plants orient their leaves in a manner so that most of the leaves get exposed to the sun's rays. The plant possesses an open canopy for the efficient capture of sunlight for photosynthesis. In this respect, Felker (2009) studied photosynthesis, water relations, root systems and nitrogen fixation,

reporting a high photosynthetic capacity for *Prosopis*.

26.9.3 Crown shape

Prosopis shows large variation in crown shape, depending on the population density. In closely spaced forest the crown is mostly globose to subglobose with close branchlets. In open space it may be flat, conical or irregular in shape. The crown shape has plasticity in *Prosopis*.

26.9.4 Branching pattern

The branching pattern in *Prosopis* shows high plasticity, depending on the habitat: wide spacing, close spacing. The stem branches twice or thrice, either right from the base or somewhat from the top of the main trunk. Primary or secondary branches show dichotomy, thereby extending vertically upwards as well as horizontally, depending on the space available. *Prosopis* in general does not like to grow in association with other dicotyledons, except some under shrubs and herbs occupying the spaces between plants. In a forest of mesquites each individual grows in harmony with its neighbours by adjusting its branches without interfering with the growth of neighbouring plants. In close spacing the branching pattern may be conical or vertically upwards, thereby exposing their leaves to sunlight for efficient photosynthesis. Sometimes they bend the branch in a zig zag pattern to suit the space available. In open space, the primary branch grows horizonally in a zig zag manner and extends to occupy open space in its territory, revealing its aggression. From these horizontal primary stems secondary branches arise and grow upwards, showing continuous dichotomy (quantitative data are needed to determine the branching pattern of *Prosopis*). It occupies a large area. The stems and branches are not straight but show curvature. No information is available on the branching pattern of *Prosopis*.

Adaptive characteristics: *Prosopis* possesses several morphological and eco-physiological characters for adaptation to xeric conditions: thick velvet leaves with a waxy coating for spectral reflectance, open leaf canopy for efficient photosynthesis, adapted to a low water potential, deep root system, adapted to saline condition, abundant blooming habit with showy yellow colour for attracting pollinators (Villagra et al., 2010).

26.9.5   Habitat

Prosopis grows in diverse habitats, starting from highly xeric/arid, subhumid to humid regions, on plain valleys, on both sides of roads, on hillocks, uphills or on mountain peaks. It is adapted to diverse soil types and shows large variability in its growth characteristics; and it is very aggressive in nature. These observations coincide with other authors (Villagra et al., 2010).

26.9.6   Biodiversity

Though the present study is directed at studying biodiversity, we want to mention here a few aspects on the flora of Arizona.

We did not have an opportunity to know the flora of Arizona. A book written by the Arizona Native Plant Society is available. This book mentions that Arizona possesses an extremely rich flora due to its diversity of altitudes and climates. It contains floristic associations ranging

from subtropical to alpine with transition zones between the Sonoran, Chihuahuan, Mohave and Great Basin deserts. This rich flora, almost 4000 species of native plants, is uncommon in other regions of the United States. For a complete discussion, read Hendricks (1985).

Arizona Flora. SENet Southwest Environmental Information (Project Managers: Arizona State University Vascular Plant Herbarium) reported that Arizona is the third or fourth most floristically rich state in the US with perhaps as many as 3900 species of vascular plants. Over the last 60 years an average of 12 new species records have been reported annually. The following species are reported in desert regions of Arizona: *Acacia angustissima, A. constricta, A. constricta, A. greggii, Agave ocahui, A. palmeri, A. vilmoriniana, Ambrosia deltoidea, Arbutus arizonica, Arctostaphylos pungens, Argemone pleiacantha, Baccharis sarothroides, Berberis haematocarpa, Carnegiea gigantea, Celtis pallida, Chamaebatiaria millefolium, Chilopsis linearis, Commelina dianthifolia, Cowania Mexicana* var. *stansburiana, Dasylirion wheeleri, Echinocereus fasciculatus, Encelia farinosa, Echinocereus nicholii, Eschscholzia californica, Ferocactus cylindraceus, F. emoryi, F. herrerae, F. wislizeni, Fouquieria columnaris, F. splendens, Juglans majo, Juniperus deppeana, J. monosperma, J. osteosperma, Kallstroemia grandiflora, Larrea tridentata, Nepeta cataria, Nolina microcarpa, Olneya tesota, Opuntia acanthocarpa, O. arbuscula, O. bigelovii, O. clavata, O. engelmannii, O. leptocaulis, O. engelmannii* var. *linguiformis, O. fulgida, O. santa-rita, O. spinosio, Parkinsonia aculeata, P. florida, P. microphylla, Peniocereus greggii, Phoradendron juniperinum, Picea pungens, Pilosocereus alensis, Pinus aristata, P. cembroides, P. edulis, P. engelmannii, Platanus wrightii, Pinus strobiformi, P. ponderosa, P. leiophylla* var. *chihuahuana, P. flexilis, P. ponderosa* var. *arizonica, Populus tremuloides, Prosopis glandulosa, P. velutina, Quercus gambelii, Q. hypoleucoides, Rhus trilobata, Sapindus drummondii, Simmondsia chinensis, Stenocereus eruca, S. thurberi, Yucca brevifolia, Y. elata* and *Y. madrensis.*

Prosopis shows large diversity in growth characteristics, branching habit, crown shapes, plant size, abundance or scarcity depending on habitat, soil strata, xeric to humid conditions. *Prosopis* is dominant in desert although scarce in number in association with *Cactus* spp. It is the only evergreen tree in desert, a source of food for cattle and herbivores and a shelter for birds. As we travelled from Phoenix to Tucson we saw vast desert areas with scanty vegetation, mainly grasses, cactus, a few agaves, other species and *Prosopis* but only sparse populations here and there in the desert areas. From the roadside one can observe *Prosopis* with evergreen leaves, high ramification and globose crown, leaves with open canopy for the capture of solar radiation. Having a deep root, they can survive in desert environment through absorption of water deep in the soil profile. From Tucson while driving through an area called Green Valley we observed the density of *Prosopis* increase slowly, with low cactus and agave species. In Green Valley we saw luxuriant growth of *Prosopis* with a higher population. The growth and size of *Prosopis* plants depend on the availability of soil moisture. It is reported that the size and density of *Prosopis* reveals the moisture status in the soil of a particular region. As we neared Nogales we clearly observed a variation in biodiversity, population density and size of plant, depending on the soil

strata, soil nutrients and moisture availability in the soil profiles. We observed that, when we passed through an area surrounded by mountains, the *Prosopis* growth density was high, owing to the prevalence of microclimates, availability of soil moistures, prevalence of clouds, possible small showers and dews at night. Clouds often accumulate on mountain peaks. The vegetation was luxuriant in these regions of microclimate wherever we passed. On both sides of the road we saw rows of *Prosopiis* like soldiers, as if guarding the ecosystem behind it. Beyond that we saw dense forests of *Prosopis*, rarely in association with pines, oaks or others. We could not identify them at a distance.

As we drove along the zig zag road up and down we saw drastic changes in *Prosopis* population and size. They were tall and dense in population in low valleys with high moisture variability. On the other hand, the density and plant size gradually decreased from the low valley up the top of the mountains, probably due to a shallow root system on hard stony strata where roots could not penetrate further. At long distance we saw *Prosopis* like a short bushy shrub, very interesting indeed. In some areas, mesquite plants were widely spaced and produced bush shrub with globose or semi-globose crowns associated with grass-lands in between. We found cattle grazing on this grassland. Sufficient food was available to the livestock. This showed *Prosopis* at a high degree of adaptation to both extreme xeric and subhumid. On diverse soil strata, most of these observations coincided with other authors (Villagra et al., 2010). The presence of a waxy coating on velvet leaves causes the reflectance of the sun's rays, thereby reducing the loss of water by transpiration and lowering the leaf temperature. In this respect, Chávez et al. (2013) modelled spectral reflectance in relation to water relationships and water stress. High spectral reflectance reduced the leaf temperature and loss of water loss by transpiration.

Nogales where we resided, showed a predominance of luxuriant growth, tall, dense populations of evergreen tall *Prosopis* trees, probably due to the availability of high moisture in the soil profiles. This region is subhumid, probably with sufficient underground water. The trees form a dense forest with plants very close to each other, but not interrupting its neighbour's light. Plants attain different heights, forming globose crowns, thereby not competing with other individuals for light. All these trees are highly branched, mostly thinner stems and with an open leaf canopy exposed to the sun's rays for the harvest of solar energy for photosynthesis. The presence of high soil moisture, deep root system and long day length (more than 15 hours) probably contribute to the prolific growth and dense populations of *Prosopis* in the major part of this area, but the size and density of population decreases on hillocks and mountain tops, depicting an excellent landscape in the region. It has been mentioned that *Prosopis* possess a deep tap root and extended lateral roots (Villagra et al., 2010).

In this respect, various authors have discussed the impact and aggressive nature of *Prosopis* on biodiversity, its invasion on grassland, roads, dwellings and irrigation systems in various countries (Maundu et al., 2009; Gichua, 2014; Kumar and Mathur, 2014).

26.9.7 *Prosopis* as an obnoxious weed

The abundance and aggressive growth of *Prosopis* when invading pastures thereby makes them inaccessible to livestock. This poses a great danger to the economy of local inhabitants, as dangerous weeds and encroachment on residential lawns causes hazards for the inhabitants, very difficult to eradicate. Several methods have been reported for the control of mequites, leading to utter failure or of short duration. The durability of seed for years is another reason for the prevalence of weeds posing a danger to human and livestock populations.

Visit to Tomestone. We made a trip to the famous historic old city of Tomestone. Many historic incidents, gun fights and so on occurred there. Most of old houses were built prior to 1800 but are well maintained by the government. We made a city tour by horse-driven carts and enjoyed the glory and past history of this old city. Many foreigners visited this city.

On the way to Tomestone, passing through Patagonia, we observed a change in the ecosystem, starting from subhumid Nogales on the way to the desert lands of Tomestone. As we proceeded onwards the height and density of *Prosopis* decreased, but further on we reached desert lands where the *Prosopis* population decreased drastically, but the cactus population and agaves, associated with grasses, increased. We observed various species of *Agave*, gigantic cactus, *Carnegiea gigantean* like a tree, *Ephedra*, *Larrea* and other species. The same species were observed in desert nearing Phoenix. The *Prosopis* population was scarce.

26.9.8 Silent valley and uphill of the *Prosopis* forest

With sunset a complete silence prevailed in the atmosphere of the valleys and uphills of *Prosopis*, with only a rare chirping of birds. The forest was completely dark. When the leaves were exposed again to the sun's rays it was as if the plants woke up and were ready to harvest solar energy.

Our survey on a few autoecological characteristics and biodiversity of a few places in Arizona revealed several interesting features of *Prosopis*. *Prosopis* is the dominant and evergreen tree in deserts of Arizona. It serves as a good source of food for desert settlers with its nutritive pods and seeds (seeds contain 35% protein and the flour of *P. tamulas* contain 65% protein), besides wood for construction and it possesses several medicinal values. It is a good source of fodder for livestock and gives shelter and nesting for birds.

Prosopis shows a diversity in distribution, population and plant size, starting from desert environments to subhumid zones. In a subhumid region it is tall. *Prosopis* possesses several morpho-physiological characters for adaptation to arid and semiarid environments, such as thick velvet leaves with a waxy coating, open leaf canopy with a high capacity to capture sun light for photosynthesis, stiff thorns for protection against animals, deep root system, extended lateral roots, high spectral reflectance for reducing loss in transpiration, adaptation to low water potential, less stomatal capacity, high photosynthestic capacity and tolerance to drought and salinity.

Its aggressive nature, invading and replacing grassland, residential areas and irrigation channels, poses a great danger to humans and livestock. Various attempts to eradicate it in different countries have been found to be inefficient. Efficient control measures need to be evolved to eradicate this gigantic weed.

The field photographs in Figure 26.1 show the diversity of *Prosopis*, its gregarious habit and its predominance in Nogales, Arizona. The plant shows dichotomous branching habit with open canopy leaves and variability in crown architecture.

The photographs in Figure 26.1 show the distribution and growth habit of mesquite along the gradients and elevations of a mountain range. They show a decrease in height and a bushy habit with increasing elevations, probably due to shallow soil depth on a rocky substrate and laterally spreading roots. One picture shows the gregarious habit of invading residential lawns. Another picture shows the variability in pod shapes. The pods and seeds are nutritious as a source of rich protein for birds and grazing animals. The gregarious plant with stout and branched basal stems shows a dichotomous branching pattern with evergreen open canopy leaves, thereby permitting sunlight to illuminate the lower leaves for efficient photosynthesis.

26.10 Conclusions and research needs

The present study examined a few aspects of the autoecology and biodiversity of *Prosopis* in a few regions, from south to north in Arizona, It is the dominant evergreen tree in desert and shows large variability in distribution and density, depending on the availability of underground water. It show plasticity in crown

Figure 26.1 Field photographs showing the diversity of *Prosopis*, its gregarious habit and its predominance in Nogales, Arizona. See text for details. (See insert for color representation.)

shape, leaf canopy and branching patterns. It has several traits for adaptation to arid lands, such as thick leaves with a waxy coating, an open leaf canopy with a high capacity for photosynthesis, a deep root system, an adaptation to drought and salinity and long-lasting seeds with a thick seedcoat. Mesquite seed has a high nutritive value. The prevalence and aggressive nature of *Prosopis* in pastures, dwellings, irrigation channels and at road sides pose a great danger to human beings and livestock; it is a gigantic weed, very difficult to eradicate. Efficient techniques need to be evolved for the control and management of this obnoxious weed.

Research needs also to be directed to a detailed study of the autoecology, ecophysiology and physiology of *Prosopis*. It is a dominant species in desert lands, offers an evergreen landscape and offers shelters for animals, birds and food and feed. Research is needed to exploit the species for its feed and food value. Efficient control measures are needed to save pastures, prevent damage to water channels and protect against road side encroachment. It affects biodiversity for its gregarious habit. The high nutritive value with rich protein (up to 65%) needs to be harnessed for human health, medicinal values and so on. There is a great necessity to study the mechanism *Prosopis* uses to adapt to xeric habitats, such as morpho-anatomical, physiological, tolerance to salinity and so on.

Bibliography

Berhanu A., Tesfaye G. **2006**. The *Proposis* dilemma, impacts on dryland biodiversity and some controlling methods. *Journal of the Drylands* **1**(2):158–164.

Chávez R.O., Clevers J.G.P.W., Herold M., Ortiz M., Acevedo E. **2013**. Modelling the spectral-response of the desert tree *Prosopis tamarugo* to water stress. *International Journal of Applied Earth Observation and Geoinformation* **21**:53–65.

Devinar G., Llanes A., Masciarelli O., Luna V. **2013**. Different relative humidity conditions combined with chloride and sulfate salinity treatments modify abscisic acid and salicylic acid levels in the halophyte *Prosopis strombulifera*. *Plant Growth Regulation* **70**(3):247–256.

Felker P. **2009**. Unusual physiological properties of the arid adapted tree legume *Prosopis* and their applications in developing countries. In: S. Nobel (ed.) *Perspectives in Biophysical Plant Ecophysiology: A Tribute to Park.*, pp. 221–255.

Gichua M.K. **2014**. *Impacts of* Prosopis juliflora *Linnaeus emend. Burkart on Aspects of Biodiversity and Selected Habitat Conditions in Baringo Kenya.* Doctoral Thesis, Jomo Kenyatta University of Agriculture and Technology. Kenya.

Hendricks D.M. **1985**. Natural vegetation of Arizona. In: *Arizona Soils.* College of Agriculture, University of Arizona, Tucson, Arizona, 10 pp.

Kumar S., Mathur M. **2014**. Impact of invasion by *Prosopis juliflora* on plant communities in arid grazing lands. *Tropical Ecology* **55**(1):33–47.

Llanes A, Masciarelli O., Ordóñez R., Isla M.I., Luna V. **2013**. Differential growth responses to sodium salts involve different abscisic acid metabolism and transport in *Prosopis strombulifera*. *Biologia Plantarum* **58**(1):80–88.

Maundu P., Kibet S., Morimoto Y., Imbumi M., Adeka R. **2009**. Impact of *Prosopis juliflora* on Kenya's semiarid and arid ecosystems and local livelihoods. *Biodiversity* **10**:33–50.

USDA. **2010**. Root systems of mesquite, *Prosopis.* Forest Service General Technical Report RMRS-GTR-8, US Department of Agriculture, Washington, D.C.

Villagra P.E., Vilela A., Giordano C., Alvarez J.A. **2010**. Ecophysiology of *Prosopis* species from the arid lands of Argentina: What do we know about adaptation to stressful environments? In: *Desert Plants.* K.G. Ramawat (ed.). Springer, Berlin, pp. 321–340.

APPENDIX 1

Leaf Morphology

Ebenopsis ebano (Berland.) Barneby & J.W. Grimes

Cordia boissieri A. DC.

Bernardia myricifolia (Sheele). Benth. & Hook. F.

Condalia hookeri M.C. Johnst.

Autoecology and Ecophysiology of Woody Shrubs and Trees: Concepts and Applications, First Edition.
Edited by Ratikanta Maiti, Humbero Gonzalez Rodriguez and Natalya Sergeevna Ivanova.
© 2016 John Wiley & Sons, Ltd. Published 2016 by John Wiley & Sons, Ltd.

Eysenhardtia texana Scheele

Croton suaveolens Torr.

Karwinskia humboldtiana (Schult.) Zucc.

Leucophyllum frutenscens (Berland.) I.M. Johnst

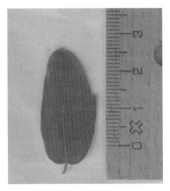

Celtis pallida Torr.

Zanthoxylum fagara (L.) Sarg.

Caesalpinia mexicana A. Gray

Havardia pallens (Benth.) Britton & Rose

Forestiera angustifolia Torr.

Leucaena leucocephala (J. de Lamarck) H.C. de Wit

Helietta parvifolia (A. Gray) Benth.

Sideroxylon celastrinum (Kunth) T.D. Penn.

Diospyros palmeri Eastw.

Parkinsonia texana (A. Gray) S. Watson

Ehretia anacua (Terán & Berland.) I.M. John.

Guaiacum angustifolium Engelm.

Amyris texana (Buckley) P. Wilson.

Sargentia greggii S. Watson

Celtis laevigata Willd.

Berberis chococo Schlecht.

Acacia wrightii Benth.

Acacia farnesiana (L) Willd.

Lantana macropoda Torrey.

Gymnosperma glutinosum (Spreng.) Less.

Diospyros texana Scheele.

Acacia berlandieri Benth.

Quercus virginiana Mitl.

Prosopis laevigata (H. & B.) Jonhst

Acacia rigidula (Benth.) Seigler & Ebinger.

Fraxinus greggii A. Gray

Acacia schaffneri (S. Watson) F.J. Herm.

Parkinsonia aculeata L.

Index

Note: Page numbers in *italics* refer to Figures; those in **bold** refer to Tables.

Autoecology and Ecophysiology of Woody Shrubs and Trees: Concepts and Applications, First Edition.
Edited by Ratikanta Maiti, Humbero Gonzalez Rodriguez and Natalya Sergeevna Ivanova.
© 2016 John Wiley & Sons, Ltd. Published 2016 by John Wiley & Sons, Ltd.